SEED HARDENING, PELLETING AND COATING

Principles and Practices

SEED HARDENING, PELLETING AND COATING
Principles and Practices

Editors:
K. Vanangamudi, K. Natarajan, T. Saravanan,
R. Renuka, N. Natarajan, R. Umarani, A. Bharathi

SSPH
SATISH SERIAL PUBLISHING HOUSE
403, Express Tower, Commercial Complex, Azadpur, Delhi - 110033 (INDIA)
Phone : 011-27672852 Fax : 91-11-27672046
E-mail : info@satishserial.com, hkjain1975@yahoo.com
Website : www.satishserial.com

Published by :

SATISH SERIAL PUBLISHING HOUSE
403, Express Tower, Commercial Complex, Azadpur, Delhi - 110033 (INDIA)
Phone : 011-27672852 Fax : 91-11-27672046
E-mail : info@satishserial.com, hkjain1975@yahoo.com

ISBN 81-89304-20-8

Printed at
SALASAR IMAGING SYSTEMS, DELHI-35

Tamil Nadu Agricultural University

Coimbatore-641 003
Tamil Nadu, India

Prof. C.Ramasamy
Vice-Chancellor

Foreword

High quality seed is the key for successful agriculture. Modern agriculture with its bias for precision farming demands that each and every seed sown must germinate and establish into a vigorous plant. Even for resource poor farmers, high quality seed is a boon as quality seed alone ensures an increase in yield upto 15-20 per cent.

Seed technology research has donated precise technologies as an answer to specific problems in agriculture. Seed hardening is one such technology. The greatest threat for Indian farmers is the onslaught of drought. Seed hardening is a simple, no-cost technology which offers an opportunity to reduce losses caused by drought condition. The simplicity of the technique assures best exploitation, once it spreads across the farming community. Large scale implementation of vast scientific knowledge depends on assembling the information on technologies for easy reference by the end users. This book has offered a scope to salvage Indian agriculture by presenting the literature on seed hardening and pelleting as a potential source of improving the quality of agricultural productivity.

Since no other book is exclusively available for seed related technologies, this book opens the gates for a new era of technology oriented farming even for small and medium farmers. The scientists and students can benefit by reading this book as it elaborates the mode of action of the seed based technologies.

Indian agricultural scenario is largely dominated by rainfed and garden land conditions. The book will be extremely useful in catering to the needs of majority of farming community.

I congratulate and commend the efforts of Dr. K. Vanangamundi and his co-authors who have brought out this unique book.

Coimbatore - 641 003

C. RAMASAMY

Preface

The increasing demand for quality seed should be fulfilled for emerging India as a major force in the global seed market. For this, there is an urgent need to refine various technologies to suit our specific requirements. Since, most of the cultivable area of our country comes under dryland and gardenland ecosystem, the regenerative and resource conserving technologies and practices should be developed for the above two eco-systems, in order to bring both environmental and economic benefits to the farmers. Seed Science and Technology is a magnanimous science which offers precise technologies to overcome acute problems of agriculture.

Seed hardening gains much more importance so as to salvage the bottlenecks of rainfed conditions and seed pelleting and coating helps in precision sowing, aids spot application of herbicides, seed protectants and fertilizers, etc., eliminating the necessary to treat the entire bulk of soil. Though these technologies have been reached extensively in the past decades, yet the practical implementation is not wide spread. The nation's crop production can be increased only, if both rainfed and gardenland ecosystems are made more productive, for which seed hardening and pelleting technologies can be efficiently harnessed.

Keeping in mind the importance of this aspect, this book on "Seed hardening, pelleting and coating technologies for rainfed / gardenland eco-system" is prepared with the prime objective of compiling all the research work carried out by different scientists in this field. This book imparts thorough knowledge of the related methodologies and underscores the mode of action within the seed to cater the needs of the people who are engrossed in seed science and enthralled by seed mechanisms.

We hope that this publication would be extremely useful to the students, academicians, seed producers, seed growers and those who are actively engaged in agriculture research of vocation.

K. Vanangamudi
K. Natarajan
T. Saravanan
R. Renuka
N. Natarajan
R. Umarani
A. Bharathi

Contributors

ANANDALAKSHMI, R. Division of Seed Technology, Institute of Forest Genetics and Tree Breeding (Indian Council of Forestry Research and Education), Coimbatore – 641 002, Tamil Nadu, India

ANSARI, N.A. Department of Plant Breeding and Genetics & Seed Science and Technology, Acharya N G Ranga Agricultural University, Rajendranagar, Hyderabad - 500 030, Andhra Pradesh, India.

BALAMURUGAN, P. Department of Seed Science and Technology, Tamil Nadu Agricultural University, Coimbatore-641 003, Tamil Nadu, India.

BALASUBRAMANIAN, T.N. Department of Agricultural Meteorology, Tamil Nadu Agricultural University, Coimbatore – 641003, Tamil Nadu, India

BALASUBRAMANI, V. Centre for Plant Molecular Biology, Tamil Nadu Agricultural University, Coimbatore-641 003, Tamil Nadu, India.

BHARATHI, A. Department of Seed Science and Technology, Tamil Nadu Agricultural University, Coimbatore – 641 003, Tamil Nadu, India

BHASKARAN, M. National Seed Research and Training Centre, Varanasi – 221 106, India.

GURUDEV SINGH, B. Division of Seed Technology, Institute of Forest Genetics and Tree Breeding (Indian Council of Forestry Research and Education), Coimbatore – 641 002, Tamil Nadu, India

HARINATH BABU, P. Department of Plant Breeding and Genetics & Seed Science and Technology, Acharya N G Ranga Agricultural University, Rajendranagar, Hyderabad - 500 030, Andhra Pradesh, India.

JAYARAJ, T. Soil and Water Management Research Institute, Thanjavur – 613 501, Tamil Nadu, India.

JEGATHAMBAL, R. Tapioca & Castor Research Station, Yethapur, Salem, Tamil Nadu, India.

JERLIN, R. Agricultural Research Station, Bhavanisagar – 638 451, Tamil Nadu, India.

KAVITHA, S. Central Institute for Cotton Research Regional Station, Coimbatore – 641 003, Tamil Nadu, India.

KRISHNARAJAN, M.V. Directorate of Seed Certification, Coimbatore, Tamil Nadu, India.

KRISHNASAMY, V. Centre for Plant Molecular Biology, Tamil Nadu Agricultural University, Coimbatore-641 003, Tamil Nadu, India.

MALLIKA VANANGAMUDI. Department of Crop Physiology, Tamil Nadu Agricultural University, Coimbatore – 641003, Tamil Nadu, India.

MANONMANI, V. Agricultural Research Station, Bhavanisagar – 638 451, Tamil Nadu, India.

MENAKA, C. Department of Seed Science and Technology, Tamil Nadu Agricultural University, Coimbatore-641 003, Tamil Nadu, India.

MURUGESA BOOPATHI, P. Department of Soil Science and Agricultural Chemistry, Tamil Nadu Agricultural University, Coimbatore – 641 00, Tamil Nadu, India

MURUGESAN, P. National Research Centre for Oil Palm, Pedavegi – 534 450, Andhra Pradesh, India.

NANTHAKUMAR, S. Department of Horticulture, Adhiparasakthi Agricultural College, G. B. Nagar, Kalavai-632 506, Tamil Nadu, India.

NATARAJAN, K. Department of Plant Breeding and Genetics, Adhiparasakthi Agricultural College, G.B.Nagar, Kalavai – 632 506, Tamil Nadu, India.

NATARAJAN, N. Department of Seed Science and Technology, Tamil Nadu Agricultural University, Coimbatore – 641 003, Tamil Nadu, India.

NATESAN, P. Department of Seed Science and Technology, Tamil Nadu Agricultural University, Coimbatore – 641 003, Tamil Nadu, India.

PADMANABHAN, G. Department of Crop Physiology, Tamil Nadu Agricultural University, Coimbatore – 641003, Tamil Nadu, India.

PALANISAMI, K. Director, CARDS, Tamil Nadu Agricultural University, Coimbatore – 641 003, Tamil Nadu, India.

PALANISAMY, V. Department of Crop Improvement, Anbil Dharmalingam Agricultural College and Research Institute, Tamil Nadu Agricultural University, Tiruchirappalli – 620 009, Tamil Nadu, India.

PARAMASIVAM, V. Oilseed Research Station, Tindivanam – 604 001, Tamil Nadu, India.

PONNUSAMY, A.S. Department of Seed Science and Technology, Tamil Nadu Agricultural University, Coimbatore-641 003, Tamil Nadu, India.

PRABAKAR, K. Department of Seed Science and Technology, Tamil Nadu Agricultural University, Coimbatore-641 003, Tamil Nadu, India.

RAJA, K. Central Institute for Cotton Research Regional Station, Coimbatore – 641 003, Tamil Nadu, India.

RAJASKARAN, R. Department of Seed Science and Technology, Tamil Nadu Agricultural University, Coimbatore-641 003, Tamil Nadu, India.

RAMAMOORTHY, K. Department of Seed Science and Technology, Agricultural College and Research Institute, Tamil Nadu Agricultural University, Madurai - 625 104, Tamil Nadu, India.

RAMANADANE, T. Department of Plant Breeding and Genetics, PAJANCOA, Karaikal – 609 603, Pondicherry, India.

RAMANATHAN, S. Former Directorate of Research, Tamil Nadu Agricultural University, Coimbatore-641 003, Tamil Nadu, India.

RAMYA, H. Department of Seed Science and Technology, Tamil Nadu Agricultural University, Coimbatore-641 003, Tamil Nadu, India.

RATHINAVEL, K. Central Institute for Cotton Research Regional Station, Coimbatore – 641 003, Tamil Nadu, India.

SABIR AHAMED, A. Soil and Water Management Research Institute, Thanjavur –613 501, Tamil Nadu, India.

SAMBANTHAMANI, S. Directorate of Seed Certification, Coimbatore, Tamil Nadu, India.

SARAVANAN, T. Department of Plant Breeding and Genetics, Adhiparasakthi Agricultural College, G. B. Nagar, Kalavai-632 506, Tamil Nadu, India.

SASTHRI, G. Department of Plant Breeding and Genetics, National Pulse Research Centre, Vamban – 622 303, Tamil Nadu, India.

SELVARAJU, P. Department of Crop Improvement, Anbil Dharmalingam Agricultural College and Research Institute, Tamil Nadu Agricultural University, Tiruchirappalli – 620 009, Tamil Nadu, India.

SHERIN SUSHAN JOHN. Department of Seed Science and Technology, Tamil Nadu Agricultural University, Coimbatore – 641 003, Tamil Nadu, India.

SHETE, D.M. Department of Seed Science and Technology, Tamil Nadu Agricultural University, Coimbatore-641 003, Tamil Nadu, India.

SIVAKUMAR, V. Division of Seed Technology, Institute of Forest Genetics and Tree Breeding (Indian Council of Forestry Research and Education), Coimbatore – 641 002, Tamil Nadu, India.

SRIMATHI, P. Department of Tree Breeding, Forest College and Research Institute, Mettupalayam – 641 301, Tamil Nadu, India.

SUNDARAJAN, P. Directorate of Seed Certification, Coimbatore, Tamil Nadu, India.

SUNDARALINGAM, K. Krishi Vigyan Kendra, Sirugamani – 636 203, Tamil Nadu, India.

SUNDARESWARAN, S. Department of Seed Science and Technology, Tamil Nadu Agricultural University, Coimbatore-641 003, Tamil Nadu, India.

THANGAVEL, P. Department of Seed Science and Technology, Tamil Nadu Agricultural University, Coimbatore – 641 003, Tamil Nadu, India.

UMARANI, R. Department of Seed Science and Technology, Tamil Nadu Agricultural University, Coimbatore – 641 003, Tamil Nadu, India.

VAKESWARAN, V. J. K. Seeds Pvt Ltd, Hyderabad, Andhra Pradesh, India.

VANANGAMUDI, K. Dean (Agri Culture), Tamil Nadu Agricultural University, Coimbatore - 641 003.

VIGNESHWARI, R. Department of Seed Science and Technology, Tamil Nadu Agricultural University, Coimbatore-641 003, Tamil Nadu, India.

VIJAYA, J. Department of Seed Science and Technology, Tamil Nadu Agricultural University, Coimbatore-641 003, Tamil Nadu, India

VIJAYAKUMAR, A. Department of Seed Science and Technology, Tamil Nadu Agricultural University, Coimbatore-641 003, Tamil Nadu, India

VILAS A. TONAPI, National Research Centre for Sorghum, Rajendranagar, Hyderabad - 500 030, Andhra Pradesh, India.

Contents

PART I

RAINFED AND GARDENLAND ECOSYSTEMS - GENERAL PERSPECTIVES

SEED RELATED TECHNOLOGIES OF DIFFERENT ECO-SYSTEMS FOR SUSTAINABLE AGRICULTURE

K. VANANGAMUDI[1], R. JERLIN[2], R. UMARANI[3],
P. BALAMURUGAN[3], A. BHARATHI[3] AND P. SRIMATHI[4]

1. INTRODUCTION

Agriculture in its present form must cater to the food, fibre, fuel and feed stock needs of the increasing world's human population which was 1.5 billion at the beginning of the last century, increased to 6.08 billion today, and is estimated to reach 8.3 billion by 2025 AD. Based on the increasing population and enhanced consumption, food grain production of 243.2 million tonnes are required by the year 2006-2007. To obtain this target of higher yield from crops without causing any adverse effect to the environment is an essential concern of today which can be achieved by the way of practicing sustainable agriculture, taking into account the prevailing different eco-systems of our country. A clear understanding of the problems encountered in each eco-system will pave way to use quality seeds by applying appropriate technologies for profitable crop production, since the quality seed is the essential input in agriculture and it plays a crucial role in boosting up the productivity and economy of the country.

2. WETLAND ECOSYSTEM

The main crop of wetland is rice which occupies a pivotal position in Indian National food and Livelihood security system and considered to

1 *Dean (Agriculture), Tamil Nadu Agricultural University, Coimbatore - 641 003.*
2 *Agricultural Research Station, Bhavanisagar – 638 451.*
3 *Department of Seed Science and Technology, Tamil Nadu Agricultural University, Coimbatore – 641 003.*
4 *Department of Tree Breeding, Forest College and Research Institute, Mettupalayam – 641 301.*

be a staple food of India. In India, nearly 80% of the total rice growing area is sown with farmers, own saved seeds without knowing its quality at the time of sowing which may lead to reduced yield compared to the usage of quality seeds of known source. A suitable pre-sowing upgrading technique will be highly useful to the farmers to maintain optimum population and to obtain higher yield. One such simple and easy low cost technology is egg flotation technique by which the seeds are upgraded before sowing. For this, 2 kg of common salt is dissolved in 10 litres of water. When a fresh egg is dropped in this salt solution, it will float in the salt water such that 25 paise coin size of egg alone will be exposed out of the water. When the bulk seeds of paddy are soaked in this solution, the immature and ill filled seeds will float in the water and this can be removed whereas the well filled dense good quality seeds are settled at the bottom, which could be utilised for sowing as they will produce vigorous seedlings.

3. GARDEN LAND ECOSYSTEM

Majority of the crops of our country are being cultivated under garden land condition. The major problem of the garden land is continuous crop cultivation which leads to depletion of nutrients from the soil. Since the fertility status of the soil is becoming poor, application of large quantity of inorganic fertilizers for cultivation alters the physical properties of the soil and also causing hazards to environment besides increasing the cost of cultivation to the farmers. Under these conditions, the crop production could be enhanced through seeds by suitable presowing treatments and this can be achieved by seed pelleting and pregermination techniques.

3.1. Seed pelleting

It is nothing but enclosing the seed with foreign material to obtain a standard size and it has its own influence on the micro environment of seed and improves the seedling growth and vigour and seed soil interface (Mucke, 1987).

In sunflower, seed pelleting with moringa leaf powder (250 g) along with DAP (100 g), *Azospirillum* (25 g) and *Trichoderma* (4 g) for one kg of seed, maximized the plant stand and yield. An increase in yield of 15% was noticed both under irrigated and rainfed conditions (Anon, 2001).

3.2. Pre-germination technique

This technique is commonly practiced in groundnut. The graded seeds are soaked in 0.5% $CaCl_2$ (50% seed volume) for 6 hours. After 6 hours of soaking, the seeds should be spread over moist gunny bags and covered with another moist gunny bag for 24 hours. After 24 hours, the seeds with sprouted radicle (just visible expression of radicle) should be separated and dried under shade. It should be repeated for 2-3 times with 2 hours interval and all the viable seeds with expressed radicle emergence should be separated and dried under shade. The viable and dead seeds are separated by this process which ensured high planting value and maintained the required plant population in the field resulting in higher yield.

4. DRYLAND ECOSYSTEM

The commonest calamity of the Indian cultivator is the failure of monsoon, which in its extreme manifestation is called drought. In some of the regions, eratic rainfall leads to drought during the vegetative phase and the water balance of a plant is upset and as a consequence, the physiological functions affecting the growth and yield are disarranged. Prevalence of drought during the initial phase of plant life is torturous. Safeguarding the seeds during this initial stage of germination will give a special impetus for the seed to overcome the moisture stress condition and develop into a vigourous plant. A simple method to alter the physiological and biochemical nature of the seed in order to induce the factors responsible for drought resistance is termed as seed hardening (Henckel, 1964).

4.1. Seed hardening

Before attempting to induce the drought tolerance externally, the potential of seed vigour should be harnessed to combat the moisture stress condition. In this technique, the seeds are soaked in lesser quantity of water for 4-16 hrs depending on the type of crop (till the moisture content of the seed raised to 30% during the imbibition process). By that period, the commencement of the radicle takes place but not allowed to protrude out. Then, the seeds are dried to original moisture content. The efficacy of the hardening process can further be improved by soaking the seeds in 1% leaf extracts of pungam, arappu and prosopis, as these plant products have growth stimulating substances which migrates into the seed that are subjected to hardening treatments (Table 1).

5. WASTELAND ECOSYSTEM

Due to reckless deforestation activities and also due to incessant rainfall by which the thick layer of the soil is washed away from the parent body which diminishes the fertility of the soil gradually, the area under wasteland is increasing day-by-day. Greening of this wasteland, is the

Table 1. Seed hardening techniques for few crop seeds

Crops	Chemical and concentration	Methodology
1. Pearlmillet	2% Potassium chloride	Dissolve 20 g of the salt in 1000 ml of water. Soak 1 kg of seed in 650 ml of this solution for 10 hrs and dry back to original moisture
2. Sorghum	2% Potassium dihydrogen phosphate	Dissolve 20 g of the salt in 1000 ml of water. Soak 1 kg of seed in 650 ml of the solution for 16 hrs and dry back to original moisture
3. Cotton	2% Potassium chloride	Dissolve 20 g of the salt in 1000 ml of water. Soak 1 kg of seed in 650 ml of solution for 12 hrs and dry back to original moisture
4. Sunflower	2% Potassium chloride	Dissolve 20 g of the salt in 1000 ml of water. Soak 1 kg of seed in 650 ml of solution for 12 hrs and dry back to original moisture
5. Pulses	100 ppm Zinc sulphate	Dissolve 100 g of the slat in 1000 ml of water. Soak 1 kg of seed in 350 ml for 3 hrs and dry back to original weight or moisture. Before soaking, the seeds are preconditioned by soaking the seeds in 1/3rd volume of water for 1 hr and then placing them in moist gunnies for a duration of ½ hr to 1hr, in order to raise the moisture content of seed to 15-18%. This is to avoid seed coat damage during rapid soaking

most effective reclamation measure which is achieved through massive afforestation programme. The seeds of the tree species involved in this programme many times lose its vigour and viability potential due to many factors like, lack of nutrients, moisture, feeding activity of insects and attack by soil borne pathogens. In afforestation programme, aerial seeding is normally practiced. In this method, the seeds will be sown in the land which cannot be managed by human force. Under these circumstances, for nursing such seed, proper method should be followed. Pelleting of seed is an effective control measure against all the above back drops.

5.1. Seed pelleting

As said earlier, it is nothing but increasing the size of seed by enclosing the seed with foreign material.

This process involves stamping, coating and rolling. Seeds are first mixed with adhesives and then coated with filler materials on rolling for uniformity. Materials used in pelleting are adhesives, coating materials and bioactive chemicals. The seed is placed in the mixer and moistened with an adhesive sticker solution such as maida solution or gum arabica. Enough treating powder is then added to dry the sticker usually in a volume relationship of 4 parts of powder to one part of sticker. The thickness of the seed coating is dependent on the amount of stickers in relation to the amount of seed. By this process, the seeds come in direct contact with the nutrients which enhances the vigour of the germinating seedlings. The seeds are protected against adverse soil effects besides pest and disease attack.

6. CONCLUSION

A closer book of the agricultural scenario of the developing country like India brings to the light of the challenging task of achieving the food grain target of about 250 million tonnes in next few years to manage the fast growing population. The task becomes more challenging is the present agricultural scenario. Use of advanced technologies including proper seed management in different prevailing eco-system, is an inevitable practice for successful crop production. To achieve uniformity and synchrony of growth of plant, these eco-friendly technologies must be followed without damaging the environment which is a step towards sustainability in agriculture. These seed management practices will be much useful to increase the agriculture productivity so as to keep pace with the increasing population of our country.

REFERENCES

1. Anonymous, (2001). Proceeding of the Oil Seed Meet, 2001. Tamil Nadu Agricultural University, Coimbatore.

2. Henckel, P.A. (1964). Physiology of plants under drought. *Ann. Rev. Pl. Physiol.*, *15*: 363-368.

3. Mucke, J.D. (1987). The regulation of water transport in pelleted sugarbeet seed. *J. Agron. and Crop Sci.*, *161*: 79-83.

AN INTRODUCTION TO CROP ECOLOGY IN RAINFED AND GARDENLAND ECOSYSTEMS

T.N. BALASUBRAMANIAN

1. INTRODUCTION

Study of plants in reciprocal relationship with their environment (climate, soil scape and biota) or study on the inter-relationship between crop plants and the environment is defined as crop ecology. Though this is the definition, exploitation of positive potential interaction between crops and environment through eco friendly technologies is very important for keeping the environment free from further damage. This is the present day task of researchers, agricultural extension workers, farmers and policy makers.In this chapter, the role of climate which is one among the ecological factors on crop production is discussed.

2. RAINFED AND IRRIGATED ECOSYSTEMS

Though the soil and rainfall pattern are same in an agricultural domain, with the introduction of irrigation in that environment to crop plants alters microclimate and there by productivity gets changed. Plants interact differently. Under dryland situation, due to moisture stress, plants fail to express their full potential.

The interaction between crop plants and climate (environment) with reference to irrigated and dryland agriculture, (though the terminology rainfed agriculture conveys somewhat different meaning) is given in Table 1.

Department of Agricultural Meteorology, Tamil Nadu Agricultural University, Coimbatore – 641 003.

Table 1. Irrigated and rainfed ecology

Particulars	Dryland agriculture	Irrigated agriculture
Climate	Arid to semi arid	Semi arid to sub humid
Temperature		
Maximum	upto 40°C	upto 34°C
Minimum	upto 21°C	upto 17.8°C
Relative humidity (%)	20 to 93	33-93
Rainfall (mm)	600-1000	800-1500
Wind speed (kmph)	2.9 to 10.5	3.3 to 12.4
Bright sunshine hours	4.8 to 10.0	4.3 to 10.3
Crop response (%)	50 – 60	100-150%

The crop response under irrigated agriculture is more positive, while it is negative in dryland situation. Enough caution must be exercised to handle the fragile dryland ecosystem. The crop response or crop's ecology changes between different seasons. In India there are two predominant agricultural seasons *viz.,* Kharif (June - October) and Rabi (November - February). During Kharif, the stress comes from moisture distribution, cloudiness, high wind speed and temperature, while during Rabi season it is from low temperature. Both seasons relatively either superior or inferior with reference to crops to be grown. This scenario could be observed in 80 per cent geographical area of India. Crop selection and selection of suitable management practices are very important for the concerned season so as to exploit positive interaction.

Productivity of crops can be enhanced through efficient use of solar radiation irrespective of the season. As reported by Kler and Dhaliwal (1997), productivity of crops like wheat, mung, raya etc., can be enhanced through efficient use of solar radiation by new planting technique of bi-directional sowing. Closer row planting, north south row orientation has also helped to capture more solar radiation for higher yields in wheat. Efficiency of solar energy utilization can be increased by increasing the surface area by planting crop species or crop varieties in different canopy structural pattern as prismatic, columnar, pyramidal etc. Inter / relay cropping systems can also help to sustain maximum productivity. Sowing on sunny slopes of ridges as in maize could manipulate soil temperature and incidence of solar radiation for better crop production. Uniform and optimum stand of crop is essential for getting higher yields due to better use of solar energy, nutrients and moisture. The future of crop productivity lies upon the management of varies ecological factors through production technology based on various scientific agrotechniques.

3. CROP PLANTS RESPONSE TO ENVIRONMENT (weather parameters)

Crop plants response to some selected weather parameters are furnished in Table 2.

Table 2. Crop plants response to environment (weather parameter)

Weather parameter	Crop response
Temperature (°C)	(i) Increased temperature with in the climatic normals of crop increases enzyme activity for carbon reduction (ii) <10°C decreases photosynthesis in maize upto 60 percent. (iii) 24.3 to 26.6°C ideal for photosynthesis. (iv) At 9°C rice exhibits yellow discoloration. (v) Germination reduces with decrease in temperature (eg. 75% reduction in germination) in groundnut. (vi) At 18°C number of grains / ear decreases in wheat. (vii) Pod length increases with temperature in soybean
Relative humidity (RH)	(i) Low RH increases transpiration, restrict the entry of CO_2 and leads to stomata closure. (ii) At 60% seed set increases in wheat (iii) Higher RH inhibits pollen dispersal and enhances germination of fungal spores. (iv) Aphids and thrips thrive well at higher RH. (v) High RH reduces maize grain yield (vi) Sorghum grain yield increases with RH.
Wind	(i) Higher wind speed increases transpiration (ii) Wind increases the enhanced supply of CO_2 (iii) NAR and RGR increases with wind speed (iv) Ethylene production in rice and barley is observed with increase in wind speed (v) Increased N fixation with increased wind speed (vi) Helps in spreading of pest and diseases

4. WEATHER PARAMETERS AND SEED QUALITY

Seed quality by its own character can influence productivity to a expected level in the absence of agricultural inputs. Hence seed quality is very important. However, the quality of seeds get damaged by unfavourable weather parameters and some examples are given hereunder.

4.1. Temperature

High temperature increases chaffyness. At a temperature range of >30°C and <27°C, sorghum yield gets reduced. Increased night temperature at silking stage in maize reduced kernel set.

4.2. Light

50 per cent shading in rice would bring the following

- Tillers from 456 to 396 / m^2
- Filled grain per panicle from 27.8 – 11.3/m^2
- Yield from 4.78 – 2.48 t/ha.
- Shade did increase abortive grains

4.3. Rainfall

Either high or low rainfall not suitable for seed production and rainfall during flowering is most dangerous for seed production.

4.4. Wind

Important for pollen dispersal for crops pollination.

5. CONCLUSION

Crop ecology differs between dryland and irrigated agriculture. According to the prevailing climate, weather based technologies have to be given top most priority. Weather based technologies can be developed successfully, when informations are known on weather sensitive crop stages and weather sensitive farm operations.

REFERENCES

1. Kler, D.S. and G.S. Dhaliwal. (1997). Management of ecological factors for higher crop productivity. In: Ecological Agriculture and Sustainable Development (ed : G.S. Dhaliwal, N.S. Randhawa; Ramesh Arrora and A.K. Dhawan). Proceedings of Intnl. Conf. on Ecological agriculture, towards sustainable development, Nov. 15-17 1997, Chandigarh, India 295-313.

WATER MANAGEMENT TECHNOLOGIES FOR RAINFED AND GARDENLAND ECOSYSTEMS

K.PALANISAMI

1. INTRODUCTION

In Tamil Nadu, the dryland area accounts for about 4 million hectares (55% of total cultivated area) in the seven agroclimatic subzones and the distribution over seven agroclimatic subzone is 25.6 per cent in North eastern zone; 24.3 per cent in North western zone; 24 per cent in Southern zone; 12.1 per cent in Western zone and 6.19 per cent in Cauvery delta zone. This paper deals with the needed strategies in terms of moisture conservation, water harvesting (watershed development), drip irrigation and tank irrigation management.

2. GENERAL RAINFALL AND SOURCES OF IRRIGATION

The pattern of rainfall distribution, irrigated area and sources of water supply in southern zone are furnished in tables 1, 2 and 3.

3. IN SITU MOISTURE CONSERVATION

3.1. Contour bunding

Contour bunding is practiced in cultivated lands to intercept the runoff, reduce the soil loss and to provide increased opportunity time for water intake. This practice is found to be useful in low rainfall areas (< 600 mm) having high infiltration and permeability rates. Generally, bunding is recommended for lands upto 6% slope and for soils having a depth of more than 7.5 cm.

Water Technology Centre, Tamil Nadu Agricultural University, Coimbatore – 641 003.

Table 1. Rainfall distribution in Southern Zone (mm)

District	South West		North East		Winter		Summer	
	Normal	Actual	Normal	Actual	Normal	Actual	Normal	Actual
Pudukkottai	345	255	392	327	55	2	125	68
Madurai	242	215	402	292	52	8	171	138
Dindigul	222	256	418	301	44	4	155	128
Ramanathapuram	135	745	498	389	75	12	109	68
Virudhunagar	162	163	432	333	59	7	164	148
Sivaganga	331	292	395	441	54	0	140	87

Table 2. Sourcewise irrigated area in Southern Zone (in ha)

	Canals	Tanks	Wells	Others (Supplementary sources)	Total	GIA/GCA
Pudukkottai	7776 (12.81)	39483 (65.04)	13436 (22.13)	8 (0.02)	60703 (100.00)	49.2
Madurai	63025 (39.60)	19362 (12.17)	76714 (48.21)	33 (0.02)	159114 (100.00)	57.7
Ramanathapuram	-	52239 (87.66)	7357 (12.34)	-	59596 (100.00)	32.4
Trinelveli	28543 (22.15)	54091 (41.98)	46175 (35.83)	55 (0.04)	128864 (100.00)	71.4
Dindigul	3376 (3.77)	7646 (8.54)	77190 (86.25)	1286 (1.44)	89498 (100.00)	38
Sivaganga	-	56627 (84.74)	10197 (15.26)	-	66824 (100.00)	63.8
Virudhunagar	-	18393 (39.91)	27695 (60.09)	-	46088 (100.00)	32

Note: GIA/GCA Gross irrigated area as percentage of gross cropped area

Table 3. Sources of water supply in Southern Zone

	Length of canals			Tanks (Nos.)			Irrigation wells		
	Govt.	PVT	TOT	>40ha	<40 ha	Total	Tube	Ordinary	Total
Pudukkottai	77	-	77	660	4791	5451	3493	40599	44092
Madurai	1696	-	1696	292	1995	2287	29	44541	44570
Theni	147	-	147	55	151	206	888	23964	24852
Dindigul	115	-	115	103	2125	2228	693	86482	87175
Ramnad	-	-	-	511	1330	1841	285	8021	8306
Virudhunagar	-	-	-	290	707	997	-	36201	36201
Sivaganga	20	-	`20	571	3830	4401	79	28804	8883
Tirunelveli	532	-	532	373	1797	2170	51	86694	86745

3.2. Compartmental bunding

Where the rainfall is low and the land slope is less than 1%, the fields can compartmentalised into 1 cent compartments by working with bund former.

3.3. Dead furrows

Dead furrows are formed by ploughing at an interval of 3.6 m across the slope between two seed rows about 20-25 days after germination on receipt of rains.

3.4. Ridges and furrows system

This is laid out in soils having less than 1% slope. Furrows serve as effective means of surface drainage and carry excess water into cutoff drains dug across the slope. This system is highly effective in medium to high rainfall areas (700-1200 mm).

3.5. Graded furrows

This system is suitable for areas receiving medium rainfall (less than 1000 mm). Productivity of upland crops can be substantially increased simply by providing graded furrows 0.2 to 0.3% slope.

3.6. Broad bed and furrow system

It essentially consists of land smoothing with less soil disturbances after primary and secondary tillages and formation of graded wide beds separated by furrows connected to grassed water ways. The beds and furrows are laid out on gradient of 0.4 - 0.8% for optimum performance. The beds are 120 cm wide and the furrows are 30 cm wide and 15 cm deep. This system is suitable for lands with a slope not exceeding 3%. This system permits safe disposal of run off and tends to conserve soil and water *in situ*.

3.7. Raised and sunken bed

In regions with assured annual rainfall of more than 1000 mm and relatively on flat topography, the crops suffer because of poor drainage during continuous and intense rainfall. On sloppy lands, run off causes severe soil, water and nutrient loss. Raised and sunken bed provides surface drainage and encourages *in situ* rain water conservation. This system consists of an array of alternating raised and sunken bends. The run off from raised bed is arrested in the adjacent sunken beds. The width of the raised bed and sunken bed vary depending upon the terrain of the land (approximately 8 m wide raised bed and 4 m wide sunken beds can be formed).

3.8. Vegetative barriers

Vegetative materials (vetiver) are planted across the slope to arrest the runoff. This is suitable for varying slope of land. For effective use if the slope is more, the distance between two rows should be less (i.e. if the slope is 57% interval may be 40 m). The use of vegetative barrier could be a permanent viable alternative to costly engineering measure for soil and water conservation.

3.9. Deep tillage

Deep tillage more than 25 cm once in a year is required for better receptiveness of precipitation. Deep tillage is also needed in areas with hard pans.

3.10. Mulching

Mulching aims at slowing the evaporation loss from the stored soil moisture using the available farm waste as vertical mulch in trenches (40 cm deep, 15 cm wide and protruding 10 cm above the ground) enhanced the available soil moisture by 4-5%.

4. WATERSHED DEVELOPMENT

4.1. Watershed management

4.1.1. Checkdam

Checkdams constructed across gullies will serve as an effective water harvesting structure like mini percolation ponds. Even though their main purpose is for controlling the gullies, series of checkdams across a gully course will help to augment the filling up of water in the existing wells nearby on both sides of the gully. The storage of water against the checkdam will be depositing silt and within 3 or 4 years the upstream side of the checkdams will be completely silted up, the checkdams will then serve as drop structures for maintaining the hydraulic gradient of the gully course. Approximate cost will be Rs.10, 000 for minor checkdam and Rs.40, 000 for major checkdam.

4.1.2. Percolation ponds

Percolation ponds are small ponds located mostly in poromboke lands and formed in order to store the run-off rain water and to allow it to percolate downwards and sideways. There are more than 7500 ponds in the State constructed by Agricultural Engineering Department. Deep ponds are preferred since evaporation of the stored water will be less. It has been observed that the percolation ponds are effective upto a distance of 1000 metres on the downstream side, and wells within this range are benefited with more replenishment of water. Approximate cost will be about Rs.1,00,000/pond.

4.1.3.Farm ponds

Farm ponds are small water bodies formed either by the construction of a small dam or embankment across a waterway or by excavating or dug out. The water is usually harvested from a small catchment area and then used for irrigation for crops during prolonged dry periods or for rabi crops. Since a farm pond is formed in the low lying area in an agricultural farm it will act as a collective drain. Approximate cost will be about Rs.25,000 - 50,000.

Achievements under DPAP watershed programme and the rise in watertable for certain DPAP watersheds in coimbatore district are given in tables 4 and 5.

5. IRRIGATION METHODS

5.1. Gradual widening technique

Gradual widening of furrows in sugarcane and gradual widening of basins in banana were efficient in water use.

Irrigation to banana (Rasthali) at 1.0 IW/CPE ratio (once in 7 days) from 0-7 months and 1.2 IW/CPE ratio (once in 5 days) from 7-14 months recorded higher mean yield of 32.7 t ha^{-1} with an increased fruit yield of 2.1

Table 4. Achievement under DPAP Watershed Programme (1995-96 to 1999-2000)

Sl. No.	Name of the activity	Unit	Unit Cost (Rs.)	Achievement	
				Physical	Financial (Rs. in lakhs)
1.	Contour bunding	Ha	1000	874	8.15
2.	Minor Checkdam	No	10000	290	27.7
3.	Major Checkdam	No	40000	104	39.86
4.	Summer ploughing	Ha	500	2300	11.50
5.	Land levelling	Ha	10000	150	14.10
6.	Percolation pond	No	100000	38	38.19
7.	Farm pond	No	25000	15	3.52

Table 5. Rise in Water table for selected DPAP watersheds (DPAP) in Coimbatore District

Sl.No.	Name of watershed	Rise in water table	Average water table rise (m)
1.	Kuppepalayam	1.5-2.0	1.75
2.	Kodangipalayam	2.0-3.0	1.50
3.	Arasur II	1.8-2.8	2.30
4.	Karumapalayam	1.5-2.5	2.00
5.	Sokkanur	1.0-1.5	1.25
6.	Ichipatti	1.5-2.5	2.00

t ha^{-1} compared to basin irrigation at 1.0 IW/CPE ratio throughout the crop growth period (higher WUE and saving of 140 mm of water). By gradual widening of ring basins from 30 x 60, 45 x 90 and 60 x 120 cm for 0-50, 51-100, 101-150 days respectively, there was a saving of 25.4 per cent of water and increased WUE of 25.9 kg ha^{-1} mm over check basin in Lower Bhavani command area.

In Periyar-Vaigai command, the gradual widening of ring basins and channels to banana saves about 45 per cent of irrigation water and in sugarcane 15 per cent saving was achieved. It was proved that using of tractor mounted cage wheel or sheep – foot roller can reduce the percolation losses by 25 per cent. Initially, the furrow width will be narrow, with the advancement of crop age, furrow widened to facilitate higher water requirement. It eliminates the use of more water during early stage of crop than in maturity stage.

5.2. Drip irrigation

In India, drip irrigated area is 2.53 lakh hectares and in Tamil Nadu it is 0.34 lakh hectares. The major crops under drip irrigation in Tamil Nadu is Coconut (21,200 ha), Mango (4000 ha), Guava (1260 ha) and Citrus (900 ha).

Drip irrigation can be employed in rainfed areas to cultivate fruit trees like mango, guava, sapota, pomegranate, amla, ber etc. During the rainy season, the rainwater can be harvested and stored in ponds and through this groundwater recharge in wells is possible. From ponds and wells the water can be applied through drip irrigation. Through drip irrigation, nearly 50 to 60% of the irrigation water can be saved in addition to saving of about 40% human labour (Table 6). Yield increase is also reported in several cases.

Table 6. Water requirement of crops in drip and conventional system

Crops	Water requirement lit/day/plant	
	Drip	Conventional
Coconut	75-100	200-300
Grapes	25-35	90-100
Mango	30-40	90-100
Guava	22-30	70-100
Sapota	20-30	70-100
Pomegranate	20-40	60-100
Banana	8-12	30-40
Lemon	10-20	20-65
Papaya	5-7	17-26
Vegetables and flowers	1-2	4-6
Tapioca	1.5-2	3-4
Cotton (hybrid)	1.5-2.5	3-5

For employing drip irrigation there is no need of land levelling in rainfed areas. The approximate drip installation cost (without motor pump set) has been worked out at Water technology Centre, Coimbatore (Table 7). For Mango Rs.24,000/ha, for Guava it is Rs.34,000/ha, for sapota it is Rs.24,0900/ha, Rs.34,000/ha for Pomegranate and Rs.34,000/ha for acid lime. The minimum requirement of pumpset is 5 HP, even with 3 HP the system can be operated satisfactorily.

Table 7. Cost and spacing of crops in drip system

Crops	Spacing (m)	Cost Rs/ha (Approximate)
Coconut	7.5 x 7.5	30,000
Grapes	3.0 x 1.5	40,000
Mango	10 x 10	24,000
Guava	5 x 5	34,000
Sapota	10 x 10	24,000
Pomegranate	5 x 5	34,000
banana	2 x 2	45,000
Acid lime	5 x 5	34,000
Vegetables	0.75 x 0.6	65,000

5.3. Affordable micro irrigation systems

Both bucket and drum kits can be used in the dry tracts using manual filling of water. The bucket kit (15 lit capacity) will cover an area of 15 m^2 (0.4 cents/unit) and the buckets should be filled in 3-4 times a day. The cost will be Rs.200/unit. In the case of drum kit (200 lit) it will cover an area of 120 m^2 (3 cents/unit). The drum will be filled in 2 times a day. The cost will be about Rs.600/unit excluding drum.

5.4. Advanced - Surface Irrigation Methods

5.4.1. Surge irrigation

Under surge irrigation, the water is applied in the border intermittently and not in continuous basis. It increases the advance rate.A relatively new concept in surface irrigation application method *viz.,* surge irrigation has been introduced and evaluated for field use. Extensive experiments covering a wide range of long furrow specifications, inflow discharges, cycle ratio and number of surges with different test crops like maize, sunflower, sorghum, groundnut have established the supremacy of surge irrigation over continuous flow. Surge irrigation was found to discernible with 40 per cent water saving, 25 per cent yield increase, 20 per cent land saving, 40 per cent labour saving (for irrigation), ease for the labour in night hours and saving of Rs.1250 ha^{-1} on tillage. The cost increase is only marginal. The use efficiency was highest (8.23 and 13.13 kg ha^{-1} cm^{-1}) under surge irrigation than check basin, short strip and long furrow with continuous flow.

5.4.2. *Cablegation*

The cablegation system is an advanced, automated surface irrigation method that can be easily adapted for different types of field crops, orchards and trees. It is designed to automatically apply a predetermined quantity of water to field at a declining rate. The water application rate is designed to closely match the infiltration characteristics of soil to minimize deep percolation and run off losses.

Compared to blocked-furrow irrigation method, atleast more than 50 per cent water savings can be achieved. Properly designed cablegation systems can attain water application efficiencies of over 80 per cent.

The system requires a person to turn on the water supply and control the rate into the stilling tank. This may take about an hour. Once started, the cablegation system will irrigate the entire field automatically. Once the plug reaches the end of the pipe, the irrigation water supply is stopped. The plug is un-hooked and the rope is wound manually. This may take another hour. Both these operations could be carried out by the farmer eliminating the need for additional labour to irrigate the field. Loam to clay-loam soils that have low infiltration rates are most appropriate for this system. The soils that reach their basic intake rate in less than 15 - 20 minutes are more suitable for this method of irrigation. The system could also be installed in soils having higher infiltration rates provided the furrow lengths are shorter than 50 m. The land slope and water flow rate to each furrow will determine the maximum length of the furrow for the particular type of soil. The aim is to minimize deep percolation losses. Cost per acre is also less (annualized cost is Rs.2500). Land that has a gentle slope between 0.1 to 0.5 per cent is more suitable. Land with higher slope can be used after the furrows are formed on benched terraces having gentle slopes.

SOIL AND NUTRIENT MANAGEMENT PRACTICES FOR RAINFED AND GARDENLAND ECOSYSTEMS

P.MURUGESA BOOPATHI

1. INTRODUCTION

Seed agronomy comprises of selection of land, preparation of field for sowing and nutrient application and their management which trigger the expression of total ability of seed. Seed production is determined by environmental, edaphic and nutrient factors. The influence of environment can be minimized by selection of proper season and area for production. Seed production needs special attention as the resultant produce need to be viable and reproducible. Hence, for rainfed and garden land ecosystem, care should be taken from seed to seed i.e. from sowing till it is marketed.

The depletion of nutrients particularly micronutrients in the soil is increasing due to adoption of modern technologies in crop production such as raising high yielding crop varieties, multiple cropping, intercropping etc., where demand for nutrients is very high and varying. The amount of nutrients depleted is not refunded to the soil in the right proportion at right time leading to wide spread deficiency of nutrients especially micronutrients.

The micronutrient delineation studies have revealed that Tamil Nadu soils are deficient in Zn by 52%, Cu by 30%, Fe by 22% and Mn by 8%. The Zn deficiency is more prevaalent in the districts of Ramanathapuram, Erode, Sivaganga, Thanjavur, Kancheepuram, while Fe deficiency is high in Ramanathapuram, Sivaganga, Erode,Kancheepuram and Coimbatore

1 *Department of Soil Science and Agricultural Chemistry, Tamil Nadu Agricultural University, Coimbatore - 641 003.*

districts. Cu deficiency is dominant in Ramanathapuram, Madurai and Tirunelveli districts whereas Mn deficiency is higher in Kancheepuram, Virudhunagar, Madurai and Coimbatore districts.

2. SELECTION OF SEED PLOT

The plot selected for seed crop must have the following characteristics.

1. Soil texture and fertility of plot should be according to the requirement of the seed crops.
2. The seed plot should be free from volunteer plants, weed plants and other crop plants.
3. The soil of the seed plot should be comparatively free from soil-born diseases and insect pests.
4. The seed plot must be levelled.
5. It should be feasible to isolate the plot as per requirements of certification standards.

3. SOIL CHARACTERS AND SELECTION OF CROPS

The crop should be selected for different soils based on their suitability (Eg. Black soil - cotton, red loamy soil - pulses, oilseeds and vegetables; alluvial soils - paddy, sugarcane) for rainfed and gardenland ecosystem.

The land for the seed crop must be prepared well. Good land preparation helps in improved germination, stand establishment and destruction of potential weeds. It also aids in water management and uniform irrigation.

Depth of sowing is extremely important in ensuring good plant stands. Small seeds should be planted shallow, but large seeds a little deeper. Seeds would emerge from greater depths in sandy soils than in clay soils, and also in warm soil as compared to cold. In dry soils, seeds should be planted slightly deeper so that they come in contact with moisture.

4. SOIL WATER POTENTIAL

Physical properties of soil have direct impact on water holding capacity. Water potential characters are viewed to identify the frequency of irrigation to the crop. Soil with different distribution pattern of different particles (sized) will have different water potentials. The water potential is important as the germination of seed consists of a numebr of phased events. The processes of imbibition, radicle emergence, plumule emergence, radicle growth and plumule growth wherein each respond to the availability of water and each has critical water potential at which the process ceases (Naylor, 1985). The problem of crop establishment relates to the sown seeds receiving an adequate supply of water over a sufficient time period to permit all these processes to occur. There is evidence that as water potential decreases (ie. as water become less available) imbibitions may be prevented. In addition, the critical values may differ between crop species, between cultivars of a single species, and also between seed lots of different quality in a single cultivar.

5. SOIL TEMPERATURE

Soil surface temperatures in tropical crop systems commonly exceed 35°C and this may inhibit the emergence of seedlings. The high soil surface temperatures may even lead to death of emerged seedlings through heat gridlines.

Soil micro organisms show maximum growth and activity at optimum soil temperature range. The biological processes for nutrient transformations and nutrient availability are controlled by soil temperature and soil water. Water transport and ion movement also depend on soil temperature.

Soil temperature has a profound influence on seed germination, root and shoot growth, nutrient uptake and crop growth. Even a difference of 1°C in soil temperature can have a significant effect on plant growth. Seed germination is very sensitive to soil temperature. Seed do not germinate below or above a certain range of temperature. During winter, seeds of vegetables have to be germinated under favourable condition and then sown in the field. Root elongation is very much dependent on soil temperature for its rapid growth and maximum yield.

Soil temperature under field conditions can be altered by suitable cultural practices such as mulching, irrigation, drainage and tillage. The underlying principle is to control the rise or fall in temperature of surface soil.

6. SOIL CRUSTING

Crusting of soil is a form of soil compaction. By beating action of raindrops or wetting with irrigation water, soil with low water table structure slakes and gets dispersed, and fine particles clog pores. When such a soil dries, a thin layer of hard crust of soil is formed. The crusts present a serious barrier for seedling emergence. High exchangeable sodium percentage, poor structure, low organic matter content of soil and also puddling during tillage operations are some of the factors responsible for crust formation. Lowering the exchangeable sodium percentage and incorporation of organic matter prevent crust formation.

Soil crusting is sometimes an important factor limiting seedling establishment. Soil crusting in red lateritic soils can be effectively managed by the application of lime gypsum along with bulky organic manure.

7. SOIL SALINITY AND SODICITY

Saline soils usually have an excess of sulphates and chlorides of sodium and calcium in the surface layers of soil with a pH < 8.5 and electrical conductivity of > 4.0 dSm^{-1}. If a large proportion of the calcium salts are washed out and sodium carbonate forms in the soil solution then sodic soil is produced with pH > 8.5. The effects of such soils are two types *viz.*, creation of moisture stress by higher osmotic potential and the lack of availability of nutrients to the plant growth and seedling establishment (Agrawal and Dadlani, 1995).

8. EVALUATION OF SOILS FOR GERMINATION

For evaluating the response of crop to soil characteristics, total germination percentage, mean germination time, intermediate counts on germination, the variation in germination around the germination time are to be considered (Hunder *et al.*, 1985).

9. SOME MANAGEMENT TECHNIQUES AGAINST THE SOIL PROBLEMS IN SEED PRODUCTION

- Crops should be selected based on soil types.
- Soil should be made into fine tilth to avoid crusting.
- Life irrigation should be given on 3rd day to avoid soil crusting and emergence problem.
- In high water potential soils, stagnation of water should be avoided. Crop varieties that can resist water logging should be selected.
- In calcareous soils, iron is not available to the growing plant. Hence, to avoid chlorosis, iron should be sprayed. The study carried out with $FeSO_4$ 0.5% spraying in sorghum CSH-5 hybrid seed production

expressed that seed setting was more in plots sprayed with $FeSO_4$ thrice (Karivaratharaju *et al.*, 1987).

- On acid soils, application of lime improved the production of quality seeds.
- In alkaline soils, where the pH is above 8.00, after the incorporation of gypsum and leaching processes, growing of green manure crops like daincha, sunhemp, sesbania and ploughing *in situ* at flowering stages reduces the alkalinity and improve the productivity of seed.
- Botanical seed pelleting with leaf powders of arappu (*Albizia amara*), vadhanarayanan (*Delonix elata*) and prosopis (*Prosopis juliflora*) favour better establishment of seeds in problem soils.
- On availability of problem soils, seed production of varieties that are resistant to the specific situation should be used.

10. EFFECT OF PROBLEM SOILS IN SEED PRODUCTION

- Poor seed set
- Flower dropping
- Infertile pollen
- Poor seed filling
- Reduced seed size
- Reduced seed yield
- Reduction in initial vigour

11. FERTILIZER RECOMMENDATION FOR SEED CROP

In general, seed crop requires higher dose of NPK compared to grain crop. A comparison of fertilizer recommendation for seed and commercial crop is given in table 1.

Table 1. A comparison of fertilizer recommendation for commercial and seed crop

Crop	NPK recommendation (kg/ha)	
	Commercial Crop	**Seed crop**
Pulses	25:50:0	25:50:20
Horsegram	12.5:25:0	12.5:30:30
Soybean	20:80:40	40:160:80
Groundnut	17:34:54	40:40:60
Sunflower	40:20:20	60:45:45
Cotton (MCU 7)	30:30:60	40:40:80
Gingelly	35:23:23	50:25:25

11.1. Nutrients recommended for foliar spray

Table 2. Foliar nutrition for crop plants

Crop	Fertilizer dosage (kg/ha)	Basal (kg/ha)	Top (kg/ha)	Foliar spray at flowering
Sorghum	100:50:50	50:50:50	25 kg N – 20 DAS	
			25 kg N – 40 DAS	
Cumbu	100:50:50	50:50:50	50 kg N – 35 DAS	
Maize	150:75:75	40:75:40	50 kg N – 20 DAS	
			60 kg N – 40 DAS	
			35 kg N – 40 DAS	
Cotton variety	75:35:35	35:35:35	35 kg N – 45 DAS	
Hybrid	75:50:25	50:50:25	25 kg N – 30 DAS	
Sunflower	60:45:45	60:45:45		
Groundnut	40:40:60	40:40:60	200 kg Gypsum	
Gingelly	50:25:35	50:25:25	5 kg $MnSo_4$	
Onion	60:60:30	30:60:30	30 kg N - __ DAS	
Tomato	60:40:40	20:40:40	20 kg N – 45 DAS	DAP 1%
			20 kg N – 60 DAS	
Chillies	60:30:30	0:30:30	20 kg N – 15 DAS	DAP 1%
			20 kg N – 45 DAS	
			20 kg N – 15 DAS	
Brinjal	40:30:30	20:30:30	20 kg N – 40 DAS	DAP 1%
Gourds	6:12:12	6:12:12	10 kg N – 30 DAS	
Bhendi	16:20:12	8:20:12	8 kg N – 10 DAS	DAP 1%
Pulses	25:50:0	25:50:20	-	DAP 1%

In addition the nutrients preferably micro nutrients are used as foliar spray to supplement the basal nutrition. Some of the chemicals recommended for foliar spray for effective seed set in crop are tabulated as below:

1. Pulses

Nutrient combination	Redgram	Blackgram	Greengram
Urea (kg)	10.0	7.5	10.0
DAP (kg)	2.6	1.95	2.6
MOP (kg)	1.75	1.31	1.75
K_2SO_4 (kg)	1.4	1.05	50
Succinic acid (g)	40	40	40
Teepol (ml)	125	125	125

Note L In 250 litres/ha : Spray schedule and 40 days after sowing

1.	Soybean	:	2% DAP spray and 0.5% $ZnSO_4$ / 1.0% / $MnSO_4$ 0.2% Boric acid spray at 20, 30 and 40 days after sowing depending on need.
2.	Sunflower	:	0.5% $ZnSO_4$ spray on 30, 40 and 50 days after sowing.
3.	Cotton	:	0.5% urea at 45 and 65 days after sowing.

12. CONCLUSION

A closer look of the agricultural scenario of the developing country like India brings to light the challenging task of achieving food grain target of 250 million tonnes of present to meet the demand of around one billion people. Adequate and quality seed production is also required to be achieved in the available area. The task becomes all the more challenging in the present trend of diminishing land and water resources. The major focus has to be on rainfed areas, which are potentially productive but for effective input of scientific techniques. Development of dryland technologies involving multi-disciplinary approach in the context of integrated technologically package would ensure soil and moisture conservation and enhance productivity of crops, especially for seed production. This type of technology should be of eco-friendly without degradation of the natural resources in the large tract of dryland areas and is bound to play a vital role of usher prosperity for millions of the people toiling hard in these areas.

SEED TREATMENTS FOR MAXIMISING THE PRODUCTIVITY IN RAINFED AND GARDEN LAND ECOSYSTEMS

A.S.PONNUSAMY[1] AND T.RAMANADANE[2]

1. INTRODUCTION

Pre-sowing treatments are given to the seed just before sowing to achieve the needy objectives. Bioactive chemicals are effective in improving the performance of seed under field conditions. These bioactive chemicals are applied to the seed by different methods such as fortification (aqueous solution), suspension treatment, infusion technique through organic solvents, pelleting, etc.

2. SEED FORTIFICATION

Seed fortification is impregnation of required substances for invigorating the seed for improved production where the seeds are soaked either in water or dilute solutions of bioactive chemicals such as micronutrients, growth regulators, vitamins and seed protectants.

2.1. Methodology

Seeds are soaked in solutions in equal volume for a duration of 6-24 hours depending on the crop to increase the moisture content of the seed to 20-25 per cent just enough for endogenous impregnation of the chemicals by exogenous application. The choice of chemicals, its concentration and duration of soaking vary with species and decide the success of the treatment.

1 *Department of Seed Science and Technology, Tamil Nadu Agricultural University, Coimbatore-3.*
2 *Department of Plant Breeding and Genetics, PAJANCOA, Karaikal – 609 603.*

2.2. Nutrients

The macro and micro nutrients are used for fortification as they are essential for the

- Nitrogen - constituents of amino acids, nucleotides and also acts as a coenzyme
- Phosphorus - constituents of phospholipids, nucleotides and sugar phosphate
- Potassium - required by all higher plants for active of enzymatic process
- Sulphur - constituent of amino acids like cystine, methionine and cystenine.
- Magnesium - constituent of chlorophyll. It interfers with the food reserve policy of the plant. It is an activator of enzymes and their metabolism. $MgSO_4$ and $MgCl_2$ are the widely used Mg forms.
- Iron - integral part of protein and constituent of phytochromes and thereby it involves in photosynthesis and respiratory action of the system. It also a consistent of enzymes such as catalase, peroxidase and dehydrogenase enzyme.
- Zinc - constituent of dehydrogenase, alcohol dehydrogenase and lactic dehydrogenase. Zinc is very much required for pollen production.

Others : Growth regulators, vitamins, botanics and fungicides.

3. SEED INFUSION

It is a method of impregnation of seeds with bioactive chemicals through organic solvents instead of water. This technique of infusion helps to avoid the damage caused to the seed due to water soaking. Hence, highly suitable to seeds that suffer with soaking / seed coat injury. Simultaneously it also provides protective, regulatory and selective functions of the chemicals to improve the performance of seed with the help of organic solvents.

3.1. Methodology

Desired solutes are dissolved in the organic solvent at desired concentration. Seeds are immersed in solvent with desired solute for 5 to 24 hrs depending on species. After the desired period, the solvent is evaporated by air and by vacuum desiccator for 30 minutes to 1 hour or under ambient condition for 24 hours for complete expiry of solvent from the seed. Then the seeds are taken for sowing.

The depth and amount of chemicals penetrated into the seed vary with (1) duration of treatment (amount of chemical reaching the embryo increased with increase in time), (2) concentration of solutes (higher the concentration more will be the penetration) and (3) type of chemical; compared to IAA, GA_3 reaches the embryo lesser in quantity in a given time due to higher molecular weight.

3.2. Bio-active chemicals used for seed infusion

1. *Seed protectants:* fungicide, bacteriocide.
2. *Herbicide antidotes:* To prevent the ill effects of herbicide, herbicide antidotes are infused.
3. *Growth regulators:* GA_3, IAA and kinetin are incorporated to nullifying the endogenous deficit of the growth substances.
4. *Nutrients:* Micro and macro elements.
5. *Substances to resist adverse stress conditions:* Substances like fusicoccin alleviate the adverse effect in soil.

3.3. Organic solvents used for infusion

Acetone, Petroleum ether, Dichloro methane, Ethanol, Dichloro ethane, etc.

4. SEED HARDENING

Seed hardening is conditioning the seed to withstand adverse environmental and edaphic conditions. It is a creation of resistance in the seed for better outgrowth of seedling. Pre-sowing hardening of seed is one of the technique where the physiological reorganisation is induced by hydration and dehydration process. Hardening process brought out physico-chemical changes by modifying the physico-chemical characters by increasing the physiological activity of the embryo and associated structure. This results in absorption of more water due to increase in the elasticity of cell wall and development of a stronger and efficient root system.

4.1. Methodology

Seeds are soaked in required quantity of water (seed to solution ratio) depending upon crop seeds and allowed to absorb moisture upto 35 per cent of their weight and spread out in thin layer for drying in shade for 2 to 3 days. During this period, the seed gets dried almost to the original weight. The soaking duration vary with crop and the hardening efficacy can be improved by use of bioactive chemicals.

4.2. Chemicals used for hardening

(a) Water

(b) Aqueous solutions of salts - sodium chloride, sodium sulphate, potassium chloride, potassium nitrate, potassium sulphate, calcium chloride, ammonium sulphate, potassium dihydrogen phosphate, zinc sulphate, sodium molybdate, manganese sulphate.

(c) Growth promoters like gibberellic acid, Chloro-choline Chloride, kinetin, 2-chloroethyl phosphonic acid, Ascorbic acid, succinic acid.

(d) Vitamins: Vit K, Nicotinic acid, Pantothenic acid, Adenin.

(e) Plant products: Garlic extract, coconut water, leaf extracts of Arappu (*Albizia amara*), prosopis (*Prosopis juliflora*) etc.

(f) Osmoticants: D-Mannitol, Polyethylene glycol.

5. OSMOTIC PRIMING

Osmotic priming is one of the treatments where the seeds are physiologically conditioned to face the adverse environment in the field when it is sown. Osmotic priming is a very expensive but required process particularly for large seeded legumes like peas, beans etc. They have high protein and large embryo and are susceptible to soaking injury. High protein seeds are hydroscopic and hydrophilic. Osmotic conditioning begins when a seed is hydrated in low water potential osmotic solution. After the moisture equilibrium is achieved, further uptake of water by the seed is prevented. Osmo conditioning can be effectively achieved by adjusting the concentration of osmoticum in the conditioning solution to a level just preventing germination.

5.1. Chemicals used

Chemicals used for osmotic priming are 1). PEG 6000: High molecular compound, acts as coating material. It absorbs water and releases ½ of its water potential to seed, 2). Mannitol : Low molecular compound; highly toxic materials and 3). NaCl, KH_2PO_4, Glycerol, KNO_3, $MgSO_4$.

5.2. Methodology

Large seeded legume seeds are soaked in 20-30% of PEG 6000 solution with or without other chemicals and incubated at 15-20°C for 7-21 days. The seeds are washed with distilled water and air dried at 25°C. Then osmoprimed seeds can be used for sowing.

6. SEED PELLETING

Seed pelleting is enclosing the seed with foreign material to obtain a standard size and it has its own influence on the micro environment of the seed and influences on seedling growth at seed soil interface.

6.1. Methodology

Pelleting process can be broadly classified into three (1) stamping, (2) coating, (3) rolling. Seeds are first mixed with adhesives and then coated with filler material on rolling for uniformity. Materials used in pelleting are (a) Adhesives, (b) Coating materials and (c) Bioactive chemicals.

6.2. Adhesives

The common adhesives are gum Arabic, methyl ethyl cellulose 5%, gelatin, casein, Carboxy methyl cellulose (CMS) etc. CMC 3% (w/v), Methyl ethyl cellulose 5% (w/v) and gum arabic upto 45% are commonly used.

6.3. Coating materials or the filler materials

Lime, gypsum, dolomite or rock phosphate and charcoal are the filler materials commonly used for pelleting.

6.4. Organic materials for pelleting

1. Dried blood : 15 parts in 85 parts of dolomite
2. Milk powder: 3 parts in 97 parts dolomite
3. Yeast extract : 1 part in 99 parts of dolomite

Since these protect the rhizobia, on the seed they are called as nutrient pellets. The size of the organic materials should pass through 150 mesh sieve.

Biological material: Leaf powder of neem, arappu, notchi and pungam etc.

6.5. Types of coating

6.5.1. Inoculant coating

Rhizobia, AM (Arbuscular Mycorrhizal fungi), and *Bacillus* or *streptomyces* are used.

6.5.2. Protecting coating

Fungicides, pesticides, rodenticides (Mestrenol), methiocarb, endrin (To prevent bird attack).

6.5.3. *Herbicide coating*

Antidote (108 naptithetic anhydride) and absorbant (2,4-D and alachlor).

6.5.4. *Nutrient coating*

Seeds coated with macronutrients such as N, P, K and S and micronutrients such as Mo, Mg, Zn and Mn, increase germination and seedling growth.

6.5.5. *Hydrophilic coating*

Use of starch graft polymers can absorb water upto 1000 times of their original weight of water and coating of $MgCO_3$ increase the movement of air and water. These fine particles act as wick (or) moisture attracting material. So it improves seed soil contact. Coating with peroxides of Zn and Ca increase the O_2 supply to seed and makes it to germinate faster.

7. CONCLUSIONS

Seed management is an inevitable practice for successful crop production. To achieve uniformity and synchrony of plant growth, pre-sowing seed management system must be sequencely followed. Seed fortification with fortifying nutrients and growth regulators are the primary steps of pre-sowing seed treatment. To get an active and uniform growth of seedlings and plants, seed pelleting technique may be helpful. Stress management in the field upon drought can be managed by seed hardening and seed infusion techniques.

PART II

SEED HARDENING

GENESIS AND IMPROVEMENT OF THE CONCEPT OF SEED HARDENING

R.UMARANI[1] AND R.JERLIN[2]

1. INTRODUCTION

Normal events do not invite many questions. A shift from normalcy unleashes search for the causes of the complications and methods to overcome the problems. After all "Necessity is the mother of invention". Drought is the most common calamity that affect farmers severely. Methods had to be developed to combat this calamity.

2. GENESIS OF THE CONCEPT OF SEED HARDENING

Research on methods to tackle drought started as early as 1883 when Will, H. made the first report that repeated soaking and drying of the seeds result in increased drought and frost resistance by the vegetative plant. This age old technology of seed hardening can thus be defined as "training the seeds under lab conditions to withstand and continue productive plant growth even under adverse field conditions of drought, heat and salt".

Intensive research on this technology was pursued by the Russian Scientist P.A.Henckel and his Co-workers whose experiments commenced in 1930's. Henckel published a review on the development of a method for increasing drought resistance and he was the first to refer this technique as "presowing drought hardening" in his review entitled "Physiology of Plants under Drought" published during 1964 in "Annual Review of Plant Physiology". The excerpts of this paper is quoted as follows.

---

1 *Department of Seed Science and Technology, Tamil Nadu Agricultural University, Coimbatore- 641 003.*

2 *Agricultural Research Station, Bhavanisagar – 638 451.*

According to many authors drought resistance is a property which is formed and developed in the process of ontogenesis and is based on the whole preceding phylogeny of the plant. Drought resistant plants are those which, in the process of ontogenesis are able to adopt to the effect of drought and which can normally grow, develop and reproduce under drought conditions, because of a number of properties acquired in the process of evolution under the influence of environmental conditions and natural selection. It is therefore clear that drought resistant property is not unchangeable but is markedly affected by the individual development of the plant. This concept that drought resistance is a property developed during ontogenesis, made possible the development of a method for increasing drought resistance termed as 'pre-sowing drought hardening' (Henckel, 1946; Henckel, 1960).

Pre-sowing hardening is done by soaking seed in water for two days and then air drying. The final water content varies: 45 per cent of the air-dry weight for wheat, 60 per cent for sunflower and 30 per cent for millet. The temperature during hardening is between 15° and 25°C. This confers a high drought-resistance without interfering with germination, growth and yield.

Henckel (1964) enumerated the following as the basis of pre-sowing seed hardening.

2.1. Physiological reorganisations

(i) Significant increase in hydrophilic property of the protoplasmic colloids namely the viscosity and elasticity.
(ii) Quality of the proteins are changed
(iii) Osmotic pressure increases
(iv) These overall changes leads to increase in water holding capacity of plants.

2.2. Metabolic changes

(i) Increased respiration
(ii) High level of synthetic reactions even during drought; leaves of hardened plants have more starch
(iii) Increased phosphorylation activity of their mitochondria
(iv) Higher rate of photosynthesis because of increase in the bound water
(v) Higher organic phosphorus and nucleoproteins.

2.3. Anatomical and morphological Changes

(i) Hardened plants have a more xeromorphic morphology than unhardened ones
(ii) More extensive and denser network of veins and ribs
(iii) Epidermal and stomatal cells are smaller
(iv) Number of stomata per unit leaf area is greater
(v) Foliage area is increased
(vi) Faster recovery from atmospheric drought
(vii) Greater total and absorbing surface in the root system, as well as more number of primary roots
(viii) Leaves of hardened plants have more of starch

Henckel acclaimed that presowing drought hardening is very efficient to enable the plant to adapt itself to soil and atmospheric drought.

3. IMPROVEMENTS IN SEED HARDENING RESEARCH

In the following years, research on seed hardening was conducted specific to each crop and results were published in English enabling wide readability.

3.1. Hydro hardening

May *et al.* (1962) recommended different amounts of water for different species. Best results are claimed for seeds subjected to two or three cycles of wetting and drying, although for some species one cycle is sufficient. For spring wheat, water equivalent to 30% of the grain weight is added at each cycle, with Caesium, 30%, 20% and 15% are added in three successive cycles, and with leucurrum, 40%, 30% and 30%. Pea and vetch seeds are given 75% water in one cycle only, the water being added in small amounts at a time.

As germination and growth proceed the resistance to drying of the embryo decreases. At the same time, the effect of treatment is claimed to be greater if the embryo is more advanced at the time of drying. Therefore, (i) the timing of the initial imbibition period(s) is a compromise between the above two conflicting tendencies. Besides (ii) the temperature of soaking and (iii) drying cycles and (iv) the rate of drying may all be important.

Increase in the resistance of the vegetative plant to drought and to dry wind also increased yields, particularly under conditions of moisture deficiency were claimed by many authors. Workers have also registered failures of drought hardening technology. Many researchers recorded inconsistencies in the responses of different species of cultivar following treatment. This suggests that the responses of plants following wetting and drying cycles at the seed stage are very variable or that the optimal conditions have not always been established. Thus, it seems that for every species and cultivar harvest must be investigated individually under a variety of climatic conditions to determine if presowing treatments are a useful proposition.

The number of physiochemical changes within the cytoplasm suggested by Henckel, were questioned by Bewley and Black (1982). How these changes, which presumably must occur within the cytoplasm of cells of the germinating seed, are able to be maintained in all of the newly formed cells and tissues of the developing hardened plants. Also there has been limited support for the possibility that hardened plants develop a more extensive root system, thus enabling them to survive better under field drought conditions (Carceller *et al.*, 1972; Woodruff, 1969). Therefore, Bewley and Black (1982) concluded that "it is possible that early radicle emergence and seedling establishment on planting in the field following dehydration – rehydration treatments simply gives the better start than non-hardened plants".

The possibility, suggested by Simon and Raja Harun (1972) is related to the possibility of reduction in the solute leakage by making the seeds imbibe water slowly, because during drying the cell membrane ceases its integrity which is re-established during imbibition, thus causing slow imbibition. This could possibly help to minimize solute leakage and therefore store the essential metabolites, to be utilized in effective turn over during germination and seedling growth even under stress conditions, which cannot be ruled out.

Sen and Osborne (1974) observed that when the seed is hydrated and then dehydrated embryos can resume a rate of protein and RNA synthesis that approaches the level at which it was interrupted by the dehydration treatment; they do not show again the low rates of RNA and protein synthesis that characterizes the first period of imbibition. An important feature of pre-hydration treatment is the shortening of the time of DNA replication in the second hydration period. In rye, embryos DNA replication initiated at 9th hour in normal germination. When a prehydration of 9 h was followed by dehydration, DNA replication commenced at once when embryos were next imbibed. It is assumed that the biochemical events that are pre-requisites to the normal onset of DNA replication at germination were achieved during the first hydration period, were stable during dehydration and were operative on rehydration of the first period of hydration is extended into phase of DNA replication, however, subsequent hydration results in cell death (Sen and Osborne, 1974).

Srivastava and Randhawa (1983) conducted hardening studies in *Triticale*. Seed hardening improved the germination accompanied by increase in synthesis of protein which is the pre-requisite to start the chain reactions leading to germination. The increase in protein content in the initial stages of seedling growth after an increase in moisture stress suggests the likelihood of the stimulation of long lived mRNA under moisture stress conditions, after hardening treatments. This increase in protein in response to hardening of seeds appears to be the key metabolic event which needs to be tested further. Proline increase was also visible after seed hardening and was reflected in the over all germination. The study revealed simultaneous increase in protein and proline content of seedlings after hardening under moisture stress conditions. So proline accumulation as reported in the present study does not seem to be due to protein synthesis but possibly due to its *de novo* synthesis. It appears that the two components specific to hardening, protein and proline which from the basis for hardening against moisture stress, have different metabolic path way, independent of their respective synthesis.

3.2. Chemical hardening

Apart from mere water soaking as a method to harden seeds, chemical solutions have also long been considered for increasing crop production under rainfed conditions. Perusal of literature reveal that first report on chemical hardening of seeds was made by J.J.Chinoy, way back in 1947. Wood Stock (1969) reported that germination rate and percentage could be increased in pepper and tomato seeds by soaking in solutions of 2% KNO_3 and 2% KH_2PO_4. From 1970 onwards many reports have been made about benefits of chemical hardening of seeds. The chemicals which are employed in such studies are K, KNO_3, KCl, KH_2PO_4, K_2SO_4, CCC, calcium chloride, sodium chloride, vitamin B6, vitamin C, Ethrel, Resistine, Kinetin, $ZnSO_4$, $Fe_2(SO_4)_3$ and Urea.

In many of the studies K_2HPO_4 and KCl were found to have augmental effects of hardening. $CaCl_2$ and Kinetin were also reported to be effective in few studies. The beneficial effects were attributed to the nutritional aspect (Hegarty *et al.*, 1970). Kinetin was reported to facilitate efficient root system with better extracting power (Karivaratharaju *et al.*, 1973). Increased chlorophyll content was possible after hardening with

CCC which leads to enhanced mobilization of phosphorus (Karivaratharaju, 1978).

In the serial of studies on chemical hardening the consequences *viz.*, seedling growth, vigour, crop yield etc., have alone been reported however, the physiological causes for the improved crop response have not been investigated.

Few seed hardening studies have correlated between the accumulation of free proline and drought tolerance. Sashidar *et al.* (1977) reported calcium chloride hardened plants accumulated more proline under stress and ability of such plants to tolerate drought also seems to be better.

3.3. Biological hardening

As an increased orientation towards organic farming, utility of biological materials as seed hardening substance have also been investigated. Kamalam Joseph and Rajappan Nair (1989) soaked paddy in water, cow's urine (5%), cowdung extract (10%) and nitric acid for 12 h and dried in the shade to original moisture level. The treatment with cowdung extract exhibited a significant superiority.

Devarani and Rangasamy (1998) used goal's urine, cowdung extract, garlic extract, calotropis leaf extract, moringa leaf extract and cow's urine for hardening sorghum seeds. Among them, calotropis leaf extract (2%) and cowdung extract (15%) were found to excel. The reasons were attributed to the physiologically active substances present in the calotropis leaf extract which might have activated the embryo and other associated structures resulting in the absorption of more water due to elasticity of cell wall leading to development of stronger and efficient root system and higher vigour index.

4. FUTURE PERSPECTIVES

Of late, proline has been recognized as an important component of drought resistance. Matysick *et al.* (2002) in his review, reports about the molecular mechanisms of drought tolerance enabled by proline. In many plants under stress, the concentration increases upto 80% of the amino acid pool. Proline may not be just a by-product of stress defence, but a chemically active compound, crucially involved in the physiology of stress protection. All stresses such as cold, heat, salt, drought, UV and heavy metal induce the production of reactive oxygen species and free radicals. Free radicals are highly reactive and rapidly oxidizes amino acids, lipids and DNA. The direct involvement of singlet oxygen in photobleaching of photosynthetic pigments, D1 protein degradation and protein cross linking has been reported.

Free radicals produced by reaction with O_2 react aggressively. Reactive oxygen species cause stress to biological systems. Stress inturn induces the production of reactive oxygen species. Therefore, a mechanism to interrupt such an auto-catalytic process is required. As mentioned before, proline accumulates in huge amount in several plants under stress. This accumulation of proline has been shown to protect plants against damage by reactive oxygen species. Proline has been shown to be an excellent quencher of singlet oxygen, salvaging the plants from the adverse effects caused by the free radicals.

Proline in its pure form is a colourless substance, highly soluble in water unlike all other amino acids. It is therefore hypothesized that seed fortification with proline can serve as an efficient seed hardening technique. Proline rich biological substances or botanicals available at lower cost can be discovered and experimented to evolve an efficient drought management mechanism.

This technology can indeed give the juvenile, tender seedlings the necessary initial vigour to compete with the weeds and establish well even under adverse environmental stress more easily. Seed hardening methodology if optimized for each crop can undoubtedly provide an efficient low cost technology to make the seed more prudent against adversities.

REFERENCES

1. Bewley, J.D. and M.Black. (1982). Environmental control of germination physiology and biochemistry of seeds II Springer-Verlag, Berlin. Heidel berg, New York.
2. Carceller, M.S. and A. Soriano. (1972).*Can J. Bot., 50* : 105-108
3. Chinoy, J.J. (1947). *Nature (London), 164* : 879.
4. Devarani, N. and A.Rangasamy. (1998). Effect of seed hardening on the germination and vigour index in sorghum. *Madras Agric. J., 85(10-12)* : 674-675.
5. Hegarty, T.N. (1970). The possibility of increasing field establishment by seed hardening. *Hort. Res., 10* :59-64.
6. Henckel, P.A. (1960). Plant water relationships in Arid and Semi-Arid conditions. *Proc. Madrid Symp., 16. Paris: Unesco,* : 167-174.
7. Henckel, P.A. (1964). *Ann. Rev. Plant Physiol., 15* : 363-386.

8. Kamalam Joseph and N.Rajappan Nair. (1989). Effect of seed hardening on germination and seedling vigour in paddy. *Seed Research., 17(2)* : 188-190.

9. Karivaratharaju, T.V. (1978). Effect of chemical hardening of seeds in cumbu (*Pennisetum typhoides* Stapt and Hubb) on ^{32}P uptake efficiency and translocation pattern. *Andhra agric. J., 15*: 115-118.

10. Karivaratharaju, T.V., R.Vaithialingam and J.Sakharam Rao - (1973). Effect of water stress and hardening of seeds on certain physiological aspects in *Pennisetum typhoides* Stapt and Hubb. *Madras Agric. J., 60(9 to 12)* : 1266-1972.

11. Khan, A.A. (1977). The physiology and biochemistry of seed dormancy and germination. Elsevier / North-Holland Biomedical Press.

12. Matysik, K., Alia, Bhalu, B. and Mahanty, P. (2002). Molecular mechanisms of quenching of reactive in oxygen species by proline under stress in plants. *Current Science, 82(5)* : 525-532.

13. May, L.H., E.J. Milthrope and F.L. Milthrope. (1962). *Field Crop Abstr., 15* : 93-98.

14. Sen, S. and D.J. Osborne. (1974). *J. Exp. Bot., 205*: 1010-1019.

15. Simon, E.L. and Raja Harun. (1972). *J. Exp. Bot., 23* : 1076-1085.

16. Srivastava, K. and A.K.Randhawa. (1983). Eco physiological exploitation of triticale seeds with presowing treatments to develop hardiness against moisture stress. Acta agronomica academial scientiarum Hungaricae, 32.

17. Will, H.Landwirisch. (1883). *Versuchsstn, 28* : 51-79.

18. Woodruff, D.R. (1969). *Aust. J. Agric. Res., 20* : 13-24.

19. Woodstock, L.N. (1969). Biochemical tests for seed vigour. *Proc. Int. Seed Test. Assoc.,*: 253-263.

PRINCIPLES AND PRACTICES OF SEED HARDENING

P.BALAMURUGAN AND R.RAJASEKARAN

1. INTRODUCTION

High quality seed is the key to successful agriculture. Modern agriculture with its bias for technology and precision, demands that each and every seed should readily germinate and produce a vigorous seedling ensuring high yield. Uniformity of growth and synchrony in development are highly desirable characters for mechanised cultural operations. As such, only seed of high quality - genetically pure and morphologically, pathologically and physiologically sound is needed to increase the productivity.

Seeds are exposed to changing and often adverse environments in the soil for a considerably long period beginning with sowing and ending with emergence. The period of imbibition is extremely sensitive to changes in the environment and slight and sudden changes appear to profoundly affect the seedling emergence. Numerous efforts have been directed, therefore, to decrease the period between sowing and emergence on the assumption that quick germination or emergence will spare the seed, the 'agony' of prolonged exposure to hostile environment during imbibition and seedling establishment and thus improve crop stands. In dryland agriculture, drought resistance of plant is one of the very important factor to get the higher yield. Though this is largely depends on genetic make up of the variety, pre-sowing treatments like 'hardening' also practiced to defy the ill effects of drought on emergence and growth of crop.

Department of Seed Science and Technology, Tamil Nadu Agricultural University, Coimbatore-641 003.

2. SEED HARDENING

Hardening of seeds is one of the methods of pre-sowing treatments which results in modifying the physiological and biochemical nature of seed so as to get the characters that are favourable for drought resistance.

During hardening process, a number of physicochemical changes occur modifying the protoplasmic characters and increasing physiological activity of the embryo and associated structures. This results in the absorption of more water due to increase in the elasticity of cell wall and development of a stronger and efficient root system.

2.1. Physiological Basis

- Greater hydration of colloids
- Higher viscosity and elasticity of protoplasm
- Increase in the physiological activity of the embryo and associated structures
- Increase in bound water content
- Increase in photosynthetic activity
- Increase in hydrophobic colloids and decrease in hydrophilic colloids
- Increase in temperature requirement for protein coagulation
- More intensive respiration
- Lower water deficits
- Increase in water balance of plants
- More efficient root system

2.2. Principle involved

Dry seed
↓
Imbibition
↓
Permits initial process of germination → Dehydration → Prevents radicle emergence
↓
Hardened seed ← Shade and sun drying (drying back to original water content)
↓
Sowing
↓
Remember the germination process where it had stopped
↓
Rapid germination and establishes its seedling quickly
↓
Increased seed yield and production of good quality seed

2.3. Methodology

Seeds are soaked in water and allowed to absorb moisture up to 35 per cent of their weight and kept in swollen condition for about 6 hours at about 25°C. These are then spread out in a thin layer for drying in shade for 2 to 3 days. During this period, the seed gets dried almost to the original weight. This treatment is repeated 3 or 4 times depending upon kind and variety in crops. This treatment is not followed for proteinaceous seeds. Because these seeds are more hydrophilic in nature. It absorbs more water i.e. 4 to 5 folds which lead to internal damages to membrane system. However, preconditioning of seeds between moist gunny bag for shorter initial period, reduce the rate of imbibition and also damage to seeds.

This hardening is done with water or with various chemicals to improve the physiological quality of seeds. Recently botanicals are also used.

2.4. Chemicals used for hardening

2.4.1. *Water*

2.4.2. *Salts*

Sodium chloride, sodium sulphate, potassium nitrate, calcium chloride, ammonium sulphate, potassium chloride, etc.

2.4.3. *Growth regulants*

Gibberellic acid, chlorocholine chloride, kinetin.

2.4.4. *Vitamins*

Vitamin K3, Nicotinic acid, Pantothanic acid

2.4.5. *Plant products*

Garlic extract, coconut water, leaf extracts of *Pongamia pinnata, Albizia amara, Prosopis juliflora.*

2.5. Advantages

- Accelerate rapid germination and growth rate of seedling
- Plants from the treated seeds recover quickly from wilting when compared to plants from untreated seeds
- Flowering is slightly accelerated in treated plants
- Induces resistance to drought and salinity
- Seeds also withstand higher temperature (80-105°C) for prolonged periods (24-48 h) without loss of viability

- By emerging early, seedlings will be able to compete more effectively with weeds
- Treated plants are generally better in growth and yield.

2.6. Limitations

- Leguminous seeds are less responsive to wetting-drying especially when long term soaking durations are employed they result in soaking injury
- Vigour of the seed at the time of treatment would greatly determine the efficacy of the treatment
- Prolonged soaking may show adverse effects because of anaerobic respiration leading to accumulation of toxic metabolites.
- Drying of seed following hydration should be slow and compatible with the degree of hydration. Often the embryo enlarges to such an extent, that drying is injurious.
- Often the chemicals do not give significant additional effects over that obtained with water alone.

3. CONCLUSION

Pre-sowing seed management is an inevitable practice for successful crop production. To achieve uniformity and synchrony of growth of plant under drought or salt stress condition seed hardening can be adopted as a regular practice to boost the yield of rainfed crops.

SEED HARDENING FOR DROUGHT RESISTANCE

S. NANTHAKUMAR[1], T. SARAVANAN[2],
A. BHARATHI[3] AND P. MURUGESAN[4]

1. INTRODUCTION

Drought is very important factor, which affects the Indian cultivator, the water balance of a plant is upset by drought and as a consequence of this, the physiological function responsible for growth and yield are disarranged.

2. AMOSPHERIC DROUGHT AND SOIL DROUGHT

In the former case, high temperature and low humidity during certain critical phases of the plants affect its metabolism. Whereas in the later case, the moisture of soil falls to such a low level that the plant is unable to avoid its water loss (caused by transpiration) from the soil and enters into a state of permanent wilting from which it can not recover its turgor even during the cool hours of light.

Drought resistance of a plant is a very complex phenomenon conditioned by many internal (physiological and genetic) and external (environmental) factors. Among them, the important factor is the capacity to withstand the enduring long spells of wilting with little or no dislocation of the vital functions. Because of the complexity of the problem, so far no method tried has given successful results.

Hence, the need for securing the varieties of crops capable of escaping the injurious effects of drought becomes apparent. Because, nearly 60-80%

1 *Department of Horticulture*
2 *Department of Plant Breeding and Genetics, Adhiparasakthi Agricultural College, G. B. Nagar, Kalavai – 632 506*
3 *Department of Seed Science and Technology, Tamil Nadu Agricultural University, Coimbatore - 641 003*
4 *National Research Centre for Oil palm, Pedavegi – 534 450, AndhraPradesh*

of the cultivated land has to endure season and drought. So we have to seriously think about the means by which we can overcome the effect of drought. For which, pre-sowing seed hardening treatment can be effectively used to mitigate the drought situation during the initial phase of seedling growth and establishment.

3. SEED HARDENING

Pre-sowing hardening of seeds is one of the best method that results in modifying the physiological and biochemical nature of seed so as to get the characters that are favourable for drought resistance. Pre-sowing hardening is the result of extensive physiological reorganisation induced by dehydration process.

It can be done with water and dilute solutions of growth regulating compounds and chemicals to induce early germination, better root and seedling growth and also to enhance the yield potential of the crop varieties.

4. STEPS IN SEED HARDENING

Dry seed

↓

Soaking in water and/or dilute solutions of growth regulator and chemicals for 1-12h at 15-25^{0}C

↓

Shade drying (1 to 24h)

↓

Sun drying (1 to 2 days) to bring back to its original water content or weight

↓

Hardened seed

5. PRINCIPLE

During hardening process, a number of physicochemical changes occur and modifies the protoplasmic characters and increases the physiological activity of the embryo and associated structures. Eventually, due to more absorption of water due to increase in the elasticity of cell and development of a stronger and efficient root system.

Principle involved

Dry seed

↓

Imbibition

↓

Permits initial process of germination (An advancement in germination processes)

↓

Dehydration

↓

Prevent cell and radicle emergence

↓

Shade and sun drying to bring back to its original water content or weight

↓

Stops the germination process

↓

Hardened seed

↓

Sowing

↓

Remember the germination process where it had stopped

↓

Rapid germination and quick establishment of seedling (survival) with available soil moisture

↓

Increased growth and productivity

6. PHYSIOLOGICAL BASIS FOR SEED HARDENING

- Greater hydration of colloids
- Higher viscosity and elasticity of protoplasm
- Increase in bound water content
- Increase in photosynthetic activity
- Increase in hydrophobic colloids and decrease in hypophilic colloids
- Increase in temperature requirement for protein coagulation
- More intensive respiration
- Lower water deficit
- Increase in water balance of plants
- More efficient root system

7. METHOD OF SEED HARDENING

Seeds are soaked in water and allowed to absorb moisture upto 30-35 per cent of their weight and kept in swollen condition for 1-12 h depending upon the crop species at 25° C (lower temperature of 15° C is found more beneficial).

Then, the seeds are spread in a thin layer for shade drying for 1 to 2 days. After shade drying, they are sundried for 1 to 2 days to bring back to the original moisture content or weight. The hardened seeds are used for sowing. The treatment is repeated for more times depending upon the kind and variety of crops. This can also be tried by using various concentrations of number of chemicals. High temperature treatment of seed can also be attempted for inducing resistance to drought.

8. CHEMICALS USED FOR SEED HARDENING

In addition to water, the following chemicals are recommended.

Chemicals

Aqueous solution of sodium chloride, sodium sulphate, potassium nitrate, calcium chloride, ammonium sulphate, potassium chloride, zinc sulphate, potassium dihydrogen phosphate.

Growth regulators	:	Gibberellic acid, CCC, kinetin ascorbic acid, riboflavin, biotin phosphonic acid, ascorbic acid succinic acid.
Vitamins	:	Vit.K_3, nicotinic acid, pantothenic acid, adenine etc.

Plant products : Garlic extract, coconut water, leaf extracts of prosopis, moringa, pungam, neem.

The concentration of this chemical, duration of soaking and seed to solution ratio varies depending upon the crop species.

Table 1. Seed hardening techniques recommended for crops

Crops	Chemicals and concentration	Methodology
1. Pearlmillet	2% Potassium chloride	Dissolve 20 gm of the salt in 1000 ml of water. Soak 1 kg of seed in 650 ml of this solution for 10 hrs. and dry back to original moisture.
2. Sorghum	2% Potassium dihydrogen phosphate	Dissolve 20g of salt in 1000 ml of the solution for 16 hrs. and dry back to original moisture or weight
3. Cotton	2% Potassium chloride	Dissolve 20g of the salt in 1000ml of water. Soak 1kg of seed in 650 ml of solution for 10 hrs and dry back to original moisture or weight.
4. Sunflower	2% Potassium chloride	Dissolve 20g of the salt in 1000 ml of water. Soak 1 kg of seed in 650 ml of solution for 12 hrs. and dry back to original moisture.
5. Pulses	100 ppm Zinc sulphate	Dissolve 100 mg of the salt in 1000 ml of water. Soak 1 kg of seeds in 350 ml for 3 hrs. and dry back to original weight or moisture. Before soaking them in water or solution, precondition the seeds for 1 hr by keeping them in between two moist gunny bags

9. BENEFICIAL EFFECTS OF SEED HARDENING

- Accelerates rapid germination and growth rate of seedling.
- Plants recover much more quickly from wilting than those from untreated plants
- Flowering is slightly accelerated in treated plants
- Induces resistance to salinity as well as drought conditions
- Seeds are able to withstand higher temperature (80-105° C) for prolonged periods (24-46 h.)
- By emerging early, seedling will be able to compete more efficiently with weeds
- Plants are generally better in growth and productivity.

10. RESEARCH HIGHLIGHTS

The Department of Seed Science and Technology conducted network experiments with rainfed sorghum at four locations *viz.,* Coimbatore, Aruppukottai, Kovilpatti and Paiyur during 2000-01. It has been concluded that seed hardening with 2% KH_2PO_4 improved the yield of sorghum in all the centres except Aruppukkottai where seeds hardened with 1% prosopis leaf extract excelled others (Table.2).

Table 2. Influence of seed hardening and pelleting in rainfed sorghum at different locations of Tamil Nadu during September 2000.

Treatment	Grain yield (kg ha^{-1})				
	CBE	PYR	KPT	APK	Pooled Analysis
Control	877	1020	1705	1240	1211
KH_2PO_4 2% hardening	1170	1379	1810	1447	1452
Prosopis 1% hardening	897	1310	1730	1323	1315
DAP 100 g pelleting	860	1257	1780	1332	1308
Prosopis + DAP	897	1329	1850	1530	1402
Prosopis + DAP+ $ZnSO_4$ 100 mg	827	1207	1554	1620	1302
CD(P=0.05)	52.3	48	95	35.5	152

CBE - Coimbatore, PYR-Paiyur, KPT-Kovilpatti, APK-Aruppukkottai

Another experiment conducted at Coimbatore and Aruppukkottai during 2001-02 expressed that seed hardening with 0.1 ppm brassinolide + 2% KH_2PO_4 +1% *Prosopis* and seed coating with DAP+$ZnSO_4$ improved the seed yield by 24.2 per cent in Coimbatore. Whereas in Aruppukkottai, seed hardening with 2% KH_2PO_4 followed by pelleting with DAP @ 100g kg^{-1} of seed recorded an yield of 1950 kg/ha compared to control. The percentage increase was 33.1. The pooled analysis also showed the superior performance of this treatment (Table 3).

Table 3. Effect of seed hardening and pelleting on seed yield in sorghum

Treatment	Coimbatore	Aruppukottai	Pooled analysis
Control	874	1458	1116
2% KH_2PO_4 16 soaking + 5h drying	903	1617	1260
0.1 ppm brassinolide 6 h	980	1735	1358
200 ppm salicylic acid 6 h	882	1370	1126
2% KH_2PO_4 16 h + seed coating with DAP 100 $^{g\ kg-1}$	931	1950	1440
1% Prosopis leaf extract 16 h + DAP 100g + $ZnSO_4$ 100 mg kg^{-1} of seed	1002	1849	1425
Brassinolide + prosopis and seed coating with DAP 100g+$ZnSO_4$ 100 mg kg^{-1} of seed.	947	1420	1184
Brassinolide + KH_2PO_4 and seed coating with DAP 100g+$ZnSO_4$ 100 mg kg^{-1} of seed.	965	1446	1206
Brassinolide + KH_2PO_4 +*Prosopis* + seed coating with DAP 100g+$ZnSO_4$ 100mg kg^{-1} of seed	1086	1558	1322

In finger millet cv CO 13, 0.1 ppm brassinolide exhibited higher yield (Table 4).

Table 4. Influence of seed hardening and pelleting on the productivity of finger millet cv. CO 13

Treatments	Yield (kg ha^{-1})
Control	942
Calicum chloride (0.5%)	1139
Brassinolide (0.1 ppm)	1148
Salicylic acid (200 ppm)	1015
KH_2PO_4 (2%) + DAP (100g) pelleting	1124
Prosopis leaf extract 1% +DAP (100g) and $ZnSO_4$ (100mg) pelleting	1120
Brassinolide (0.1 ppm) and KH_2PO_4 (2%) + DAP (100 g) $ZnSO_4$ (100 mg) pelleting	1131
Brassinolide (0.1 ppm) and prosopis leaf extract (1%) + DAP (100g) and $ZnSO_4$ (100 mg) pelleting	1136
Brassinolide (0.1 ppm), KH_2PO_4 (2%) and prosopis leaf extract (1%) DAP (100 g) and $ZnSO_4$ (100 mg) pelleting.	1142
CD (P=0.05)	86.8

The seeds of blackgram hardened with prosopis (1%) and pelleted with DAP (40g), $ZnSO_4$ (100 mg), $FeSO_4$ (100 mg) and Ammonium molybdate (250 mg) recorded higher field emergence, survival, chlorophyll content, seed number and seed yield compared to $ZnSO_4$ (100 ppm), hardening and control. The seed yield ha^{-1} in the best treatment was 789 kg ha^{-1}, which was 12.2 per cent increase over the control and 4.6 per cent increase over $ZnSO_4$ hardening.

12. STORAGE OF HARDENED AND PELLETED SEED

Hardened seeds store better for atleast 4-6 months with minimum loss in vigour and viability. If hardened seeds are not used immediately for sowing, that can be stored for 4-6 months. One such experiment conducted with blackgram seeds expressed that the reduction in germination, seedling growth and vigour index was minimal in $ZnSO_4$ hardened seed compared to prosopis hardened and pelleted seeds and control (Plate 1;Table 5).

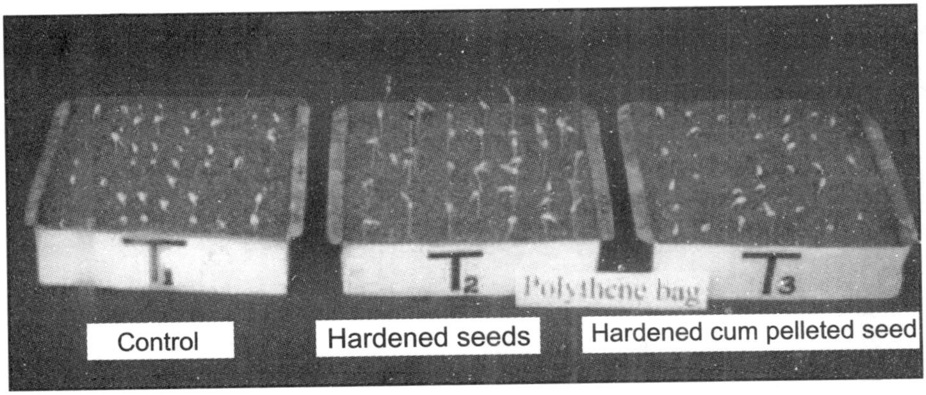

Plate 1. Effect of hardening and pelleting and storage container on germination and seedling growth on 4 DAS at 6 months after storage

Table 5. Effect of seed hardening and pelleting on storability of blackgram seeds

Treatments	Initial			After 4 months storage		
	Germination (%)	Dry matter production	Vigour Index	Germination (%)	Dry matter production	Vigour Index
Control	94	0.25	3365	75	0.23	2422
$ZnSO_4$ hardening	95	0.27	4047	80	0.25	2951
Prosopis hardening and DAP + $ZnSO_4$ + $FeSO_4$ + Ammonium molybdate pelleting	98	0.28	4445	73	0.22	2301

SEED HARDENING WITH CHEMICALS AND BOTANICALS

N. NATARAJAN

1. INTRODUCTION

In general, the low productivity is due to the reason that the crops are grown in marginal and rainfed areas of inadequate soil moisture and poor fertility status. To overcome these adverse conditions like low rainfall, low soil moisture which prevent the germination and establishment of seedlings, seed hardening given as presowing seed treatment is a boon, in dryland agriculture. The effect of presowing hardening on crop plants is explained primarily on changes occurring in the physico chemical properties of the cytoplasm such as the greater hydration of colloids, the higher viscosity and elasticity of the protoplasm, the greater amount of water and the more intense metabolism, variations in xeromorphil structure, more intense transpiration, the ability to retain greater quantity of water associated with a more efficient root system are the results of these changes. This seed hardening accelerates rapid germination and growth rate of seedlings and also it is an important morphological adaptation in plants that helps in drought resistance without loosing productivity.

For the purpose of presowing treatment, number of chemicals such as KCl, KH_2PO_4, succinic acid, ascorbic acid etc., growth regulators like NAA, IAA, IBA, GA, CCC etc., and major plant nutrients like nitrogen, phosphate in the form of urea, ammonium sulphate, potassium, hydrogen orthophosphate, minor nutrients like $ZnSO_4$, $MnSO_4$ etc., and botanicals such as prosopis leaf extract, pungam leaf extract, arappu leaf extract etc., are used.

Department of Seed Science and Technology, Tamil Nadu Agricultural University, Coimbatore – 641 003.

2. METHODOLOGY

The methodology is slightly varied depending upon the seed type. Hydration can be accompanied by spraying the seed with fixed quantity of water, dipping the seed in water or by using the moist gunny bags.

First the seeds should be preconditioned by soaking in water. Then the seeds should be soaked in hardening solution at recommended volume for a particular duration. This volume of solution and duration of soaking varied depending on the type of seed. Then the seeds should be air dried in ambient temperature to get the original moisture content of the seed.

2.1. Hardening with chemicals

Presowing seed hardening with growth regulators influenced the yield components of finger millet. Seeds of one kg soaked in one litre solution of Na_2HPO_4 100 ppm for 12 h followed by 2 h drying at room temperature before sowing recorded highest grain yield, harvest index, the highest uptake of nutrients in the field experiment. This was closely followed by 100 ppm of KH_2PO_4 treatment (Table 1) (Maitra *et al.*, 1997).

Table 1. Effect of seed hardening on crop productivity and nutrient uptake by finger millet

Treatments	1000 grain wt (g)	Grain yield (kg/ha)	Harvest index (%)	Nutrient uptake (kg/ha)		
				N	P	K
Control	2.78	856	34.1	16.5	7.8	25.9
Water	2.91	1037	32.0	21.0	10.5	34.1
$CaCl_2$ 0.25%	3.00	1139	33.0	22.8	11.2	36.1
KH_2PO_4 100 ppm	3.03	1259	32.8	25.5	13.2	40.2
Na_2HPO_4 100 ppm	3.05	1331	33.5	26.9	13.8	41.7
CD (P=0.05)	0.13	104	NS	2.3	1.8	3.5

(Maitra *et al.*, 1997)

Seeds of five sweet corn cultivars were hardened with water by soaking it for 16 hrs and drying back and seeds of one carrot cultivar was given three cycles of soaking and drying back in water or in solutions of vitamin B_6 or K_2HPO_4 increased field emergence. Reduction in germination of carrot seeds due to low temperature could be counteracted by seed hardening (Table 2).

In carrot, this increase in field emergence is suggested that the carrot embryo may be after ripened when the seed is hardened in water (Hegarty, 1970).

Table 2. Percentage of field emergence of the hardened seeds and control of carrot and sweet corn

Sweet corn			Carrot (chantenay I)	
Cultivars	**Untreated**	**Hardened**	**Treatments**	**Field emergence**
Northern Belle	41.3	46.7	Untreated	58.0
Prima`	28.0	51.3	Water hardened	65.0
Seneca Beauty	68.0	70.7	Vitamine B6	69.8
Seneca Daybreak	73.3	66.7	K_2HPO_4	72.5
Sugar Daddy	20.0	55.3		

The seeds of sorghum, pigeonpea, cowpea and groundnut were soaked for 12, 4, 4, 6 hours respectively in calcium chloride 0.4% and CCC 0.2%, registered maximum germination percentage, vigour index, root and shoot length (Table 3) (Rangaswamy *et al.*, 1993).

Table 3. Effect of seed hardening on physiological parameters of crops

Chemicals	Sorghum		Groundnut		Pigeonpea		Cowpea	
Concentration	Germination (%)	Vigour index	Germination (%)	Vigour index	Germination (%)	Vigour index	Germination (%)	Vigour index
KH_2PO_4 10%	79.5	936	83.3	1756	89.5	3969	86.80	3931
NaCl 0.6%	88.8	1370	83.3	2333	91.0	4465	90.5	4814
$CaCl_2$ 0.4%	97.3	1668	94.8	2769	97.8	5007	95.0	5246
$CaCl_2$ 0.6%	90.5	1496	91.5	2603	94.3	4677	93.0	4922
Succinic acid 0.04%	80.5	966	72.3	1517	84.3	3780	83.8	4074
CCC 0.2%	97.3	1675	95.5	2791	97.6	5001	97.8	5359
CCC 0.3%	90.8	1424	91.8	2603	91.5	4630	91.8	4966
CD	2.26	40.8	1.88	63.32	1.49	89.31	1.46	86.06

Seeds of moong soaked in solutions of ascorbic acid, succinic acid, cycocel for 4 h and then shade dried were significantly increased in yield as compared to control and soaking in water. Among this, soaking seeds in 0.002% succinic acid recorded highest yield closely followed by 0.004 % ascorbic acid (Table 4) (Bhatnagar *et al.*, 1975)

Seed hardening of ragi seeds with chemicals like calcium chloride, ascorbic acid and benzyl adenine recorded superiority over the control in high ESP (29.6) red sandy loam soil. They recorded better germination percentage and more vigourous seedling growth under high ESP as against

Table 4. Effect of hardening with chemicals on development and yield of moong var T.44

Treatments	No. of pods/plant	1000 grain weight (g)	Yield (q/ha)
Unsoaked	18	28.16	5.54
Soaked in water	18.5	28.00	5.54
Ascorbic acid 0.002%	24.4	28.87	7.78
Ascorbic acid 0.004%	24.5	28.83	8.52
Succinic acid 0.002%	27.4	30.37	9.01
Succinic acid 0.004%	22.1	28.83	7.70
Cycocel 0.4%	25.3	28.00	5.83
Cycocel 0.8%	26.1	28.33	6.67
CD (P=0.05)	3.00	1.08	0.28

untreated seeds (Viswanath *et al.*, 1972). Here seeds were soaked for 18 hr using a solution to seed ratio of 1:0.5 (w/v). Then the seeds were shade dried for 48 h (Table 5).

Table 5. Effect of seed hardening on ragi seeds in high ESP (29.6) soil

Treatments	Germination (%)		Root length (cm)		Shoot length (cm)	
	Normal soil	High ESP soil	Normal soil	High ESP soil	Normal soil	High ESP soil
Control	85	76.25	64.65	60.65	17.90	12.68
Water	100	90.00	83.25	77.80	25.50	22.90
Calcium chloride (1%)	98.75	97.50	85.00	84.00	25.95	25.40
Ascorbic acid (50 ppm)	100.00	98.75	85.45	90.95	25.10	26.60
Benzyl adenine (20 ppm)	97.50	96.00	53.10	56.25	20.05	20.45

2.2. Hardening with growth regulators

Cumbu seeds of HB3 variety were subjected to presowing hardening using kinetin (5 ppm), CCC (5 ppm), Ethrel (5 ppm) Resestine (10 ppm) for

18 h recorded an increased drought resistance by promoting deep and extensive root system, although other biochemical influences are involved (Karivaratharaju *et al.*, 1973)

Seed hardening of ragi seeds with CCC 100 ppm, succinic acid 0.15% and sodium chloride 1% with the soaking duration of 12 h and then air drying recorded better results in improving the germination per cent, root:shoot ratio and vigour index (Shanmugasundaram and Kannaiyan, 1989).

2.3. Hardening with nutrients

2.3.1. Potash as hardening material

Under rainfed conditions, the wheat variety kalyansona hardened with potash with 2% and 3% concentrations found increased yields over the other concentrations and control due to an increase in ear length, fertile spikelets per ear, grain number per ear and grain weight per ear. Seeds were soaked in 60% of seed weight of solution for 12 hrs. Then air dried for original moisture content (Table 6) (Subramaniyam and Misra, 1980)

Table 6. Effect of presowing seed treatment with potash on yield components of wheat

Treatments	Grain wt/ear (g)	No. of ears/ha (in thousand)	Grain yield (kg/ha)
Control	1.06	1230.33	1092.83
Water	1.22	1755.16	1366.83
K 0.5%	1.27	1716.33	1449.50
K 1%	1.29	1783.33	1550.00
K 2%	1.42	1910.16	1666.66
K 3%	1.50	2039.50	1754.50
K 4%	1.18	1556.50	1383.00
K 5%	1.16	1474.66	1303.33
K 6%	1.09	1203.00	1086.16
K 7%	1.04	1178.50	1046.16

Different K salts like KCl, KH_2PO_4 and K_2SO_4 are used for the presowing seed hardening of wheat seeds. Among these under field conditions, the germination, root length, shoot length, vigour index was significantly higher in KH_2PO_4 1.0% over KCl, K_2SO_4 and control with 18 h of soaking duration (Paul and Choudhury, 1993).

2.3.2. *Phosphate as hardening material*

Another experiment conducted by Mehrotora *et al.* (1968) reported that wheat seeds hardened with higher concentration of potassic phosphate solution of 7.5-10% with the soaking duration of 12 to 18 h gave better germination and yield parameters (Table 7).

Table 7. Effect of soaking seeds in KH_2PO_4 solution on wheat

Treatments	Tillers/plant no. of grains/ear	No. of grains/ ear	1000 grain wt (g)	Yield of grain (q/ha)
Unsoaked	5.3	36	38.88	28.54
Soaked in water	5.3	40	39.40	30.86
KH_2PO_4 2.5%	5.5	39	39.34	30.59
KH_2PO_4 5.0%	5.6	39	38.73	30.13
KH_2PO_4 7.5%	6.3	43	39.91	31.24
KH_2PO_4 10%	6.2	42	38.24	28.61
CD 5%	NS	NS	NS	NS

Analysis on phosphorus transport in individual organs such as leaves, stem showed that seeds hardened with CCC, and resistine treatment greatly influenced the translocation efficiency.Ethrel and resistine hardening caused markedly increased uptake of phosphorous (Karivaratharaju, *et al.,* 1973).

2.3.3. *Chemical, Hormones and Nutrients*

Wheat seeds were hardened with water, IAA, NAA, Ascorbic acid, Succinic acid, $ZnSO_4$, $MnSO_4$, Urea, KH_2PO_4 for 6 h in the ratio of 1:2.5 seed: solution. Then the seeds were air dried for their original moisture content. Among these treatments IAA 20 ppm, NAA 20 ppm, 40 ppm, ascorbic acid 40 ppm, urea 2% and KH_2PO_4 2% were significantly increased the grain yield and nutrient uptake of wheat than the water soaking (Table 8).

2.4. Hardening with botanicals and animal waste

Paddy seeds hardened with cow's urine, cowdung extract and N/10 nitric acid for 12 h and dried in shade found to be better than the control. The treatment with cow dung extract 10% exhibited a significantly higher superiority in the earliness in germination, this being followed by water soaking (Kamalan Joseph and Rajappan Nair, 1989). The beneficial effect of cow dung extract may be due to the contents of physiologically active substances and since it is an easily available one (Table 9).

Table 8. Effect of hardening on wheat seeds

Treatments	Grain yield (kg/plot)	Nutrient uptake (g / plot)		
		N	P	K
Unsoaked	0.700	18.83	2	2.48
Water	0.830	19.30	2.48	2.98
IAA 40 ppm	1.083	24.43	3.92	4.46
NAA 20 ppm	1.200	30.06	3.70	4.49
NAA 40 ppm	1.206	27.20	3.50	5.14
Ascorbic acid 40 ppm	1.200	27.86	4.21	5.13
Urea 2%	1.083	24.43	3.37	3.32
KH_2PO_4 2%	1.200	27.40	3.43	4.91
CD (P=0.05)				11.07

Table 9. Effect of treatments on germination percentage, vigour index in rice

Treatments	Germination (%)	Vigour index
Dry (control)	70.12	383
Water	75.99	526
Cow's urine 5%	73.67	465
Cow's urine 10%	76.4	413
Cow dung extract 10%	77.57	693
N/10 nitric acid	66.60	377
CD (P=0.05)	-	101.44

Presowing hardening of sorghum seeds with water, goat urine, cow dung extract, garlic extract, calotropis leaf extract, cow urine, moringa leaf extract, KH_2PO_4 were recorded higher germination and vigour index over control. One kg of seeds was soaked in 600 ml of hardening solution and then air dried. Among these treatments, cow dung extract 15% and calotropis leaf extract 2% recorded highest germination percent and vigour index. The physiologically active substance present in the calotropis leaf extract might have activated the embryo and other associated structure which resulted in the absorption of more water due to elasticity of cell wall and development of strong and efficient root systems which would have ultimately resulted in higher vigour index (Table 10) (Devarani *et al.,* 1998).

Table 10. Germination per centage and vigour index of the different seed hardening treatment on sorghum (Devarani *et al.,* 1998)

Treatment	Germination (%)	Vigour Index
Control	63	587.3
Water	65	697.2
Goat urine whole	46	468.1
Goat urine 15%	66	708.4
Goat urine 10%	62	503.4
Goat urine 5%	61	582.6
Cowdung extract 15%	76	895.0
Cowdung extract 10%	73	810.4
Cowdung extract 5%	75	843.7
Garlic extract 5%	70	752.6
Garlic extract 2%	70	775.1
Garlic extract 1%	66	653.7
Calotropis leaf extract 2%	76	969.1
Calotropis leaf extract 1%	72	810.1
Cow urine whole	24	386.5
Cow urine 15%	56	525.0
Cow urine 10%	51	470.5
Cow urine 5%	58	548.2
Morinda leaf extract 2%	70	765.0
Morinda leaf extract 1%	65	635.6
KH_2PO_4 2%	70	761.6
CD (P=0.05)	2.8	37.3

Seeds of ragi hardened with pungam leaf extract at 1% concentration recorded higher germination per cent, root length, shoot length, vigour index and field emergence compared to control (Table 11) (Punithavathi, 1997).

Table 11. Effect of seed hardening on seed quality of ragi

Treatments	Germination (%)	Vigour index	Field emergence (%)
Control	82.0	1074	62.0
Water	87.0	1149	71.0
KCl 1%	86.0	1240	71.0
$CaCl_2$ 1%	82.0	1125	71.0
KCl + $CaCl_2$ 0.5%	82.0	1137	72.5
Prosopis leaf extract 1%	82.0	1155	71.0
Pungam leaf extract 1%	86.0	1259	74.0

3. CONCLUSION

Improvement of seed quality by seed hardening is a simple, easy and inexpensive approach to enhance the seed performance and agricultural productivity especially in the dry lands and marginal lands of resource poor farmers. This will maintain higher water balance in the tissue and enhanced photosynthetic activity which might have ultimately contributed to an increase in the drought tolerance capacity and increased plant vigour.

REFERENCES

1. Bhatnagar. O.K., J.P.Srivastana and J.K. Garg. (1975). Effect of seed soaking in ascorbic acid, succinic acid and cycocel on the yield of moong. *Indian J.Agron., 20(3)* : 240 - 241.
2. Devarani, N. and Rangasamy. (1998). Effect of seed hardening on the germination and vigour index in sorghum. *Madras. Agric. J.,85 (10-12)* : 674 - 675.
3. Hegarty. T.W. (1970). The possibility of increasing field establishment by seed hardening. *Hort. Res., 10* : 59 -64.
4. Kamalam Joseph and N.Rajappan Nair. (1989). Effect of seed hardening on germination and seedling vigour in paddy. *Seed Res.,17 (2)* : 188 - 190.
5. Karivaratharaju, T.V. (1973). Effect of chemical hardening of seeds in cumbu (*Pennisetum typhoides* stape and Hubb) on 32p uptake efficiency and translocation pattern. *Andhra Agric. J.,*: 115 - 118.
6. Karivaratharaju,. T.V., R.Vaithialingam and J. Sakharam Rao. (1973). Effect of water stress and hardening of seeds on certain physiological aspects in *Pennisetum typhoides* stapt and Hubb. *Madras. Agric. J., 60 (9 to 12)* : 1266 - 1272.

7. Maitra, S. Sounda. G., D.C. Ghosh. and P.K. Jana. (1997). Effect of seed treatment on finger millet (*Eleusine coracana*) Varieties in rainfed upland. *Indian Journal of Agricultural Sciences, 67(10)* :678 - 80.
8. Mehrotra. O.N. J.P.Srivastana and Shina Nath. (1968). Germination, Growth and yield of wheat as influenced by presowing seed treatment in phosphate solution. *Indian Journal of Agronomy, XIII* ; 51 - 56.
9. Padole. V.R. (1979). Effect of presoaking seed treatment of wheat seed with chemicals and harmones on yield and uptake of NPK. *Journal of MAU, 4(1)*: 85 - 88.
10. Paul, S.R. and A.K.Chowdhury. (1993). Effect of seed hardening with potassium salts at different concentrations and soaking durations on germination and seedling vigour of wheat. *Ann. Agric. Res., 14 (3)* : 357 - 359.
11. Punithavathi, N. 1997. Seed hardening and pelleting studies in Ragi cv. CO 13. M.Sc. (Ag.) Thesis, Tamil Nadu Agricultural University, Coimbatore.
12. Rangaswamy. A, S. Purushothaman, and P. Devasenapathy. (1993). Seed hardening in relation to seedling quality characters of crops. .
13. Shanmugasundaram. V.S. and M. Kannaiyan. (1989). Effect of concentration of seed hardening chemicals on physiological characters of pearl millet. (*Pennisetum typhoides stapf and Hubb*). *J. Agronomy and Crop science, 163*: 174 - 176.
14. Subramanyam. M.V.R. and Misra. N.M. (1998). Effect of potash seed treatment on yield and yield components of high yielding wheats under rainfed conditions. *Seeds and Farms, 6(10)*: 21 - 24.
15. Viswanath. P.P., K.S. Krishana sartry and B.V. Venkata rao. (1972). Effect of seed treatment on germination and growth of *Eleusine coracana* Gaertn in soils with high exchangeable sodium. *Mys. J.Agric Sci., 6* : 24 - 28.

SEED HARDENING RESEARCH AROUND THE WORLD FOR DRYLAND AND GARDEN LAND ECO SYSTEM

P. SRIMATHI[1] AND K. NATARAJAN[2]

1. INTRODUCTION

Rapid and uniform field emergence are the essential prerequi sites to increase the yield and quality, ultimately profits in annual crops. Uniformity and percentage of seedling emergence of direct seeded crops have a direct impact on final yield and quality and are decided by the quality of seed used for sowing and the environmental conditions prevailing during seedling emergence.

Rainfed agriculture is one of the adverse agroecological situations where the success of the crop depends on proper management practices to resist against existing drought. Water balance of a plant is upset by drought and as a consequence of this the physiological functions are affected resulting in poor growth and yield. The urgent need for securing varieties of crops capable of escaping this deleterious effect of drought becomes increasingly important and one has to seriously try to defy drought. Pandey and Sinha (1999) defined drought resistance as the capacity of the plant to develop normally in dry habitat yielding maximum crop or with endureness of drought without injuring the crop.

2. SEED HARDENING

Presowing seed hardening with water and solutions of growth regulating compounds to induce early germination, better root and seedling growth and increased yield has been employed by many workers. Seed hardening has been investigated by those concerned with problem of seed

1 *Department of Tree Breeding, Forest College and Research Institute, Mettupalayam – 641 301.*
2 *Department of Plant Breeding and Genetics, Adhiparasakthi Agricultural College, G.B.Nagar, Kalavai – 632 506.*

germination and seedling emergence in the field. In ancient time placement of seeds of vegetables in lukeworm water overnight had been practiced to increase the germination velocity. In 17th century pre sowing seed treatment with salt solution was the common practice among the Russian farmers. It is also reported that ancient Greek farmers soaked cucumber seeds in water or milk or honey before sowing to increase the germination rate and emergence. The Russian work on seed hardening reviewed by May *et al.*(1962) was concerned with inducing a physiological resistance to water stress in growing plants by exposing the seeds to water stress before they were sown i.e. by soaking and then drying back the seeds one or more times.

Henkel (1964) opined that wetting and drying treatment could create drought resistance in plants and promoted the yield. Heydecker (1972) also advocated that seed invigouration treatment could promote vigour, viability and field performance of crop seeds. Abdul Hafeez and Hudsan (1967), Evenari (1964) and Hegartz (1970) also reported beneficial effects of presowing hardening or wetting and drying treatments for invigourating seeds that were just at the threshold of decline in vigour and viability either for prolonging shelf life or for immediate sowing to increase yield.

Researchers used water as a hydrating agent and reported that presowing seed hardening (wetting and drying) modified the seed and seedling quality characters and thereby enhanced the yield. Hence seed hardening is one of the physiological presowing seed management practice given to seeds to resist drought and to boost up the yield and is being in practice from time immemorial owing to the better performance among the agriculturists.

3. RESEARCH HIGHLIGHTS ON THE INFLUENCE OF SEED HARDENING WITH CROP PRODUCTIVITY

3.1. Paddy

In paddy Kamalam and Nair (1989) recommended soaking of seeds in 10 per cent cowdung extract to enhance the seedling quality characteristics that could boost up yield. Ros *et al.* (2000) in paddy also reported that soaking the seeds in potassium dihydrogen phosphate and sodium dihydrogen phosphate improved the seedling emergence and yield attributes.

3.2. Wheat

In wheat, Mehrotra *et al.* (1968) reported that the seeds hardened with 7.5 per cent monobasic potassium phosphate recovered quickly from wilt on rewatering. Padole (1979) of Akola reported that wheat seed soaked in 40 ppm of IAA, ascorbic acid, succinic acid, 0.04 % of $ZnSO_4$, $MnSO_4$ and 2% of urea and KH_2PO_4 in the seed to solution ratio of 1:2.5 for 6 h and dried under shade improved grain and straw yield of kalyansona. Subrahmanyam and Misra (1980) of Varanasi reported that wheat seed soaking in potash as KCl at 2% and 3% concentration (Treatment : 500 gm seed soaked in 300 ml (1:0.6 ratio) for 12 h and dried back to original moisture content) increased the yield due to increase in ear length, fertile spikelets per ear, grain number per ear and grain weight per ear under rainfed conditions. While, Asgarjallis (1986) reported that seed hardening with 0.5 per cent $CaCl_2$ increased the grain yield, 100 seed weight and number of grains per ear in wheat. Paul and Chowdhury (1993) of Assam reported that seeds treated with 1% KH_2PO_4 for 18 h at 1: 0:6 ratio improved the seed and seedling vigour of treated seeds.

3.3. Sorghum

In sorghum, seed hardening with potassium dihydrogen phosphate (KH_2PO_4) was recommended by Balasubramanian (1976), Anon (1999a) and Periyathambi (1980). But the concentration differed with the authors where 10 per cent was recommended by Balasubramanian (1976) while 2 per cent was recommended by Anon (1999a). However, Alexander (1985) and Vanangamudi and Kulanthaivelu (1989) recommended seed soaking in 0.5 per cent KH_2PO_4 for the increased seedling emergence and vigour. But Rangasamy (1986) recommended seed soaking in CCC and $CaCl_2$ for the increased efficacy of hardening treatment that reflected in increasing the phenotypic observations pertaining to increased seed yield. Pawar and Kadam (1981) reported that soaking the seeds (12 h) in chemical solutions (NaCl 3%, $MgSO_4$ 1%) improved the uptake of N, P and K from soil and boosted the yield of wheat (under rainfed conditions).

Devarani and Rangasamy (1998a) evaluated different low cost seed hardening agents *viz.*, cows urine (5-100 per cent), goats urine (5-100 per cent), cow dung (5-15 per cent), calotropis leaf (1-2 per cent), garlic (1-5 per cent), moringa leaf (1-2 per cent) and KH_2PO_4 (2 per cent) and reported that cows and goats urine could not be recommended for seed hardening as they reduce the seedling quality characters even immediately after the treatment imposement. However Devarani and Rangasamy (1998b) also reported that seed hardening with 2 per cent calotropis leaf extract improved the growth attributes and resulted in higher grain yield compared to other hardening agents.

The researchers have also evaluated the synergistic effect of beneficial presowing seed management practices with pelleting on hardening treatment. Jegathambal (1996) in sorghum reported that, seed hardening with *prosopis* + pungam leaf extract mixed at the ratio of 1:1 for 16 h in 1:0.6 seed to solution ratio and pelleted with pungam leaf powder @ 200-300 g kg^{-1} of seed improved the seed yield under the dryland conditions of Coimbatore and Aruppukottai of Tamil Nadu.

3.4. Pearlmillet

In pearlmillet, Sandhu and Hussain (1961) recommended 250 -500 ppm of gibberellic acid (GA_3) for seed hardening, while CCC was recommended by Ramachandran (1972), and Kannaiyan (1987). Karivaradaraju (1987) reported that in cumbu, seed hardening with kinetin, chlorocholine and ethrel improved dry weight of seedlings and its protein per cent which was expressed to be due to the stronger root system with more absorption of nutrients under dryland conditions. Karivaratharaju *et al.* (1973) reported that cumbu seeds soaked in kinetin, CCC, Ethrel and resistive for 18 h in equal volume of solution improved the drought resistance in plants by promoting deep and extensive root system through modification of biochemical system of seed. But Manoharan and Mathur (1966) recommended soaking seeds in 0.09 per cent succinic acid for 3 h to promote the germination and plant growth besides increasing survival of plants. Ramachandran and Sakharamrao (1975) also reported that seed hardening with cycocel improved the water retaining capacity of the plants in desiccation and increased the diffusion pressure deficit of cells compared to ethrel, kinetin and resistine in pearlmillet.

Shanmugasundaram and Kannaiyan (1989) of Coimbatore reported that seed hardening with (12 h) different chemicals *viz.*, CCC, (100 ppm), succinic acid (0.15%) and sodium chloride (1%) improved the germination percentage, root and shoot ratio and vigour index of pearl millet.

3.5. Maize

Hardening of seed improved the field germination of corn seeds. Asthana and Srivastava (1978) also reported that presowing soaking treatment with ascorbic acid and salicylic acid either individually or in combination, hastened the flowering and ear emergence under drought condition.

3.6. Ragi

In Ragi, Viswanath *et al.* (1972) reported that soaking of seeds in $CaCl_2$ (1%) and ascorbic acid (50 ppm) for 18 h in seed to solution ratio of 1:0.5

(v/v) and dried back to original moisture content for 48 h under shade germinated better in soils with higher exchangeable sodium percentage (30) recording higher value for seedling growth characters. Narayanan (1981) recommended seed soaking and drying in 20 per cent Dibasic potassium phosphate (KH_2PO_4) for improved grain and straw yield while Karivaratharaju and Ramakrishnan (1985) recommended soaking in 2.5 per cent calcium chloride ($CaCl_2$) increased drymatter production and yield. Palanisamy and Punithavathi (1998) recommended potassium chloride KCl, $CaCl_2$ and aqueous leaf extract of one per cent *Prosopis* and pungam for 12 h to improve the seed quality characters and they found these botanical treatments were on par with seed treatment with inorganic chemicals like KCl and $CaCl_2$. While Kalarani *et al.* (2001) reported that hardening of seeds with KCl one per cent + $CaCl_2$ one per cent for 6 h increased grain yield in ragi under late sown condition at Coimbatore. But Punithavathi (1997) advocated the seed to solution ratio of 1:0.7 for better results. Maitre *et al.* (1997) reported that presowing hardening treatment (12 h in equal volume ie., 1 kg in 1 litre of solution) with 100 ppm solution of Na_2HPO_4 or KH_2PO_4 were useful for further improvement in growth and productivity of 2 varieties of *Elucine coracana*. Palanisamy and Punithavathi (1998) also reported that in fingermillet, seed hardening with 1 per cent KCl in the seed to solution ratio of 1:0.7 for 12 h and pelleting with pungam leaf powder improved the yield potential of the crop at harvest. Vijaya and Ponnusamy (1998) also reported that blackgram seeds fortified with $ZnSO_4$ + $MnSO_4$ + Sodium molybdate and pelleted with DAP (120 g kg^{-1} of seed) registered higher yield of 1.37 kg $plot^{-1}$. However Vigneshwari (2002) reported that seeds hardened with brassionoloid 0.1 ppm had positive influence on drought resistance which was evident from its increased physiological and biochemical aspects of seed that resulted in increased yield of seed. She also reported that seed hardening followed by DAP pelleting was harmful for improving the yield of ragi under rainfed conditions.

3.7. Groundnut

In groundnut, Arjunan and Srinivasan (1989) recommended 1 per cent $CaCl_2$ for seed hardening, for increased germination, yield and yield attributes. Sandhya and Gopalsingh (1994) reported that soaking groundnut seeds in $Ca(OH)_2$ 0.15% and $CaCl_2$ 1.5% in double the volume of solution for 8 h and drying back to original moisture content improved the emergence index in groundnut. Similar hardening combined with pre sprouting is recommended by Anon (1999a) in groundnut. Pandey and

Sinha (1999) reported that drought resistance in plants can be induced by presowing soaking of seeds with 0.25 per cent solution of calcium chloride for 20 h and drying back to original moisture content. They also recommended soaking the seeds in a solution of boron for drought tolerance.

3.8. Sunflower

In sunflower soaking in GA_3 and vitamin C were recommended by Karivaratharaju *et al.* (1974) for improved seed set under dry conditions. Kathiresan *et al.* (1984) reported that seed hardening with potassium nitrate (KNO_3) 500 ppm was better to alleviate water stress on germinability than other chemicals *viz.*, ascorbic acid, $CaCl_2$, KH_2PO_4, NaCl, succinic acid and zinc sulphate ($ZnSO_4$). Anon (1998) also reported that moist hydration for 24 h and drying followed by slurry treatment with thiram improved the seed yield of sunflower.

3.9. Cotton

Jhorer *et al.* (1982) reported that soaking of cotton seeds in 100 ppm gibberellic acid for 24 h improved the plumule length and root length in addition to enhancement in seed germination. Pothiraj and Sankaran (1984) also reported that seed hardening with 1000 ppm CCC for 6 h improved the vigour of the seedlings and that was possible as it maintained increased turgidity and efficient root system under the dryland condition. Nirmala *et al.* (1994) soaked cotton seeds in 2% KCl for 8 h and dried for 5 h and found that germination per cent was improved at lower moisture content of 50 to 75%. While Rathinavel and Dharmalingam (1999a) recommended seed hardening with botanical leaf extracts of one per cent pungam (*Pongamia pinnata*) and 0.5 per cent *prosopis* which was expressed to be better than KCl hardening to augment the productivity of rainfed cotton. Rathinavel (1997) in cotton also reported the efficacy of hardening cum pelleting treatments in enhancing the productivity and seed quality characters which could be attributed to the additive effect of these two pre-sowing seed treatment for the betterment of productivity as detailed elsewhere and the performance had been found to be better with dryland.

3.10. Pulses

The pulses are most susceptible for imbibitional injury due to the delicate seed coat character, which is the genetic hinderance in seed management practices through hardening. For these seeds also several scientists recommended seed hardening. Bhatnagar *et al.* (1975) reported that soaking of seeds in succinic acid (0.002%) and drying back under shade improved the grain yield under 4.5 irrigations given to crop as and when needed under 2 year trial of arid season.

Dharmalingam and Basu (1989) in greengram, Kohli *et al.* (1992) in gardenpea, Gopalsingh (1995) in soybean, Sahulhameed *et al.* (1988) in rice fallow pulses and Vijaya (1996) in blackgram and cowpea reported that hydropriming or water soaking of seed in limited water improved the germination percentage by quickening the germination process and enhanced the seed yield by improving the yield attributing characters. Vijayakumar (1982) in blackgram, Rajendran (1982) in redgram, Basha (1982) in greengram and Vanangamudi and Karivaratharaju (1986) in blackgram and greengram recommended fortification of seeds with succinic acid (20 ppm), cobalt nitrate (1 per cent), ascorbic acid (20 ppm), sodium molybdate (100 ppm), potassium sulphate (100 ppm), manganese sulphate (100 ppm), zinc sulphate (100 ppm), potassium chloride (100 ppm), and IBA (200 ppm) using $1/3^{rd}$ volume of seed to solution ratio and drying back for increasing the initial quality yield and the maintenance of the same at seed storage. However Singh and Kumar (1992) with chickpea reported that 24 h of water soaking and drying back to original moisture content activated the enzyme activity and enhanced the seed quality characters. They also reported that priming of seeds with zinc and copper enhanced the protein synthesis and enzyme activity of amylase and peroxidase.

Not only mear water, inorganic chemicals were also elucidated to improve the seed quality characters on seed hardening by different researchers. Subbian *et al.* (1987) in greengram reported that $MnSO_4$ at 10 ppm improved the seed quality characters. While Sabir- Ahamed (1999) recommended the hardening of seed with 50 ppm of Na_2HPO_4 and 20 ppm of succinic acid improved seed yield in greengram. In blackgram Vijaykumar (1982) recommended seed hardening with sodium molybdate and $ZnSO_4$ for enhancing initial seed quality characters while Renugadevi *et al.* (2001a) reported that seed hardening with botanicals enhanced the seed quality in a better way than chemical hardening with $ZnSO_4$ 100 ppm. Similarly sodium molybdate (Iliev *et al.*, 1989) for *Vicia fabae*, ammonium molybdate 500 or 100 ppm (Xia *et al.*, 1989) for redgram, one per cent sodium molybdate (Khan and Hedge, 1989), and one per cent $FeSO_4$ (Renugadevi *et al.*, 2001b) for pigeon pea were identified as the best hardening agents by the workers who elaborated that yield improvement was by improving the yield parameters and nodulation from seedling to harvesting stage. Rangasamy *et al.* (1993) scored $CaCl_2$ 0.4% and CCC 0.2% as the best chemicals for hardening sorghum, pigeon pea groundnut and cowpea seeds but soaking duration adopted were 12, 4, 6 and 4 h respectively. Sasikala (1997) also reported that seed hardening for 6 and 4 h respectively in blackgram and cowpea increased the field emergence and yield potential.

Palanisamy and Jeyaseelan (1998) reported that seed fortification (seed dried to original moisture content after fortification) with the $ZnSO_4$ + $MnSO_4$ + sodium molybdate in 1/3rd volume of seed to solution ratio for 5 h followed by DAP pelleting @ 120 g kg^{-1} of seed improved the seed yield in redgram by increasing the yield attributes. Begam (2001) in blackgram cv CO 5 reported that seed hardening with one per cent pungam leaf extract and pelleting with DAP (120 g kg^{-1} of seed) + *Trichoderma* (4 g kg^{-1} of seed) + *Rhizobium* improved seed yield under restricted irrigation condition.

Anon (2000) also reported that blackgram seeds respond well to hardening and pelleting treatment under rainfed condition.

Begam (2000) reported that seed hardening with ZnSO4 100 ppm followed by two cycles of pelleting with DAP, *Rhizobium* and *Trichoderma* improved the seed yield under rainfed and restricted conditions. Prakash and Varadarajan (2001) of Tirupathi reported that soaking of seeds in 500 ppm of ethrel for 6 h also promoted growth, nitrate reductase activity and yield at moisture stress conditions in blackgram. Shenbaganathan (2002) and Kavitha (2002) reported that seed hardening using 1% *Prosopis*

Plate 1. Effect of seed hardening and pelleting of blackgram on plant growth and productivity

leaf extract followed by pelleting with DAP (40 g) + $ZnSO_4$ + $FeSO_4$ + NH_4 MO_4 (each @ 300 mg) using gypsum as filler material and 10% maida as adhesive improved the seed yield of greengram and blackgram respectively under dryland conditions of Coimbatore (Plate 1).

3.11. Tomato

Tomato seeds kept for 1-4 days in 0.5-2 per cent solution of K_3PO_4 and KNO_3 germinated five days earlier and had a higher germination rate than control (Ellis, 1963). Woodstock (1969) reported that germination rate and percentage could be increased in tomato seeds by soaking in solution of 2 per cent KNO_3 and 2 per cent KH_2PO_4 and then drying back. Kargopolova (1976) reported that not only germination but also yield potential could be increased by hardening treatment. He revealed that tomato seeds treated with 0.2 per cent solution of copper sulphate and 0.02 per cent solution of boric acid followed drying increased fruit maturation, ripening and fruit yield. Yield increase in tomato and capsicum by soaking the seeds in 1-1.5 per cent KNO_3 was reported by Dimov *et al.* (1978). Kuperitskaya (1978) recorded an increase of 35 per cent yield with total sugar content by soaking tomato seeds for 24 h in solution of manganese, copper and zinc. The increase in sugar content of the resultant seed of tomato and cucumber was observed by soaking the seeds in 0.02 per cent solution of $ZnSO_4$ (Viasova and Kartasheva, 1981). But Doijode an Raturi (1987) reported that soaking tomato seeds in plain water increased the germination and vigour.

3.12. Carrot

Austin *et al.* (1969) reported increased rates of germination and field emergence in hardened carrot seeds although there was no increase in final percentage emergence. In both cases the effect of hardening was to promote embryo growth in the seed. In carrot, Austin *et al.* (1969) have suggested that the hardening treatment was equivalent to a natural pre-harvest ripening process that would occur in the field if it were not terminated prematurely in dry seed growing areas by water stress. Hegarty (1970) reported an increase in field emergence when seed hardening was carried out with water alone and a significant additional increase when K_2HPO_4 was included in the solution.

The laboratory germination also showed the better performance of the hardened seeds at low temperature.

Alekseeva and Rasskezaov (1976) also reported increase in yield by 13 per cent and increase in sugar, carotene content and drymatter production by soaking the seed in solution of boron, manganese, cobalt and molybdenum. An increase in germination and field emergence was observed by Bevilaqua *et al.* (1998) in carrot by soaking the seeds in GA_3 and ethrel for 15 hr at 25°C.

3.13. Onion

Ozerova and Ozerova (1960) reported that treatment of onion seeds with 0.05 and 0.1 per cent and 0.5 per cent solution of microelements ($MnSO_4$) for 12 h increased their germination per cent. Soaking onion seeds in solution of NAA and IBA @ 10 mg/lit for 7 h ensured a germination of 89 to 93 per cent against 69 per cent in control.

3.14. Sugarbeet

Namedov (1971) stated that pre sowing treatment of seeds of sugarbeet with solutions containing 0.05 per cent KNO_3, 0.01 per cent boron and 0.03 per cent $ZnSO_4$ for 24 h markedly increased plant growth, decreased leaf spot and sugarbeet weevil incidence and increased the yield and sugar content. Soaking the seeds of sugarbeet in 5 and 10 per cent solution of succinic and glutamic acid increased growth, vigour and germination.

3.15. Other vegetables

Higher germination and vigour in onion, garden pea and cluster bean seeds was reported by Doijode and Raturi (1987) by soaking their seeds in water before sowing Kohli *et al.* (1992) recorded increased pod yield in garden pea by soaking the seeds before sowing for 6 h in water.

Omran *et al.* (1980) observed that seeds of Okra showed increase in germination and vigour by soaking in aqueous solution of GA, IAA and NAA for 12 h.

In Fenugreek, increase in germination and seedling growth were reported when the seeds were soaked in sulphates of ammonium, manganese and boric acid @ 0.1 to 1 per cent for 1 h (Saraswathamma and Jayachandra,1979).

Motta De and Leon (1978) reported that soaking the seeds of cauliflower for 30 min in solution of molybdenum at 16-32°C improved the field emergence.

Veeraraghavathatham *et al.* (1980) reported improved germination in papaya seeds after soaking in water, mono and di sodium phosphate and EDTA for 12 h.

3.16. Seed hardening in forestry

The regeneration of most of the forestry species is through seeds. To increase the coverage under forestry, sowing of seeds for afforestation is being mainly done by aerial sowing in hillocks and valley and to ensure the better establishment of seedling, pelleting is the most common practice of pre-sowing seed management in forestry. Physiological procedures for inducing tolerance to environmental stress have been scarcely if at all explored. Umarani *et al.* (1999) reported that *Acacia nilotica* seeds after soaking for 90 h in water followed by 3 h air drying recorded higher germination at low soil water holding capacity of 30-50 per cent.

3.17. Hardening of seed to assess temperature stress

Hardening treatments using seed hydration dehydration techniques are physiological methods for improving plant behaviour under drought conditions (Henkel 1964). In addition, other hardening methods that have been proposed consists of exposing seedlings to high temperature (Altschuler and Mascarenhas 1982; Cooper and Ho, 1983). It has been reported that a brief exposure to sublethal temperatures (heat stok) is capable of inducing heat thermotolerance in wide range of organisms (Vierling, 1991).

Combination of a cycle of partial hydration in water with heat shock (36°C or 38°C) for 1 h produced the optimum increase on germination and thermotolerance in tomato and pepper seedlings. For cucumber seeds, partial hydration in water upto 2 h before the onset of germination was the best procedure for inducing heat tolerance (Sanchez *et al.,* 2001).

3.18. Storability of hardened and hardened cum pelleted seeds

Storability is one of the important quality characteristics of seed. On adoption of management technique with imbibition and drying, maintenance of seed quality is still a question, seed treatment normally used as presowing seed management, which were advised just before sowing. Effective storage of these treated seed is also said to be essential for extended use of seed in case of unforeseen situations that hinder with the usage of seed with specificity. Researchers have worked on the storage potential of treated seeds in several crops.

In redgram, blackgram and greengram seeds fortified with IBA (200 ppm) and pelleted with peat based *Rhizobium* culture and dusted with thiram 5 per cent wettable dust powder @ 2 g kg^{-1} of seeds before storage found to maintain their vigour and viability upto 48 months by

Vanangamudi and Karivaratharaju (1986). Selvaraju (1992) evaluated the storability of hardened sorghum seed and found that upto 35 days, the seed quality was maintained and thereafter it declined. Jegathambal (1996) in sorghum, reported that hardened and pelleted seeds of sorghum with botanicals could be maintained upto 2 weeks without any deterioration of seed quality, while Punithavathi (1997) found that in ragi, hardened (KCl) and pelleted seed (Pungam leaf powder) could be stored well upto 5 months, while Nagaraj (1996) in rice also found that seeds hardened with KCl 1per cent + $CaCl_2$ 1per cent and pelleted with pungam leaf powder could be stored well upto 3 months under ambient conditions. Vijaya (1996) reported that fortified and pelleted seeds of cowpea and blackgram could be stored upto 2 months in cloth bags under the ambient conditions. Bashyam (1999) with fenugreek, coriander and fennel reported that hardened and pelleted seed can be stored upto 3 months in storage. Begam (2001) with blackgram cv. CO 5 also reported that hardened cum pelleted seed could be stored upto 3 months without much deleterious effect on seed and seedling quality characteristics. Shenbaganathan, Kavitha and Vigneswari (2002) highlighted container variation in storability of hardened and hardened cum pelleted seed. Vigneswari (2002) stored seeds hardened with 1% brassinoloid for 6 months in 700 gauge polyethylene bag while the storability was only for 2 months in cloth bag upto seed certification standard. Kavitha (2002) stored seeds hardened with $ZnSO_4$ 100 ppm in 700 gauge polyethylene bag for 6 months while Shenbaganathan (2002) upto 12 months upto the level of seed certification standard.

3.19. Causes for the influence of seed hardening techniques

The better efficacy of seed hardening, under the given situation was attributed to the seedling characteristic which was speeded up at the germination process and subsequent germination. It is also pointed out that maintenance activities for germination could take place in soaked tissue and repair mechanism was activated during initial phase of presoaking. During the first phase of germination, synthesis of protein and mechanism of repair and replacement started. The second phase is characterized by DNA synthesis and cell division which was stopped by dehydration without affecting the embryo. Mitra and Basu (1979) also opined that with presoaking, metabolic changes of the 1st phase of germination were activated before sowing and it has the further advantage of better growth establishment of the crop of higher yield. Henkel (1964) also reported that presowing hardening or imbibition and drying of seed is one of the methods which results in modifying physiological and biochemical nature of seed so as to get the characters that are favourable for drought resistance. Singh and Kumar (1992) also reported that hardening treatment enhances the enzyme activities and thereby improving the seed yield of chickpea, while Sabir-Ahamed (1999) discussed that the physical and chemical treatments could have triggered the biosynthesis of nucleic acids, proteins and the consequential enhancement of cell division besides the enhanced metabolic activity of the plant resulting in the increased uptake of nutrients which are associated with improved crop growth.

There is considerable evidence to show that drought resistance of plants can be increased by subjecting seeds to a cycle of wetting and drying prior to sowing. Seed hydration treatments included seed moisturing, seed hardening and seed priming which could be compared to a dry seed control. Moisturing and hardening treatments improved early emergence. The early advantage to stand establishment from seed hardening also resulted in a more uniform emergence. The practical value of this technique over a wide range of condition for different crop seed, need exploration. However, Anon (1999b) recommended hardening treatments as follows for improving the crop productivity made dryland conditions (Table 1).

Table 1: Hardening technology recommended for major crops of Tamil Nadu under dryland condition

Crop	Chemicals used	Concentration	Duration of soaking (h)	Seed to solution ratio
Cholam	Potassium dihydrogen phosphate	0.55%	10	1:0.3
Cumbu	Potassium chloride	2%	16	3:1.0
Cotton	Potassium chloride	2%	16	1:0.5
Sunflower	Potassium chloride	2%	12	1:0.6
Paddy	Potassium chloride	1%	16	1:0.6
Ragi	Potassium chloride	1%	12	1:0.7
Redgram	Zinc sulphate	100 ppm	3	1:0.3
Blackgram	Zinc sulphate	100 ppm	3	1:0.3
Greengram	Manganese sulphate	100 ppm	3	1:0.3
Bengalgram	Potassium dihydrogen phosphate	1%	3	1:0.3

REFERENCES

1. Abdul Hafeez , A.T and J.P.Hudson. (1967). Effect of hardening radish seeds. *Nature, London, 216*: 688.
2. Alekseeva, A.M. and M.A. Rasskazaov. (1976). The effect of boron, manganese, cobalt and molybdenum on the yield, quality and storability of carrots. Nanchnye Trudg voronezh S-Kh Int, 85: 5-13. *Seed Abstr., 1978 1 (3)*:688.
3. Alexander, L. (1985). Seed quality studies in Kovilpatti tall hybrid sorghum. M.Sc.(Ag.) Thesis, Tamil Nadu Agricultural University, Coimbatore.
4. Altschuler, M and J.M.. Mascarenhas. (1982). Heat shock proteins and effects of heat stock in plant. *Molecular Biology, 1*: 103 - 115.
5. Anonymous, (1999a). Crop production guide. Office of the Commissioner of Agriculture, Chepauk, Chennai and Tamil Nadu Agricultural University, Coimbatore.
6. Anonymous. (1998). Annual Report, All India Co-ordinated National seed project (Crops) 1997 - 1998. Project Coordinator Unit, NSP (Crops), Indian Agricultural Research Institute, New Delhi-110012.
7. Anonymous. (1999b). Annual Report, All India Co-ordinated National Seed Project (Crops). Project Coordinator Unit, NSP (Crops), Indian Agricultural Research Institute, New Delhi-110012.
8. Anonymous. (2000). Seed hardening in blackgram: Project final report, Department of Seed Science and Technology, TNAU, Coimbatore
9. Arjunan, A. and P.S. Srinivasan. (1989). Pre-sowing seed hardening of groundnut. *Madras Agric. J., 76 (9)* : 523-524.
10. Asgarjalis, M. (1986). Effect of pre-sowing seed treatment on the yield of wheat under drought conditions. *J. Agrl. Res. Pakistan, 24(2):* 305-312.
11. Asthana, J.S. and H.S. Srivastava. (1978). Effect of presowing treatments of maize seeds with ascorbic acid salicylic acid on seed germination, seedling growth and nitrate assimilation in the seedling. *Indian J. Pl. Physiol., 21(2)*: 150-155.
12. Austin, R.B., P.C.Longden and J.Hutchinson. (1969). Some effects of hardening of carrot seeds. *Ann. Bot. 33,* : 883-95.
13. Balasubramanian, S. (1976). Note on the effect of pre-sowing seed treatment on total and reducing sugars of two sorghum hybrids. *Indian J. Agric. Sci., 48*: 346-347.
14. Basha, M.(1982). Effect of chemical fortification of seeds on storability, field performance and quality of the resultant seeds in greengram. M.Sc. (Ag.) Thesis, Tamil Nadu Agricultural University, Coimbatore
15. Bashyam, S. (1999). Seed quality testing and management procedures in three important condiments of coriander (*Coriandrum sativum* Linn.) fennel (*Foeniculum vulgare* Mill) and fenugreek (*Trigonella feonum graecum* Linn.). M. Sc. (Ag.) Thesis, Tamil Nadu Agricultural University, Coimbatore.

16. Begam, J. (2001). Seed hardening cum pelleting studies in Blackgram cv. CO5. M.Sc. (Ag.) Thesis, Tamil Nadu Agricultural University, Coimbatore

17. Bevilaqua, G.A.P., S.T. Peske., Santos Filho, B.G. Dos., Santos and D.S.B.Dos. (1998). Effect of treatment of carrot seeds with growth regulators. *Efeito do tratamento de sementes de cenours wm reguladores de crescimento pesquisa Agropecuaris Brasileira, 33 (8)*:1271-1280.

18. Bhatnagar, O.K., J.P. Srivastava and J.K. Garg. (1975). Effect of seed soaking in ascorbic acid, succinic acid and cycocel on the yield of moong. *Indian J. Agron. 20(3)*: 240-241.

19. Cooper, P and Ho, D.T. (1983). Heat shock proteins in maize. *Plant Physiology, 71:* 215-222.

20. Devarani, N. and A. Rangasamy. (1998a). Effect of seed hardening on the germination and vigour index in sorghum. *Madras Agric. J., 85(10-12)* : 674-675.

21. Devarani, N. and A. Rangasamy. (1998b). Nutrient and drought management practices for rainfed sorghum under vertisol. *Madras Agric. J., 85 (7-9)* : 391-393.

22. Dharmalingam, C. and R.N. Basu. (1989). Invigouration treatment for increased production in carried-over seeds of mung bean. *Seeds & Farms, XV (5)* : 34-36.

23. Dimov, I., L. Iranov and K.Milliev. (1978). Preplanting seed treatment for direct sown tomatoes and capsicum. *Gradinarsto, 59 (5)*:11-14.

24. Doijode, S.D. and G.Raturi. (1987). Effect of hydration dehydration on germination and vigour of certain vegetable seeds. *Seed Res., 15 (2)*:156-159.

25. Ells, J.E (1963). The influence of treating tomato seeds with nutrient solutions on emergence rate and seedlings growth. *Proc. Amer. Horti. Sci., 83*: 684-687.

26. Evenari, M. (1964) Hardening treatment of seeds as means of increasing yields under conditions of inadequate moisture. *Nature, London. 204* : 1010-11.

27. Gopalsingh, B. (1995). Effect of seed size in relation to pre-sowing treatments on growth and productivity of soybean. *Annals Agric. Res., 16(2)* : 254-255.

28. Hegartz, T.W. (1970). The possibility of increasing field establishment by seed hardening. *Hort. Res., 10*: 59-64.

29. Henkel, P.A. (1964). Physiology of plants under drought. *Ann. Rev. Pl. Physiol., 15* : 363-386.

30. Heydecker, W. (1972). Interrelated effects of imbibition temperature and oxygen on seed germination. In: Seed Ecology (ed.) Heydecker, W. Butterworths. London. pp. 157-169.

31. Iliev, V., P. Ivanov and M. Mamkova. (1989). Ways and terms of molybdenum and Wuxal (suspension) fertilization of bean. *Rastenievud. Nauk., 26(6)* : 5-9 *Micronutrient News and Information. 1991. 11(3)* : 0629.

32. Jegathambal, R. (1996). Pre-sowing seed treatment to augment productivity of sorghum cv. CO26 under rainfed agriculture. Ph.D. Thesis, Tamil Nadu Agricultural University, Coimbatore.

33. Jhorar, B.S., S.K. Varma and R.P. Agarwal. (1982). Effect of gibberellic acid on the seedling emergence and early growth of cotton (*Gossypium hirsutum* L.). *Indian J. Plant Physiol., XXV* (4): 423-426.

34. Kalarani, M.K., R. Sivakumar, Malliga Vanangamudi and K.B. Sujatha.(2001). Effect of seed soaking treatments in late sown finger millet under rainfed condition. *Madras Agric. J., 88(7-9)*: 509-511.

35. Kamalam, J. and N.R. Nair. (1989). Effect of seed hardening on germination and seedling vigour in paddy. *Seed Res., 17(2)*: 188-190.

36. Kannaiyan, M.(1987). Effect of dates of sowing and seed hardening on yield of pearlmillet (*Pennisetum typhoides* Stapf and Hubb) under rainfed condition. M.Sc. (Ag.) Thesis, Tamil Nadu Agricultural University, Coimbatore .

37. Kargopolova, E.N.(1976). Speedy ripening of tomato and potato under influence of micro elements.

38. Karivaratharaju, T.V. (1987). Seed production manual. Department of Agriculture. Tamil Nadu, India.

39. Karivaratharaju, T.V. and V. Ramakrishnan. (1985). Seed hardening studies in two varieties of ragi (*Eluesine coracana). Indian J. Pl. Physiol., 28(3):* 243-248.

40. Karivaratharaju, T.V., R. Vaithialingam and J. Sakhoram rao. (1973). Effect of water stress and hardening of seeds on certain physiological aspects in *Pennisetum typhoides* stapf and Hubb. *Madras Agric. J., 60(9-12)*: 1266-1272.

41. Karivaratharaju, T.V., V. Ramakrishnan and M. Thangaraj. (1974). Pre-sowing hardening studies in sunflower EC 68415 (*Helianthus annuus*) (unpublished). , Tamil Nadu Agricultural University, Coimbatore..

42. Kathiresan, K., V.K. Alyani and J.L. Gnanarathinam. (1984). Seed treatment for alleviating water stress on germinability of sunflower. *Seed Res., 12(2)*: pp. 92-94.

43. Kavitha, S. (2002). Seed hardening and pelleting for maximising the productivity of blackgram (*Vigna mungo* (L) Hepper) cv. vamban 3 under rainfed conditions M.Sc (Ag.) thesis. Tamil Nadu Agricultural University, Coimbatore.

44. Khan, A.A. and S.V. Hedge. (1989). Effect of molybdenum seed treatment on nodulation and growth of pigeon pea (*Cajanus cajan* (L.) Mill sp.). *Indian J. Exp. Biol., 27(10)*: 919-920.

45. Kohli, V.K., I.V. Thakur and Y.R. Shukla. (1992). Studies on improving field emergence in the autumn crop of garden pea (*Pisum sativum* L.) *Ann. Agric. Res., 13(4)*: 394-395.

46. Kuperitskaya, M.A. (1978). The effect of seed treatment with minor elements on the yield and fruit quality of green house grown tomatoes. *Trudy Dal' nevost NII S-Kh. 24*:149-150.

47. Lin, J.M. and Sung J.M. (2001). Pre-sowing treatments for improving emergence of bitter gourd seedlings under optimal and sub-optimal temperatures.
48. Maitra, S.G., Sounda, D.C. Ghosh and P.K. Jana. (1997). Effect of seed treatment on finger millet *Eleusine coracene* varieties in rainfed upland. *Ind. J. Agric. Sci., 67(10)*: 478-80.
49. Manohar, M.S. and M.K. Mathur. (1966). Effect of succinic acid treatment on the performance of pearl millet. *Adv. Front. Pl. Sci., 17* : 133-148.
50. May, L.H., E.J.Milthorpe and F.L.Milthorpe (1962). Pre-sowing hardening of plants to drought. *Fld. Crop Abstr., 15* :93-98
51. Mehrotra, O.N., J.P. Srivatsa and S.Sivanath. (1968). Germination, growth and yield of wheat as influenced by pre-sowing seed treatment in phosphate solution. *Indian J. Agron., 13*: 53-56.
52. Mitra, R. and R.N. Basu. (1979). Seed treatments for viability, vigour and productivity of tomato. *Scientia Horticulturae, 11* : 365-369.
53. Motta, De, M.B. and S.A.Leon. (1978). The effect of two methods of fertilization with molybdenum on the yield of the cauliflower cultivar snowball growing in the mosquera soil. *Seed Abstr., 1980, 3(3)*: p.99:830.
54. Muhyaddin, T and H.J. Wiebe. (2001). Influence of PEG seed treatment on emergence we under stress conditions. *Acta Horticulture,* 215.
55. Nagaraj, S. (1996). Seed hardening and coating studies in upland rice (*Oryza sativa* L.). M.Sc. (Ag.) Thesis, Tamil Nadu Agricultural University, Coimbatore.
56. Namedov. N.M.(1971). Effect of the micro elements on the growth, development and the yield of sugarbeet on irrigated light chestnut seeds. Boron in Agrl. July.PP.12.
57. Narayanan, T.R. (1981). Pre-treatment – A new aid for improving crop yields. *Madras Agric. J., 38*: 100-103.
58. Nirmala, M.S., J. Radhakrishnan and U. Bangarusamy. (1994). Effect of seed hardening and moisture regimes on germination of sorghum and cotton in vertisols. *Madras Agric. J., 81(12)*: 695-696.
59. Omran A.F, E.L.Bakry and R.A Gawish. (1980). *Seed Sci. & Technol., 8*: 161-168.
60. Padole, V.R. (1979). Effect of presoaking seed treatment of wheat seed with chemicals and hormones on yield and uptake of NPK. *Journal of MAU, 4 (1)* : 85-88.
61. Palanisamy, V. and K. Jayaseelan. (1998). Effect of pre-sowing seed treatments on seed quality in redgram. *Madras Agric. J., 85(10-12)*: 612-613.
62. Palanisamy, V. and N. Punithavathi. (1998). Effect of seed hardening and pelleting on seed quality in ragi. *Madras Agric. J., 85 (10-12)* : 583-585.
63. Pandey, S.N. and B.K. Sinha. (1999). Mineral nutrition. In: Plant Physiology, Vikas Publishing house Pvt. Ltd. New Delhi. pp. 100-119.

64. Paul, S.R. and A.K.Chowdhury.(1993). Effect of seed hardening with potassium salts at different concentrations and soaking duration on germination and seedling vigour of wheat. *Ann. Agric. Res.,14(3)*: 357-359.

65. Pawar, H.K. and R.M. Kadam. (1981). Soaking of seeds in solutions of different chemicals in relation to uptake of nutrients and protein content in wheat under rainfed conditions. *Journal MAU., 6(1)*: 73-76.

66. Periyathambi, C. (1980). Effect of pre-monsoon sowing, depth of seed placement and seed hardening on rainfed sorghum. M.Sc. (Ag.) Thesis, Tamil Nadu Agricultural University, Coimbatore.

67. Pothiraj, P. and S. Sankaran. (1984). Effect of seed hardening on the physiological vigour of the cotton seedlings. *Indian J. Plant Physiol., 27(3)* : 300-302.

68. Punithavathi, N. (1997). Seed hardening and pelleting studies in ragi cv. CO 13. M.Sc. (Ag.) Thesis, Tamil Nadu Agricultural University, Coimbatore.

69. Rajendran, P. (1982). Effect of chemical infusion of seeds on storability field performance and quality of the resultant seeds in redgram cv. CO 4. M.Sc. (Ag.) Thesis, Tamil Nadu Agricultural University, Coimbatore.

70. Ramachandran, K. (1972). Inducing drought tolerance in bajra (*Pennisetum typhoides* Stapf and Hubb) by pre-sowing seed treatment. M.Sc. (Ag.) Thesis, Tamil Nadu Agricultural University, Coimbatore.

71. Ramachandran, K. and J. Sakharamrao. (1975). Inducing drought tolerance in Bajra (*Pennisetum typhoides* Stapf. and Hubb) by pre-sowing seed treatment. *Madras Agric. J., 62*(3): 127-130.

72. Rangasamy, A. (1986). Evaluation and management of sorghum and groundnut based intercropping system in a rainfed Alfisol. Ph.D. Thesis, Tamil Nadu Agricultural University, Coimbatore.

73. Rangaswamy, A., S. Purushothaman and P. Devasenapathy. (1993). Seed hardening in relation to seedling quality characters of crops. *Madras Agric. J., 80(9)*: 535-537.

74. Rathinavel, K. (1997). Seed management, reproduction, physiology and control of seed deterioration in cotton. Ph.D Thesis, Tamil Nadu Agricultural University, Coimbatore.

75. Rathinavel, K. and C. Dharmalingam, (1999a). Seed hardening to augment the productivity of cotton cv. LRA 5166 (*Gossypium hirsutum* L.). *Madras Agric. J., 86*(1-3): 68-72.

76. Renugadevi, J., P. Sumathi, A. Nirmalakumari, K. Natarajamoorthy and P. Gnanamurthy. (2001a). An inorganic ingredients on seed hardening of blackgram (*Vigna mungo* L.) National symposium on pulses and oil seeds for sustainable Agriculture, July 29-31, 2001. (Ed.) S. Kannaiyan and M. Subramanian. Tamil Nadu Agricultural University, Coimbatore. pp. 84.

77. Renugadevi, J., P. Sumathi, A. Nirmalakumari, K. Natarajamoorthy and P. Gnanamurthy. (2001b). Seed hardening management on seed yield quality in cowpea (*Vigna unguiculata* L. Walp). National symposium on pulses and oil seeds for sustainable Agriculture, July 29-31, 2001. pp. 98.

78. Ros, C., R.W. Bell and P.F. White. (2000). Phosphorus seed coating and soaking for improving seedling growth of *Oryza sativa* (rice) cv. IR 66. *Seed Sci. & Technol., 28*: 391-401.

79. Sabir-Ahamed, A.(1999). Field performance of hardened greengram seeds. *Legume Res., 22(3)* : 207-208.

80. Sahulhameed, M., R. Murugesan and S. Anthoniraj. (1988). Bacterization of soaked seeds of blackgram with *Rhizobium* sp. *Madras Agric. J., 75 (7-8)* : 281-282.

81. Sanchez, J.A, B.C Munoz and Fresneda J. (2001). Combined ffects of hardening hydration-dehydration and heat shock treatments on the germination of tomato, pepper and cucumber. *Seed Sci. & Technol., 29:* 691-697.

82. Sandhu, A.S. and A. Hussain. (1961). Effect of seed treatment with gibberellic acid on germination and growth of bajra. *Indian J. Agron., 5*: 269-272.

83. Sandhya, K.P. and B. Gopalsingh. (1994). Effect of pre-sowing seed treatment with calcium containing compounds on emergence index in groundnut (*Arachis hypogae* L.). *Seed Res., 22(2)*: 170-171.

84. Saraswathamma, D.N. and Jayachandra.(1979). The influence of pre-sowing soaking with mineral solution on the seedling growth in methi *Trigonella foenum-gracum* L. *Curr. Sci., 48(7)*:786-787.

85. Sasikala, K. (1997). Management of aged seeds and their progenies in Bhendi (*Abelmoschus esculentus* L.) cv. Parbanikranti and cowpea (*Vigna sinensis* L.) cv. CO 6. Ph.D. Thesis, Tamil Nadu Agricultural University, Coimbatore.

86. Selvaraju, K. (1992). Studies on certain aspects of seed management practices for seed production under moisture stress condition in sorghum cv. CO 26 (*Sorghum bicolor* (L.) Moench). M.Sc. (Ag.) Thesis, Tamil Nadu Agricultural University, Coimbatore.

87. Shanmugasundaram, V.S. and M. Kannaiyan. (1989). Effect of concentration of seed hardening chemicals on physiological characters of pearlmillet (*Pennisetum typhoides* stapf and Hubb.). *J. Agronomy and Crop Science, 163*: 174-176.

88. Shenbaganathan, A. (2002). Seed Management techniques for quality seed production of greengram (*Vigna radiata* (L). cv. CO 6 under dryland condition. M.Sc (Ag) Thesis, Tamil Nadu Agricultural University, Coimbatore.

89. Singh, R.A. and Kumar. (1992). Effect of water stress and hardening on metabolic changes in chickpea (*Cicer arietineum* L.). *Indian J. Pl. Physiol., 35(3)* : 252-257.

90. Subbian, P., D.S. Pandey, and S. Palaniappan. (1987). Effect of seed soaking on the performance of greengram (*Vigna radiata* L.). *Madras Agric. J., 74(3)* : 175-176.

91. Subramanyam, M.V.R. and N.M. Misra. (1980). Effect of potash seed treatment on yield and yield component of high yielding wheats under rainfed conditions. *Seeds and Farm, 6(10)*: 21-24.

92. Umarani, R., A.Bharathi, K.Vanangamudi, T.V.Karivaratharaju and R.S.V.Rai. (1999). Seed hardening studies in *Acacia nilotica* (Linn.) willd. Ex. Del. *Indian Journal of Forestry, 22(4)*:387-390.

93. Vanangamudi, K. and R. Kulanthaivelu. (1989). Presowing seed treatment for dryland farming. *Seeds & Farms, 15* (9-10) : 22.

94. Vanangamudi, K. and T.V. Karivaratharaju. (1986). Effect of pre-storage chemical fortification of seeds on shelf life of redgram, blackgram and greengram seed. *Seed Sci & Technol., 14*: 477-482

95. Veeraraghavathetham,D. K.K.Vaidvelu and T.B.Ranganathan (1980). Seed invigouration in CO2 papaya. *South Indian Hort., 28 (3)*:69-71.

96. Viasova, V.I. and I.A.Kartasheva, (1981). Results of experiments on the effects of trace elements and growth substances of vegetable crops. Nauchn. Tr.-Stavrop. S-Kh. Inst., 44 (1):76-83.

97. Vierling, E (1991). The roles of heat shock proteins in plants, *Annual Review of Plant Physiology and Plant Molecular Biology,42* :579-620.

98. Vigneswari, R. (2002). Seed hardening and pelleting in ragi cv. CO 13 to induce drought resistance and to improve productivity under rainfed condition. M.Sc. (Ag.) Thesis, Tamil Nadu Agriculture University, Coimbatore.

99. Vijaya, J. (1996). Standardization of pre-sowing seed management techniques for pulses. M.Sc.(Ag.) Thesis, Tamil Nadu Agricultural University, Coimbatore.

100. Vijaya, J. and A.S. Ponnusamy. (1988). Studies on seed fortification and pelleting in blackgram. *Madras Agric. J., 85 (10-12)* : 549-551.

101. Vijayakumar, A. (1982). Effect of chemical impregnation of seeds on storability, field performance and quality of the resulting seeds on blackgram (*Vigna mungo* (L.) Hepper; syn; *Phaseolus mungo* L.) cv. CO. 5. M.Sc. (Ag.) Thesis, Tamil Nadu Agricultural University, Coimbatore.

102. Viswanath, D.P., K.S. Krishna Sastry. and B.V. Venkata rao. (1972). Effect of seed treatment on germination and growth of *Eleusine corocana* Galrin, in soils with high exchangeable sodium. *Mysore J. Agric. Sci., 6*: 24-28.

103. Xia, M.Z., Y. Tang and H.Y. Bai. (1989). Effect of some mineral nutrients on physiological functions and yield of faba bean. In: Faba bean production and research in China: Proceedings of an International Symposium, Hangzhou, China, 24-26 May (edited by Saxena, M.C., S. Weigland, L.J. Lang).

104. Zubenko. V.KH. (1959). The effect of preplanting of seeds against drought and the grain harvest of corn in late plantings. *Pl. physiol., 6*: 341-351.

PAST AND PRESENT STATUS OF RESEARCH ON SEED HARDENING IN HORTICULTURAL AND FORESTRY CROPS

P.BALAMURUGAN[1] AND P.SRIMATHI[2]

1. INTRODUCTION

The commonest calamity of the Indian agriculture is the failure of the monsoon which in its extreme manifestation is called drought. Water balance of a plant is upset by drought and as a consequence of this the physiological functions are affected resulting in poor growth and yield. The urgent need for securing varieties of crops capable of escaping this deleterious effect of drought becomes increasingly important and one has to seriously try to defy drought.

Presowing seed hardening with water and solutions of growth regulating compounds to induce early germination, better root and seedling growth and increase yield has been employed by many workers. Seed hardening has been investigated by those concerned with problem of seed germination and seedling emergence in the field. The Russian work on seed hardening reviewed by May *et al.* (1962) was concerned with inducing a physiological resistance to water stress in growing plants by exposing the seeds to water stress before they were sown i.e. by soaking and then drying back the seeds one or more times.

2. SEED HARDENING IN HORTICULTURAL CROPS

Whatever the merits of the treatment for this purpose (Abdul Hafeez and Hudson, 1967; Evenari, 1964) other effects of seed hardening have been reported. Zubenko (1959) stated that the hardening of seed increased

1 *Department of Seed Science and Technology, Tamil Nadu Agricultural University, Coimbatore – 641 003.*
2 *Department of Tree Breeding, Forest College and Research Institute, Mettupalayam – 641 301.*

the field germination of corn and Austin *et al.* (1969) reported increased rates of germination and field emergence in hardened carrot seeds although there was no increase in final percentage emergence. In both cases the effect of hardening was to promote embryo growth in the seed. In carrot, Austin *et al.* (1969) have suggested that the hardening treatment was equivalent to a natural pre-harvest ripening process that would occur in the field if it were not terminated prematurely in dry seed growing areas by water stress.

Tomato and several vegetables which are widely employed for human consumption being grown under the most diverse ecological conditions. However, physiological procedures for inducing tolerance to environmental stress have been scarcely exploited. The available information on research activities on this line is presented below.

2.1. Tomato

Tomato seeds kept for 1-4 days in 0.5-2 per cent solution of K_3PO_4 and KNO_3 germinated five days earlier and had a higher germination rate than control (Ells, 1963).

Germination rate and percentage could be increased in tomato seeds by soaking in solution of 2 per cent KNO_3 and 2 per cent KH_2PO_4 and then drying back.

The treatments of tomato seeds two times with 0.2 per cent solution of copper sulphate and 0.02 per cent solution of boric acid followed drying increased fruit maturation, ripening and fruit yield (Kargopolova, 1976). Yield increase in tomato and capsicum by soaking the seeds in 1-1.5 per cent KNO_3 was reported by Dimov *et al.* (1978).

Kuperitskaya (1978) recorded an increase of 35 per cent yield with total sugar content by soaking tomato seeds for 24 hr in solution of manganese, copper and zinc.

The increase in sugar content of the resultant seed of tomato and cucumber was observed by soaking the seeds in 0.02 per cent solution of $ZnSO_4$ (Viasova and Kartasheva, 1981).

Soaking tomato seeds in plain water increased the germination and vigour as reported by Doijode and Raturi (1987).

2.2. Carrot

Hegarty (1970) reported an increase in field emergence when seed hardening was carried out with water alone and a significant additional increase when K_2HPO_4 was included in the solution (Table.1)

Table 1. Field emergence (%) of untreated and hardened carrot seed

Treatment	Field emergence
Untreated	58.0
Water hardened	65.0
Vitamin B hardened	69.8
K_2HPO_4 hardened	72.5

The laboratory germination also showed better performance of the hardened seeds at low temperature (Table.2)

Table 2. Germination at 20°C and 10°C of untreated and hardened seed

Cultivar	Germination at 20°C			Germination at 10°C		
	Untreated	Water	K_2HPO_4	Untreated	Water	K_2HPO_4
Nantes	89.0	88.3	90.0	89.0	88.3	88.5
Chantenay I	88.0	86.0	81.8	70.3	77.3	84.8
Chantenay II	80.5	74.8	75.8	67.5	76.0	76.3

The results indicated that better performance may be achieved through seed hardening treatment.

Alekseeva and Rasskezaov (1976) also reported increase in yield by 13 per cent and increase in sugar, carotene content and drymatter production by soaking the seed in solution of boron, manganese, cobalt and molybdenum.

An increase in germination and field emergence was observed by Bevilaqua *et al.* (1998) in carrot by soaking the seeds in GA_3 and ethrel for 15 hr at 25°C.

2.3. Onion

Treatment of onion seeds with 0.05, 0.1 per cent and 0.5 per cent solution of microelements ($MnSO_4$) for 12 hr increased their germination per cent.

Soaking onion seeds in solution of NAA and IBA @ 10 mg/lit for 7 hr ensured a germination of 89 to 93 per cent against 69 per cent in control.

2.4. Sugarbeet

Namedov (1971) stated that presowing treatment of seeds of sugarbeet with solutions containing 0.05 per cent KNO_3, 0.01 per cent boron and

0.03 per cent $ZnSO_4$ for 24 h markedly increased plant growth, decreased leaf spot and sugarbeet weevil incidence and increased the yield and sugar content.

Soaking the seeds of sugarbeet in 5 and 10 per cent solution of succinic and glatamic acid increased growth, vigour and germination .

2.5. Other vegetables

Motta De and Leon (1978) reported that soaking the seeds of cauliflower for 30 min in solution of molybdenum at 16-32 improved the field emergence.

Increase in germination and seedling growth were reported in Fenugreek when the seeds were soaked in sulphates of ammonium, manganese and boric acid @ 0.1 to 1 per cent for 1 hr (Saraswathamma and Jayachandra,1979).

Omran *et al.* (1980) observed that seeds of okra showed increase in germination and vigour by soaking in aqueous solution of GA, IAA and NAA for 12 hr.

Veeraraghavathatham *et al.* (1980) reported improvement in germination of papaya seeds after soaking in water, mono and di sodium phosphate and EDTA for 12 hr.

Higher germination and vigour in onion, garden pea and cluster bean seeds was reported by Doijode and Raturi (1987) by soaking their seeds in water before sowing. Kohli *et al.* (1992) recorded increased pod yield in garden pea by soaking the seeds before sowing for 6 h in water.

3. COMBINED EFFECT OF HARDENING AND SHOCK TREATMENTS

Hardening treatments using seed hydration and dehydration techniques are physiological methods for improving plant behaviour under drought conditions (Henckel 1964). In addition, other hardening methods that have been proposed consists of exposing seedlings to high temperature (Altschuler and Mascarenhas 1982; Cooper and Ho, 1983). It has been reported that a brief exposure to sublethal temperatures (heat stok) is capable of inducing heat thermotolerance in wide range of organisms (Vierling, 1991).

Combination of a cycle of partial hydration in water with heat shock (36°C or 38°C) for 1 hr produced the optimum increase on germination and thermotolerance in tomato and pepper seedlings. For cucumber seeds, partial hydration in water upto 2 hr before the onset of germination was the best procedure for inducing heat tolerance (Sanchez *et al.*, 2001).

Table 3. Analysis of variance for the effects of pregerminative treatment, temperature and its interaction on germination and vigour of seedlings of vegetable species

Dependent Variable	Source of variance	F-test		
		Tomato	**Pepper**	**Cucumber**
Germination	Pregerminative treatment (A)	44.7 ***	7.7 *	4.9 **
	Temperature (B)	131.7 ***	116.5 ***	17.5 ***
	A x B	5.6 ***	3.8 ***	4.3 **
Rate of Germination	Pregerminative treatment (A)	6.5 ***	5.0 **	12.6 ***
	Temperature (B)	16.9 ***	10.4 ***	32.3 ***
	A x B	5.2 ***	13.6 ***	30.0 *
Root length	Pregerminative treatment (A)	28.3 ***	➢ ***	38.3 ***
	Temperature (B)	121.5 ***	1. ***	1. ***
	A x B	20.1 ***		28.2 ***
			10.8 ***	

*, **, *** Significant at the 0.05, 0.01 and 0.001 probability levels, respectively.

3.1. Influence of PEG under stress conditions

Muhyaddin and Wiebe (2001) treated the seeds of carrot and tomato with PEG 4000 to determine its effect on emergence in soil with high salinity. The result indicated the earlier and higher percentage of emergence of PEG treated seeds compared to control. The relatively highest effect of seed treatment was found under conditions with increasing salinity as a result of decreasing soil water content.

For improving emergence of bittergourd seedlings under optimal and sub-optimal temperature Lin and Sung (2001) gave seed priming with vermiculite and warm water soaking and then air dried to the original moisture level. The seedling emergence was determined at 25, 20 and 15°C (Table.4)

The results indicated that both priming and warm water soaking improved the vigour response of bittergourd seed under sub-optimal (20°C) condition.

4. SEED HARDENING IN FORESTRY

The regeneration of most of the forestry species is through seeds. To increase the coverage under forestry, sowing of seeds for afforestation is

Table 4. Presowing seed treatment on dry weight accumulation (mg seedling-1) of 14 days old seedling

Seed Treatment	Temperature		
	25°C	20°C	15° C
Check	32.3	-	-
Priming	52.4	38.9	-
Warm water soaking	48.8	35.3	-
Priming-warm water soaking	55.2	40.5	13.6
(CD P=0.05)	5.1	3.7	

being mainly done by aereal sowing in hillocks and valley and to ensure the better establishment of seedling, pelleting is the most common practice of pre-sowing seed management in forestry. Physiological procedures for inducing tolerance to environmental stress have been scarcely if at all explored.

Umarani *et al.* (1999) reported that *Acacia nilotica* seeds after soaking for 90 h in water followed by 3 h air drying recorded higher germination at low soil water holding capacity of 30-50 per cent.

4.1. Hardening Plants

Hardening is the process of altering the quality of plant growth to withstand the change in environmental conditions which occurs when plants are transferred from a greenhouse or home garden. A severe check in growth may occur if plants produced in the home are planted outdoors without a transition period. Hardening is most critical with early crops, when adverse climatic conditions can be expected.

Hardening can be accomplished by gradually lowering temperatures and relative humidity and reducing water. This procedure results in an accumulation of carbohydrates and a thickening of cell walls. A change from a soft, succulent type of growth to a firmer, harder type is desired.

This process should be started at least 2 weeks before planting in the garden. If possible, plants should be moved to 45 to 50°F temperature indoor shady location. A coldframe is excellent for this purpose. When put outdoors, plants should be shaded, then gradually moved into sunlight. Each day, gradually increase the length of exposure. Don't put tender seedlings outdoors on windy days or when temperatures are below 45°F. Reduce the frequency of watering to slow growth, but don't allow plants to wilt. Even cold-hardy plants will be hurt if exposed to freezing temperatures before they are hardened. After proper hardening, however, they can be planted outdoors and light frosts will not damage them.

The hardening process is intended to slow plant growth. If carried to the extreme of actually stopping plant growth, significant damage can be done to certain crops. For example, cauliflower will make thumb size heads and fail to develop further, if hardened too severely. Cucumbers and melons will stop growth if hardened.

There is considerable evidence to show that drought resistance of plants can be increased by subjecting seeds to a cycle of wetting and drying prior to sowing. Seed hydration treatments included seed moisturing, seed hardening and seed priming which could be compared to a dry seed control. Moisturing and hardening treatments improved early emergence. The early advantage to stand establishment from seed hardening also resulted in a more uniform emergence. The practical value of this technique over a wide range of condition for different crop seed need exploration.

REFERENCES

1. Abdul Hafeez , A.T and J.P.Hudson. (1967). Effect of hardening radish seeds. *Nature, London. 216* : 688.
2. Alekseeva, A.M. and M.A. Rasskazaov, (1976). The effect of boron, manganese, cobalt and molybdenum on the yield, quality and storability of carrots. *Nanchnye Trudg voronezh S-Kh Int, 85*: 5-13. *Seed Abstr., 1978 1 (3)*:688.
3. Altschuler, M and Mascarenhas, J.M. (1982). Heat shock proteins and effects of heat stock in plant. *Molecular Biology, 1*: 103 – 115.
4. Austin, R.B., P.C.Longden and J.Hutchinson (1969). Some effects of hardening of carrot seeds. *Ann. Bot., 33*: 883-95.
5. Bevilaqua, G.A.P., S.T. Peske., Santos Filho, B.G. Dos., Santos and D.S.B.Dos. (1998). Effect of treatment of carrots seeds with growth regulators. *Efeito do tratamento de sementes de cenours wm reguladores de crescimento pesquisa Agropecuaris Brasileira, 33 (8)*:1271-1280.
6. Cooper, P and Ho, D.T, (1983). Heat shock proteins in maize. *Plant Physiology, 71*: 215-222.
7. Dimov, I., L. Iranov and K.Milliev (1978). Preplanting seed treatment for direct sown tomatoes and capsicum. *Gradinarsto, 59 (5)*:11-14.
8. Doijode, S.D. and G.Raturi. (1987). Effect of hydration dehydration on germination and vigour of certain vegetable seeds. *Seed Res., 15 (2)*:156-159.
9. Ells, J.E (1963). The influence of treating tomato seeds with nutrient solutions on emergence rate and seedlings growth. *Proc. Amer. Horti. Sci.*, 83: 684-687.

10. Evenari, M. (1964) Hardening treatment of seeds as means of increasing yields under conditions of inadequate moisture. *Nature, London. 204* : 1010-11.

11. Hegartz, T.W. (1970). The possibility of increasing field establishment by seed hardening. *Hort. Res., 10*, 59-64.

12. Henckel, P.A. (1964). Physiology of plants under drought. *Annual Review of Plant Physiology,115* :363-386.

13. Kargopolova, E.N.(1976). Speedy ripening of tomato and potato under influence of micro elements.

14. Kohli, V.K., I.V.Thakur and Y.R.Shukla. (1992). Studies on improving field emergence in the autumn crop of garden pea (pisum sativum L.). *Ann. Agric. Res., 13 (4)*: 394-395.

15. Kuperitskaya, M.A. (1978). The effect of seed treatment with minor elements on the yield and fruit quality of green house grown tomatoes. *Trudy Dal' nevost Nll S-Kh. 24*:149-150.

16. Lin, J.M. and Sung J.M. (2001). Pre-sowing treatments for improving emergence of bitter gourd seedlings under optimal and sub-optimal temperatures.

17. May, L.H., E.J.Milthorpe and F.L.Milthorpe. (1962). Pre-sowing hardening of plants to drought. *Fld. Crop Abstr., 15*: 93-98

18. Motta, De, M.B. and S.A.Leon. (1978). The effect of two methods of fertilization with molybdenum on the yield of the cauliflower cultivar snowball growing in the mosquera soil. *Seed Abstr., 1980, 3(3)*: p.99:830.

19. Muhyaddin, T and H.J. Wiebe. (2001) Influence of PEG seed treatment on emergence we under stress conditions. *Acta Horticulture*, 215.

20. Namedov. N.M.(1971). Effect of the micro elements on the growth, development and the yield of sugarbeet on irrigated light chestnut seeds. Boron in Agrl. July.PP.12.

21. Omran A.F, E.L.Bakry and R.A Gawish. (1980). *Seed Sci. & Technol., 8*: 161-168.

22. Sanchez, J.A, B.C Munoz and Fresneda J. (2001).Combined effects of hardening hydration-dehydration and heat shock treatments on the germination of tomato, pepper and cucumber. *Seed Sci. & Technol. 29* : 691-697.

23. Saraswathamma, D.N. and Jayachandra.(1979). The influence of pre-sowing soaking with mineral solution on the seedling growth in methi *Trigonella foenum-gracum* L. *Curr. Sci., 48(7)*:786-787.

24. Umarani, R., A.Bharathi, K.Vanangamudi, T.V.Karivaratharaju and R.S.V.Rai. (1999). Seed hardening studies in *Acacia nilotica* (Linn.) willd. Ex. Del. *Indian Journal of Forestry, 22(4)*:387-390.

25. Veeraraghavathetham,D. K.K.Vaidvelu and T.B.Ranganathan (1980). Seed invigouration in CO2 papaya . *South Indian Hort.*, 28 (3):69-71.

26. Viasova, V.I. and I.A.Kartasheva, (1981). Results of experiments on the effects of trace elements and growth substances of vegetable crops. *Nauchn. Tr.-Stavrop. S-Kh. Inst., 44 (1)*:76-83.

27. Vierling, E (1991). The roles of heat shock proteins in plants. *Annual Review of Plant Physiology and Plant Molecular Biology*,42 : 579-620.

28. Zubenko. V.KH. (1959). The effect of preplanting of seeds against drought and the grain harvest of corn in late plantings. *Pl. physiol.*, .6: 341-343.

CROP MANAGEMENT THROUGH SEED TREATMENT AND FOLIAR SPRAY TO IMPROVE PRODUCTIVITY

MALLIKA VANANGAMUDI[1], S. NANTHAKUMAR[2]
AND K.NATARAJAN[3]

1. INTRODUCTION

Indian agriculture is predominantly rainfed agriculture. Dryland areas are those where the annual evapotranspiration exceeds precipitation. World wide, 6150 million hectares of land are under rainfed agriculture. Out of the 143 million ha of total cultivated area in India, 107 m.ha are rainfed and encounters moisture stress at one time or other. The dryland area of the country contributes to about 44 per cent of the food grain production and support 40 per cent of India's population. Successful production in regions of frequent drought requires methods and practices for providing or maintaining sufficient available water for crop growth (Singh *et al.*, 1985). Among them, seed management and development of physiological methods that would reduce evapotranspiration are of significant importance (Singh *et al.*, 1990) to boost up yield under rainfed condition.

2. SEED MANAGEMENT AGAINST WATER STRESS

Seeds are hardened by soaking with some chemicals. To make the crop to tolerate drought and grow well under rainfed conditions, which leads to increase in osmotic potential of the cell protoplasm and thereby the water absorbing capacity is increased. Presowing hardening of seeds with few chemicals reduces the water requirement of crop at very early stages of germination. Besides, hardening of seed with chemicals make

1 *Department of Crop Physiology, Tamil Nadu Agricultural University, Coimbatore - 641 003*
2 *Department of Horticulture*
3 *Department of Plant Breeding and Genetics, Adhiparasakthi Agricultural College, G.B.Nagar, Kalavai - 632 506*

the protoplasm to become adaptable for moisture deficit condition for a period of 10 to 15 days under rainfed situations. The chemicals used for hardening technique are different for each crop as follows.

Crop	Chemicals
Paddy	Potassium chloride (1%)
Sorghum	Potassium dihydrogen phosphate (2%)
Maize	Cycocel (260 ppm)
Sunflower	GA_3 (500 ppm) or magnesium sulphate or ferrous sulphate (0.2%)
Groundnut	Calcium chloride (0.5%)
Greengram	Manganese sulphate (100 ppm)
Bengalgram	Potassium dihydrogen phosphate (1%)
Cotton	Cycocel (100 ppm) for 6 hrs. or potassium chloride (2%) for 12 hrs

Seed treatment

Crop	Treatment	Effect
Wheat, Maize and Cotton (salinity)	Soaked in 0.2% $MnSO_4$	Improved yield 15.3%, 6.2% and 29.5%
Wheat	NaCl and Na_2SO_4 @ 3%	Improved yield from 733 kg/ha to 2321 kg/ha
Barley	Gamma radiation (3000r)	Induced drought tolerance and growth
Ragi (late sown)	KCl (1%) + $CaCl_2$ (1%) 6 hrs soaking	Increased yield upto 1942 kg/ha over control 1560 kg/ha

3. CROP MANAGEMENT

Water requirement of crops vary with physiological stages of growth. Water stress at any growth stage will reduce the yield of crop depending on its duration and intensity. The most common symptoms of water deficiency are leaf rolling, leaf scorching, impaired tillering, stunting, delayed flowering and incomplete panicle exertion and grain filling.

4. PHYSIOLOGICAL IMPACT OF WATER STRESS

- Water deficits impair the nucleic acid system as a result of increasing RNAse activity.
- Degradation of RNA produced more than its synthesis and this, in turn affect enzyme production and growth (May and Milthorpe, 1962).
- Protein synthesis is very much affected by drought .

- Accumulation of certain amino acids, especially proline.
- Reducing cytokinin activity whereas increasing ABA content
- Reducing photosynthetic activity
- Increasing transpiration so as to reducing relative water content
- Increase the chlorophyllase activity and reducing the chlorophyll content and stability.

Cereals show a marked sensitivity to moisture stress during the formation of reproductive organs and flowering stages. About 70 per cent of the yield are lost due to stress at reduction division stage, which follows closely primordial initiation in rice. Moisture stress at this stage seriously reduces grain yield through reduction in the number of grains formed per panicle. Even a short period of water stress at anthesis will markedly reduced the number of flowers and seed set (May and Milthorpe, 1962). Among the different growth stages, crop plants are most sensitive to drought during reproductive growth stages (Table 1).

Table 1. Critical growth stages of crops for moisture stress

Crops	Critical stages	Changes due to water stress
Cereals		
Rice (rainfed)	Tillering, panicle initiation, flowering and milky	Reproductive structures affected.
Sorghum	Primordial initiation, booting and flowering	Decrease in grain number and weight
Maize	Knee high, tasselling, silking and maturity	Decrease in grain number and weight
Oilseeds		
Groundnut	Flowering, peg formation and pod development	Decrease in pod number
Gingelly	Flowering and maturity	Reduction in capsule number
Sunflower	Two weeks before and after flowering	Head size and weight reduced
Soybean	Blooming to seed formation	Reduction in pod number and pod weight
Pulses		
Greengram and blackgram	Vegetative and flowering	Reduction in pod number and grain
Cash crop		
Cotton	Flowering and boll development	Decrease in boll number and weight
Sugarcane	Formative stage (upto 120 days)	Reduction in cane height
Banana	All stages	Reduction in branch number

5. USE OF ANTITRANSPIRANTS

The loss of water from the living tissues of aerial parts of the plants in the form of water vapour is termed as transpiration. In transpiration almost 98 per cent of water absorbed by the plants is lost and only insignificant amount is being utilised by the plant for own purpose. Due to it, plant has to face several problems. Antitranspirant is a substance that is applied to transpiring plant surface with the aim of reducing water loss due to transpiration. In recent years, efforts on these lines have been made to find substances reducing exchange of gases during photosynthesis, respiration and plant growth. The term antitranspirant is used to designate any material applied to plant for the purpose of retarding transpiration. The widely used groups of antitranspirants are as follows.

5.1. Antitranspirants that form a film over the foliage surface

This group consists of colourless plastics, oils and low viscosity waxes that are sprayed in leaves with the object of forming a film permeable to CO_2 and O_2 but not to water. Only limited success has been achieved with this group. Kaolinite is also an antitranspirant that reflects light and reduces transpiration.

5.2. Substances that cause partial closure of stomata

A fungicidal substances Phenyl Mercuric Acetate (PMA) and aspirin are used today as one of the important and most promising stomatal closing type of antitranspirant. When applied at concentration of 10^{-4} M, PMA has been found to have little toxic effect on leaves and results in partial closure of stomatal pores for a period of 2 weeks or more.

5.3. Substances that induce complete closure of stomata

The growth regulator ABA is highly effective in causing complete closure of stomata. But the effect does not persist for more than few hours so ABA has no practical value. Carbon dioxide is an effective antitranspirant. A little increase in the natural 0.03% to 0.05% induces partial closure of stomata. Its high concentration cannot be used which results in complete closure of stomata affecting the photosynthesis and respiration. The commonly used antitranspirants for various crops are as follows (Table 2).

6. FOLIAR APPLICATION OF NUTRIENTS

Delayed flowering, spikelet sterility and incomplete emergence of panicle are the common symptoms of drought occurring at the

Table 2. Antitranspirants and their efficiency in various crops

Crop	Antitranspirants	Effect
Barley	Nonenyl succinic acid (stomatal closing type)	Decrease transpiration by 12 to 33%
Wheat	Polycot (3.6 to 9%)	Increase yield
Maize	PMA (10^{-3} M)	Decrease transpiration by 54%
Sorghum	Kaolin 3% foliar spray	Increase seed yield
Ragi (CO 13)	Salicylic acid 100 ppm at 25 and 50 DAS	Improved grain yield (3.9 tonnes/hr)
Millets	Hydroxyquinone (10^{-4} to 10^{-8})	Increase grain yield
Cotton	Kaolin (6-12%)	Increase yield
Cotton	Foliar spray of Kaolin 3% and liquid paraffin 1% spray	Increase yield by 27.4 per cent
Cotton	Salicylic acid 400 ppm	Increase the seed cotton yield, no. of bolls, ginning % (Patil, 1993)
Cotton (dry land)	PMa (10 ppm)	Increase yield by 28%
Groundnut	Salicylic acid 400 ppm	Increased the no. of branches, no. of filled pods, kernal weight and 100 seed weight (Patil and Wele, 1993)
Sesamum	Salicylic acid 100 ppm	Increased the capsule volume, number and yield (Sugumar, 2000)
Soybean	Triadimetron	Increase yield
Soybean	TPC (Tributyl phosphonium chloride)	Increase seed yield
Soybean	Kaolin 3% or liquid paraffin at 1% as foliar spray	Increase seed yield
Soybean	Foliar spray of salicylic acid @ 100 mg/litre on 30th and 45th day	Increased yield by 2.16 per cent
Greengram	Salicylic acid 100 ppm at 30 and 40 DAS	

reproductive phase. In addition, water stress will reduce the availability of nutrients further delaying panicle emergence and increasing the sterility. Under such situation, foliar spray of nutrients at critical growth stages will improve the emergence of panicle and reduce sterility. The foliar application recommended for mitigating water stress for different crops are as follows (Table 3).

Table 3. Foliar spray against moisture stress

Crops	Recommendation	Yield improvement
Rice	Foliar spray of muriate of potash 2.0% or DAP 2% at flowering initiation	400 kg ha^{-1} increased yield
Rice	Foliar spray of 0.3 ppm of Brassinolide at panicle initiation and flowering	Increased yield
Cumbu (Pearlmillet)	Combined sprays of 0.5% $ZnSO_4$ and 0.5% $CuSO_4$ together with 0.2% boric acid	16.8 per cent yield increase
Ragi (CO 13) (late sown)	Urea 2% at 40 and 50 DAS	Increased grain yield upto 23%
Groundnut	Spraying KCl at 0.5%	Improved yield
Groundnut	Foliar spray of DAP 0.5% along with planofix (40 ppm), ammonium sulphate (0.2%) and Boric acid (2%)	
Cotton (Rice fallow)	1.5% DAP and 0.5% potash at 45th and 55th DAS	Increased the seed yield
Sesamum	KCl @ 0.5% at flowering and capsule maturity stage	Increased the yield by 30% over control
Blackgram	Foliar spray of KCl at 0.5% during flowering phase	100 kg ha^{-1} increased yield
Blackgram and Greengram	Urea 2% spray	Increase the grain yield, no. of pods/plant
Blackgram (rice fallow)	DAP 2% + 0.5% $ZnSO_4$ at flowering stage	Effectively improved pod retention with higher yield
Greengram	KCl 1% at vegetative stage	Effectively improved pod retention with higher yield
Redgram	Combined application of 1% Ammonium phosphate, 0.5% KCl + 0.2% $ZnSO_4$ + 0.2% borax at flowering stage	Effectively improved the leaf area, photosynthetic rate and yield
Soybean	Urea 1% at 30 and 40 DAS	Increased grain yield (1612 kg/ha) over control (1080 kg/ha) (Kalarani *et al.*, 2000)
Soybean	1% KCl at flowering stage	Increased the yield by 19% over control

7. APPLICATION OF GROWTH HORMONES

7.1. Role of plant growth regulators in rainfed pulses

The chemical substances which are required for normal crop growth are producėd naturally in the crop and utilized for increasing growth and yield of the crop. When such substances are in low quantity the normal physiological functions are disrupted. As a result, 40 to 80% flowers shed which drastically reduce the yield. Under drought condition such growth regulators can be applied to the crop to increase the production. In rainfed crops, more over because of rain water shortage and high air temperature leaf, flower, buds, fruits shedding increases. To prevent this, according to the crop, growth regulators (Table 4) are used to regulate translocation of

Table 4. Growth regulators in agricultural crops

Crop	Chemical	Effect of spray
Rice	Kinetin and GA	Grain filling and partitioning. delayed senescence
Ragi (CO 13)	Cytokinin (10 ppm) at 40 and 50 DAS	Improved the grain yield leaf area and total drymatter production
Cotton	Triacontanol (1000 ppm)	Increased grain yield
Cotton	IAA, NAA (30 ppm), CCC	Decrease boll shedding
Cotton	NAA 40 ppm at 45, 60 and 75 DAS	Increased the seed cotton yield and reduce the boll shedding (Pothiraj *et al.*, 1995)
Sunflower	BA (250 ppm) GA + BA (150 ppm)	Increased yield, Achene weight and number
Groundnut	Mepiquat chloride (125 ppm)	Increased grain yield and chlorophyll synthesis and decreased chlorophyllase activity
Sugarcane	Ethephon	Reduced growth rate and regulate ripening
Pigeonpea	Ethrel (40 ppm) and GA (20 ppm)	Increased grain yield
	CCC (64 mM)	Respond well for RWC, Chrolophyll and stomatal conductance
Pigeonpea	NAA 40 ppm at 80, 90, 100 and 120 DAS	Improved seed yield, no. of pods and reduced pod shedding (Prakasa Rao and Narayanan, 1997)
Blackgram	NAA 40 ppm at 30th and 40th DAS	Increase the yield (Anon, 2000)
Bengalgram and soybean	NAA 100 ppm at 30th and 45th DAS	Increased the yield (Anon, 2000)
Greengram	NAA 150 ppm at 35th and 45th DAS	Increased the yield

photosynthates from leaves to developing grains. Because of this, growth regulators increase the number of floral buds, prevent flower, young fruits, matured fruits shedding, thereby increases the number of fruits and their weight and increases the final yield.

Water stress results in reduced cytokinin production in roots and its supply to shoots. This storage of cytokinin hastens the leaf senescence. Leaf senescence due to water stress can be postponed by the application of cytokinin or benzyl adenine.

Yet another approach for enhancing drought tolerance is to slow down the rate of vegetative growth by spraying specific chemicals. Growth retardants like CCC (2-dichloroethyl trimethyl ammonium chloride) have been found to slow down the vegetative growth and enhance the root growth and activity to reduce the plant susceptibility to water stress. Reduced vegetative growth enables the treated plants to survive the dry period especially at the time of ear emergence, when drought is most critical. The growth retardants (Table 5) in the treated plants results in reduced stomatal opening and hence lowered transpiration. Growth retardants like CCC also delays leaf senescence and increase the ability of the plant to survive drought.

Table 5. Growth retardants in agricultural crops

Crop	Chemical	Effect of spray
Wheat (late sown)	Foliar spray of CCC at vegetative stage	Increased the yield
Ragi	Seed treated with CCC 5 ppm solution	Yield increased by 12 and 18 per cent in Paiyur 1 and CO 11 respectively
Sunflower	MC (1000 ppm) at vegetative stage	Increased the seed yield and 100 seed weight
Pigeonpea	Cycocel (0.64 mM) at 45 and 75 DAS	Better for carbon exchange and stomatal conductance, RWC, chlorophyll and photosynthetic rate

The studies on water stress management in rice revealed that seed hardening with potassium chloride one per cent and CCC 500 ppm followed by foliar spray of 1000 ppm at critical water stress stage recorded the maximum grain yield of 5720 kg ha^{-1} as against 4080 kg ha^{-1} in control. In Brinjal, foliar spray of CCC at 5 ppm concentration improved the fruit set and fruit yield by 24 per cent under water stress condition at flowering stage.

8. CONCLUSION

Thus it can be concluded that, seed treatment foliar application of antitranspirants, nutrients, growth regulators and growth retardants are proved to be very effective in manipulating various physiological process at different growth stages from germination to maturation without affecting yield. However the selection of chemicals, optimum concentration and stage of application are very important factor for consideration before application. High concentration of chemical causes deleterious effects and less concentration may be ineffective. So we can use the optimum concentration, but it is varying with the species.

REFERENCES

1. Anon, (2000). Pulses production technology, pp.1-3
2. Kalarani, M.K., Mallika Vanangamudi and M.Thangaraj. (1999). The Hindu, Thursday, September 2.
3. May, L.H. and F.L.Milthorpe. (1962). Drought resistance of crop plant. *Field Crop Abstract, 15* : 171-179.
4. Patil, S.M. (1993). Effect of aspirin on summer groundnut. *PKV Res. J., 17(1)*: 117-118.
5. Patil, S.M. and A.D.Wele. (1992). Yield of cotton as influenced by antitranspirants and land surface modification. *PKV Res. J.,16(2)*: 265-266.
6. Pohlman, J.M. (1991). The mungbean. Oxford and IBH Pub. Co. Pvt. Ltd., New Delhi, p.375.
7. Pothiraj, P., N.T. Jaganathan, R.Venkitasamy, M.Premsekhar and S.Purushothaman. (1995). Effect of growth regulators in cotton MCU 9. *Madras Agric. J., 82(4)*: 283-284.
8. Singh, D.P., H.C.Sharma, B.D.Chaudhari, S.P.S.Karwasra, and P.Singh. (1990). Drought tolerance in closed brassicas and chickpea. Directorate of Research, HAU, Hissar, Haryana, 1-60.
9. Singh, D.P., P.Singh and H.C.Sharma. (1985). Effect of crop geometry at irrigation management on yield, evapotranspiration and water use efficiency of mustard. *Indian J. Agric., 30(1)* : 59-63.
10. Sugumar, R. (2000). Effect of different sulfur levels and plant growth regulators on seed yield and quality of sesamum. M.Sc.(Ag.) Thesis, TNAU, Coimbatore.
11. Tuner, N.C. (1979). Drought resistance and adaptation to water deficits in crop plants. In: Stress physiology in crop plants (pp.343-372). Russel, H. and R.C.Staples, Eds., Wiley, NY.
12. Turner, N.C. (1986). Adaptation to water deficits: a changing perspective. *Aus. J. Pl. Physiol., 13* : 175-190.

PHYSIOLOGICAL BASIS OF DROUGHT TOLERANCE INDUCED BY SEED HARDENING

G.PADMANABHAN

1. INTRODUCTION

Rapid and uniform field emergence are two essential prerequisites to increase the yield, quality and ultimately profits in annual crops. Uniformity and percentage of seedling emergence of direct-seeded crops have a major impact on final yield and quality. Rate and uniformity of emergence are inherent to seed quality and environmental conditions during seedling emergence. Slow emergence resulted in smaller plants and seedlings, which are more vulnerable to soil-borne diseases. Extended emergence periods predispose the planting bed to deterioration and increased soil compaction, particularly under adverse environmental conditions.

Various presowing seed treatments have been used to reduce the time between seed sowing and seedling emergence and to improve synchronization of field emergence in annual as well as perennial crops.

In the last two decades, seed priming has become a common seed treatment to increase the rate and uniformity of emergence in many vegetable and flower species. The term "priming" of seeds was used to describe, a presowing seed treatment to enhance germination and increase seedling emergence with uniformity under adverse environmental conditions. The terms such as halopriming (soaking in salt solutions) or osmopriming (soaking in other osmotic solutions) were proposed as alternatives to priming, osmoconditioning or osmotic conditioning are also used to describe the same treatment. The term "Solid Matrix Priming"

Department of Crop Physiology, Tamil Nadu Agricultural University, Coimbatore-641 003.

(SMP) was used for a presowing seed treatment in which a solid-matrix instead of an osmotic solution was used to enhance germination. SMP is a process that uses solid matrix materials, water, and seed in combinations to control water, oxygen, and temperature effects on germination. This process controls the hydration of seed to a level that allows "pre-germination" activity but that prevents radicle emergence. Matriconditioning was proposed as an alternative term to SMP, to distinguish seed conditioning by matric and osmotic forces. The word "biopriming", is used for treatment where sweet corn seeds were coated with a bacteria and soaked in warm water until the seed moisture content increased to 35-40%. The use of a salt as an osmoticum can lead to an increase in fresh weight (water) of a seed. In this case, germination is delayed through increased solute potential of the embryo.

Various reports on seed hydration and dehydration physiology and their effects on germination suggested that inhibition of germination by water stress was controlled via restriction of cell elongation processes. The membrane may play a role in controlling cell elongation and that membrane integrity was important for seed activation and repair at various seed hydration levels. The osmotic potential may be decreased during priming due to solute accumulation in the embryo, which might generate sufficient turgor pressure to overcome endosperm or seed coat restraint.

2. SEED PRIMING

Seed priming is defined as a presowing treatment in which seeds are soaked in osmotic solution that allows them to imbibe water and go through the first stages of germination, but does not permit radicle protrusion through the seed coat. The seeds can be dried to their original moisture content and stored or planted via conventional techniques. This definition can be partially extended to SMP. The main difference between the two treatments is that in SMP, matric and/or osmotic forces regular water uptake by the seed instead of the osmotic potential of a liquid soak solution.

2.1. Seed Priming and Osmotic Solutions

Several osmotica have been used for seed priming (Table 1). Inorganic salts, such as KNO_3, K_3PO_4, and a chemically inert compound, polyethylene glycol 6000 (PEG 6000) or a polyethylene glycol 8000 (PEG 8000) are materials most commonly used to adjust osmotic potential.

Various inorganic salts were first used as osmotica for seed treatments. Greater germination of pepper seeds, was reported when they were soaked in a solution containing $NaNO_3$, $MnSO_4$ and $MgCl_2$. Use of K_3PO_4 and KNO_3 solution improved the germination rate of the tomato at low temperatures (below 20°C).

Table 1. Osmoticants used for seed priming in selected species

Common Name	Species	Osmoticum
Beet	*Beta vulgaris* L.	PEG, $MgSO_4$
Cabbage	*Brassica oleracea* L. Capitata group	PEG
Carrot	*Daucus carota* L.	PEG
Cucumber	*Cucumis sativus* L.	Mannitol NaCl, Milk
Maize	*Zea mays* L.	PEG, KH_2PO_4, K_2HPO_4, $KNO_3 + KH_2PO_4$
Onion	*Allium cepa* L.	PEG
		PEG + Enriched air
		Mannitol, Nacl
		Glycerol
		$KNO_3 + K_2HPO_4$
		K_3PO_4, KNO_3
Pea	*Pisum sativum* L.	PEG
Peanut	*Arachis hypogea* L.	PEG
Pepper	*Capsicum annuum* L.	PEG, Mannitol, Sodium Poly-propionate
		KNO_4, KH_2PO_4, $KNO_3 + K_2HOP_4$, NaCl $MgSO_4$
Sorghum	*Sorghum bicolor* (L.) Moench	PEG, KNO_3, K_2HPO_4, KPO_4, KNO_3 + K_3PO_4, $KNO_3 + K_2HPO_4$
Soybean	*Glycine max* (L.)	PEG, Mannitol, PEG + GA_3 + Kinetin Sodium polypro-pionate
Sweet corn	*Zea mays* L.	PEG
Tomato	*Lycopersicon* spp	PEG, Mannitol, NaCl, Sodium polypro-pionate, KNO_3, K_2HPO_4. K_3PO_4, KNO_3 + K_2HPO_4,

2.1.1. *Effect of the salts*

In fact, the term "vigourization" is used to denote the vigour of seeds that had been improved by the priming process. The utilization of inorganic salts as osmotica has generated conflicting recommendations. Seeds of leek and carrot had a reduced germination percentage when they were primed with KH_2PO_4 as compared with PEG 6000. The primed carrot seeds had improved emergence when inorganic salts (KNO_3, K_3PO_4) were used as osmotica instead of PEG. Tomato seeds primed in KNO_3 solution germinated more rapidly at 15°C as compared to seeds primed in PEG 8000 solutions. There were no differences among germination rates of pepper seeds, when PEG, $MgSO_4$, or NaCl were used as osmotica. Pepper seeds emerged more rapidly after priming in a solution of KNO_3 + K_3PO_4 as compared to seeds primed in PEG 6000. Seedlings from 'Jalapeno' pepper seeds primed in PEG 6000 solution had lower dry weight compared with seedlings from seeds primed in salt solutions. Thus, the results from using the various osmotica are mixed. Sometimes salts are more effective osmotica, sometimes PEG, and sometimes similar results are obtained from either osmotica.

Compared to PEG, salts are easier to aerate and to remove from the seed after treatment, are less costly, and may provide a nutritional effect. Nitrogen, potassium, or phosphate concentration in embryos did not change after soaking cotton seeds in salt solutions; however, the concentration of micronutrients such as manganese and copper increased. The size, structure, biochemical constitution, position of seed protecting layers, and the type of salt and soaking time are the principal factors influencing ion penetration into the embryo. High ion concentration in the embryo may affect germination adversely when inorganic salts are used as osmotica in the seed priming process. It was reported that the lowest leek seeds recorded germination percentage when primed in KH_2PO_4 solution than in PEG solutions. The seeds of carrot, leek, and onion absorbed less water when they were primed in PEG or glycerol solutions than in KH_2PO_4 solutions. Lettuce seeds primed in K_3PO_4 solution absorbed more water than seeds primed in PEG or PEG + K_3PO_4 solution or soaked in water. These results suggest that ions from inorganic salt solutions are able to penetrate and accumulate inside the seed during priming. The accumulated salt might then reduce the osmotic potential of the seed and thereby induce more water absorption as the treatment progresses.

Polyethylene glycol (PEG) is also commonly used as a priming osmoticum. PEG has advantages over inorganic salts because it is chemically inert and does not have adverse effects on the embryo. The large size of the PEG molecule prevents its penetration into seed tissues, thus reducing toxic side effect on seeds. The disadvantages of using PEG are its relative high cost, problems in removing it from the seed coat after priming, obtaining uniform aeration during priming, and environmental hazards with PEG disposal.

In addition to inorganic salts or PEG, saccharides such as mannitol have been used successfully to regulate the osmotic potential of the priming solution. Improved germination rate and final germination percentage in pepper seeds after priming in mannitol solution was reported. Similarly, faster germination and larger seedlings from eggplant, pepper, and melon seeds primed in mannitol solution than from nonprimed seeds was reported. The germination rate and final germination percentage of leek seeds at high temperature were improved by using mannitol as a priming osmoticum. Seeds of tomato and cucumber germinated under unfavourable light conditions (continuous far-red) had faster and greater final germination compared to untreated seeds after priming in mannitol solution.

Other products such as sodium polypropionate, glycerol, and synthetic seawater also have been used as alternative osmotica to salts and PEG in seed priming. A mixture of various salts in the solution is a common recommendation for many plant species.

Osmotic agents may be combined with seed protectants or growth regulators to improve the effect of seed priming. The addition of fungicide to the priming solution prevented pathogen growth during priming. PEG or salts combined with growth regulators have improved germination under stress condition. Under suboptimal temperatures and soil crusting, greater emergence of soybean seedlings was observed when seeds were primed in a solution of PEG and gibberellic acid (GA_3). When celery seeds were primed in PEG with ethephon and GA, high temperature dormancy was overcome and germination was synchronized. Thermodormancy of lettuce was reduced by priming the seeds in a solution of K_3PO_4 and benzyladenine (BA). The incorporation of BA (100 mg/liter) in a PEG priming solution increased the germination of 'Earlybelle' celery at 25°C to 43% as compared to 27% when PEG was used alone. Priming pepper seeds in a PEG + GA solution compared to PEG alone increased germination rate of pepper slightly at low temperature. *Antirrhinum majus* seeds had more germination at low temperature (15°) after priming in PEG + GA solution compared to nonprimed or primed seed in PEG alone.

2.2. Solid Matrix Priming

The considerable volume of solution required per seed and problems with aeration of the solution can become obstacles when priming large-seeded species or large quantities of seed. Furthermore, the cost of the osmotica such as PEG is high, and the osmotica must be removed from the seeds before they are dried. Disposal of both salts and PEG are an environmental concern.

Solid matrix priming (SMP) consists of mixing seeds with an organic or inorganic carrier and water for a prescribed period of time. The moisture content of the mixture is brought to a level just below that required for radicle protrusion. Seed water uptake is regulated by the matric potential of the seed.

Different materials have been used as a solid-matrix priming agent. The ideal characteristics of these materials were (1) no phytotoxicity, (2) high water-holding capacity, (3) remain friable at different moisture contents, and (4) easy to remove from seeds after treatment. SMP has a potential application to both large-and small-seeded species and large amounts of seeds can be efficiently primed at one time. Promising results from SMP have been reported in many species. The SMP treatment reduced the time between seeding and emergence of tomato, onion, and carrot seeds compared to nonprimed seeds. After SMP treatment using calcined clay, pepper seeds had more rapid and greater final germination compared to nonprimed seeds. Primed table beet seeds via SMP had faster field emergence and better final stand compared to nontreated seeds.

The beneficial *Pseudomonas fluorescens* survived the process after dehydration giving rise to SMP as a method to add biological control agents to seeds. SMP and sodium hypochlorite also increased the rate and final germination percentage of celery at high temperatures. Improved germination rate and marketable yield of table beet was reported after SMP using a combination of synthetic calcium silicate as a solid-matrix with fungicides.

Besides growth regulators, phosphate and the fungicide Apin (metalaxyl) and thiabendazole were added to cucumber seeds during SMP and improved germination percentage and rate. SMP with captan (Carboximide) also improved emergence of cucumbers planted in soil infected with *Pythium* and *Rhizoctonia*.

2.3. Factors affecting seed priming

2.3.1. Aeration

Several factors must be controlled during the priming treatment in order to improve seed performance. Proper aeration of the osmotic solution was essential to living seeds and helped in seed germination. It was showed that the response to aeration during priming varied according to species. Onion seeds primed in an aerated PEG solution, using enriched oxygen, had greater per cent germination after treatment compared to seeds primed in nonaerated solutions.No differences in final germination of lettuce were observed after seed priming in aerated or nonaerated solutions of K_3PO_4.

Aeration of the solution, however, modified the optimal duration of priming. Regardless of the osmoticum or temperature used during lettuce seed priming, aeration of the priming solution lessened soaking time to achieve greater germination percentage at 35°C.

2.3.2. *Light*

The beneficial effects of seed priming could be modified by light quality during treatment. When seeds of endive were primed in KNO_3 solution under continuous red light, they had a greater germination percentage and more uniform germination than seeds primed in the dark. Low light levels from fluorescent and incandescent bulbs during priming did not affect germination of lettuce 'Mesa 659' as compared to seeds primed in the dark. However, with another two cultivars, 'Minetto' and 'Ithaca', the seeds had higher germination if they were primed in light. Seed priming in K_3PO_4 solution improved germination of 'Valmaine' lettuce at high temperature regardless of light conditions. However, seeds germinated more rapidly if they were primed in the dark. Celery seeds primed in a solution of PEG + BA had more germination at high temperature when the priming treatment was done under light than in the dark.

2.3.3. *Duration of the Priming Process*

The ideal duration of the priming process varies according to the type of osmotica, osmotic potential of the solution, temperature during the treatment, and species. If radicle protrusion occurs during seed priming, irreversible embryo damage could be expected during dehydration after priming. Seeds primed with PEG for 6 days has the fastest germination, but they produced almost 60% abnormal seedlings compared to 14% or 0% of seeds primed for 5 or 4 days, respectively. In some species, such as lettuce, the duration of priming might be several hours. Lettuce seeds needed only 6 h of soaking in Na_2HPO_4 solution to overcome thermodormancy. Other species required weeks of soaking to produce significant effects.

2.3.4. *Temperature*

Duration of seed priming depends on the species and seed quality. However, temperature during treatment and osmotic potential of the solution affect the length of the soaking period. If the soak temperature is maintained below the optimum range for rapid germination, radicle growth during priming may be restricted. At 20°C, a high percentage (63%) of carrot seeds germinated in the priming solution (-0.5 MPa PEG), whereas no germination was observed with solution at 15°C.

The temperature during treatment also appears to be related to the type and concentration of osmotica, and length of the treatment period. Temperature during priming may alter the soaking period in many species. Seed priming for 14 days at 15°C was more effective in improving seed germination of tomato, carrot, and onion than 14 days at 25°C.

2.3.5. *Osmotic Potential*

Osmotic potential of the solution is another factor affecting the length and effectiveness of the priming treatment. seed priming in high osmotic potential solutions may allow rapid seed germination and radicle growth because seed water contents become too great. Decreasing the osmotic potential from -0.5 to -1.0 MPa prevented the germination of carrot seeds in the priming solution. A wide range of osmotic potentials has been used with different responses. Tomato seeds germinated more rapidly when the osmotic potential of the solution was -0.58 or -0.86 MPa compared to -1.19 or -1.46 MPa.

2.3.6. *Seed Quality*

A differential response to seed priming among seed lots of the same species has been reported. Seed vigour is another factor influencing the response to seed priming. However, those seeds, which were still viable after ageing, had improved germination rate after priming. Priming accelerated the germination rate of both high and low vigour seed lots, but not to the same extent. The high vigour lots generally performed more consistently. The differential response was attributed to seed genotype rather than seed vigour. Seed companies generally conduct some type of seed vigour analysis before priming a seed lot because of the effects of seed quality on priming.

2.3.7. *Dehydration after Priming*

To achieve the maximum potential of priming, strict control of all the factors previously discussed is required. However, following priming and dehydration subsequent seed storage become crucial components of this technology for commercial application. To facilitate handling and storage, the seeds should be dried to an acceptable moisture level for long-term storage after the priming treatment. Water uptake during priming is influenced by the osmotic potential of the solution, type of osmoticum, duration of the treatment, and the physical/chemical characteristics of the seeds.

2.3.8. Seed Storage

After seed priming and dehydration, seeds usually are stored before planting. The conditions and duration of storage may alter the advantages achieved from priming. Primed spinach seed maintained high germination at stressful high temperatures after 1 month of storage.

An interaction between the osmotica used during the treatment and the ability of the seed to store well has been reported. Seeds of tomato primed in KNO_3 solution has a lower capacity to tolerate high temperature storage (30°) than seeds primed in a PEG solution. Germination rate was not affected in nonprimed tomato seeds stored for 6 months at 30°C. Following the same storage period and conditions, seeds primed in PEG or KNO_3 lost almost 50% of the germination rate gained immediately after the priming treatment. Viability and germination of primed tomato seeds were severely reduced after 6 months of storage at 30°C; however, no deleterious effects were observed in seeds stored at 4°C for the same.

2.4. Effect of seed priming on seedling growth and development

Seed priming treatment modify embryonic axis growth and subsequent seedling development. The response varies according to the species and priming conditions. Embryo volume and cell number per embryo of leek and onion were not modified by priming. In the same experiment and under the same conditions, the carrot embryo volume increased almost 50% and the number of cells increased by twofold. Progressive rupture of the endosperm after 9 h of priming was one of the potential factors leading to increased germination of 'Minetto' lettuce seeds at high temperature.

The number of basal and lateral roots and taproot length, 14 days after seeding were not modified by seed priming in pepper. Root length of primed lettuce seeds germinated at 35°C was greater than that of nonprimed seeds after 96 h. Root dry weights of primed and nonprimed bell peppers, grown with sprinkler or drip irrigation, were similar 50, 70 and 90 days after planting. Seed priming did not affect the rate of radicle growth and degree of root branching in tomato seeds after germination. Primed parsley seeds yielded 52% more shoot fresh weight compared to nonprimed seeds 24 days after sowing. The differences in root or shoot growth between primed seeds and nonprimed seeds is more evident under stressful conditions. Root growth from perennial ryegrass seeds germinated at low temperature (5°, 10°, or 15°) was greater is primed seeds than in nonprimed seeds, but no differences were observed at 25°C. Shoot dry weight was higher for primed tomato seeds germinated at stressful temperature conditions than for nonprimed seeds. Generally, the major effects of seed priming on growth has been observed as earlier more-uniform emergence and not accelerated growth, *per se* of the species.

2.5. Molecular and physiological aspects of seed priming

The physiological and biochemical effects of seed priming during or after the treatment are not understood completely. Several reports shows that seed priming treatment leads to changes in seed nucleic acids. The synthesis of RNA, proteins, and low levels of DNA occurred as a normal germination process during hydration. However, numerous metabolites were stable after dehydration and accounted for the ability of the embryo to synthesize proteins and RNA rapidly after rehydration. Lettuce seeds primed in PEG 600 had increased enzymatic activity (acid phosphatase and esterase) and reduced time for RNA and protein synthesis than nonprimed seeds. There was no change in protein quality of the embryonic axis of pea after soaking in osmotic solutions. Amylase activity during germination of primed pea and tomato, and peroxidase activity of primed tomato and spinach were higher than nonprimed seeds. High levels of nucleic acids in primed tomato seeds may account for their rapid germination. The increase in nucleic acid content amount was due to RNA synthesis during and after priming treatment. In later experiments, tomato seeds primed in distilled water at 10°C for 35 days then redried, all nongerminated seeds showed a large increase in nucleic acid content, especially rRNA. During subsequent germination, treated seeds had reduced rates of nucleic acid accumulation and reduced RNA polymerase activity per unit DNA compared to the control. It was postulated that rRNA synthesis within the primed seeds was under same type of stringent control and extensive rRNA synthesis associated with the priming treatment was unlikely to be the immediate cause of increased germination rates.

SEED FORTIFICATION, INFUSION AND OSMOPRIMING

R. JERLIN[1], R. UMARANI[2] AND K. VANANGAMUDI[3]

1. INTRODUCTION

High quality seed is the key to successful agriculture. Modern agriculture with its bias for technology and precision, demands that each and every seed should readily germinate and produce a vigourous seedling ensuring high yield. Uniformity of growth and synchrony in development are highly desirable characters for mechanized cultural operations. As such, only of high quality i.e., genetically pure and morphologically, pathologically and physiologically sound seed is capable of increasing the productivity. Apart from increased productivity, the seeds should also have better storability to produce good crop during the next season. To accomplish these characters, the seeds should be properly invigourated by way of seed fortification, infusion and osmopriming.

2. SEED VIGOUR

It is the quality of the seed responsible for rapid and uniform germination, extended storability, good field emergence and ability to perform well in a wide range of edaphoclimatic conditions.

3. SEED INVIGOURATION

Seed invigouration is an improvement in seed performance by any post harvest treatment which results in improved germinability, greater

1 *Agricultural Research Station, Bhavanisagar – 638 451.*
2 *Department of Seed Science and Technology, Tamil Nadu Agricultural University, Coimbatore – 641 003.*
3 *Dean (Agriculture), Tamil Nadu Agricultural University, Coimbatore - 641 003.*

storability and better field performance than the corresponding untreated control.

These invigouration treatments are physiological in nature. These treatments can be broadly classified as,

- Pre-sowing treatments for improved field performance
- Pre – storage treatments for better storability
- Mid-storage treatments for improving vigour, viability and productivity of old seeds.

Seed fortification, infusion and osmopriming are the important pre-sowing treatments to accomplish the seed vigour and thereby improving the plant stand in the field and finally increasing the yield and quality of the resultant seed.

4. SEED FORTIFICATION

It is an impregnation of required substances for invigourating the seed for improved production where, the seeds are soaked either in water or dilute solutions of bioactive chemicals such as micronutrients, growth regulators, vitamins and seed protectants.

4.1. Method

Seeds are soaked in solutions of equal volume for duration of 6-24 h depending upon the crop to raise the moisture content of the seed to 20-25% just enough for endogenous impregnation of the chemicals by exogenous application. The choice of chemicals, its concentration and duration of soaking vary with the species and decides the success of the treatments.

4.2. Chemicals used for fortification

The macro and micro nutrients are used for fortification as they are essential for the physiological activation of seeds.

Nitrogen : constituent of aminoacids, nucleotides and also act as co-enzymes.

Phosphorous: constituent of phospholipids, nucleotides and also sugar phosphates.

Potassium: required by all higher plants for activation of enzymatic process.

5. SEED INFUSION

It is a method of impregnation of seeds with bioactive chemicals through organic solvents instead of water. This technique of infusion helps to avoid the damage caused to the seed due to soaking in water. Hence, this method is highly suitable to seeds that suffer from soaking/ seed coat injury. Simultaneously, it also provides protective, regulatory and selective functions of the chemicals to improve the performance of seed with the help of organic solvents.

5.1 Method

Desired solutes are dissolved in the organic solvent at desired concentration. Seeds are immersed in solvent with desired solute for 5 to 24 h depending on species. After the desired period, the solvent is evaporated by air or by vacuum desiccator for 30 minutes to 1 h or under ambient condition for 24 h for complete evaporation of solvent from the seed. The seeds are then ready for sowing.

The depth and amount of chemical penetration into the seed vary with duration of treatment (amount of chemical reached the embryo increases with increase in time), concentration of solutes (higher the concentration more will be the penetration) and type of chemical compared to IAA, lesser quantity of GA_3 reaches the embryo lesser in a given time due to its higher molecular weight.

5.2. Changes caused due to infusion treatment

1. Mechanical integration of cell membrane is altered by increasing its permeability.
2. Membrane proteins are denatured.
3. Inhibitors in seed coat embryo are leached out
4. Enzyme systems are activated.
5. It enhances the polyribosome formation and increases germination of seeds.

5.3. Bio-active chemicals that can be infused through organic solvents

1. Seed protectants: fungicide, bacteriocide.
2. Herbicide antidotes: To prevent the ill effects of herbicide
3. Growth regulators: GA_3, IAA, kinetin are incorporated to nullify the endogenous deficit of the growth substances.
4. Nutrients: Macro and micro nutrients

5. Substance to resist adverse stress conditions: Substances like fusicoccin alleviate the adverse effect in soil.

5.4. Organic solvents used for infusion

1. Acetone
2. Petroleum ether
3. Dichloromethane
4. Ethanol
5. Dichloroethane

5.5. Benefits of infusion technique

1. Simple technique, easy to operate on large quantities of seed.
2. Performance is better to other methods.
3. Can be used for incorporating all chemicals into seed, since it dissolves all chemicals easily.
4. The treatment duration is short and drying back is an easier process. Normally hazards of drying do not occur.
5. Rate of deterioration of seed is much controlled
6. Toxicity of chemicals to adhering fauna and flora is prevented. Hence, environmental pollution is controlled.
7. Soaking/hydration injury does not occur in seed.
8. Bioactive chemicals are made available to the embryo directly.

6. OSMOPRIMING/OSMOCONDITIONING

It is a pre-sowing treatment in which seeds are soaked in an osmotic solution that allows the seeds to imbibe water and go through the first stages of germination but does not permit radicle protrusion through the seed coat. It is based on the principle of controlled imbibition of the seed to a level that permits pre germinative metabolism to proceed, but prevents actual emergence of the radicle (Bradford, 1986).

7. OSMOLYTES

The salts used to prepare the osmotic solution are called osmolytes or osmoticants. The most common osmolytes are inorgnic salts such as KNO_3 (Ells, 1963), K_2PO_4 (Cantiliffe, 1981) and a chemically inert compound, polyethylene glycol PEG 6000 or PEG 8000 (Ali *et al.* 1981). Use of PEG as a pre-sowing treatment has been much more successful in terms of germinability, uniformity of stand and subsequent crop performance than the simple water soaking treatments. PEG regulates the entry of water into the seed, without carrying injury in leguminous seeds, and advances germination to more or less fixed level determined by the water potential. On sowing, individual seeds of a lot subjected to osmoconditioning could start from near identical levels of metabolism and could therefore show rapid and uniform germinability, synchrony in growth and development and in many cases increased biological and agricultural yield.

Apart from PEG, other inorganic salts like KNO_3 and K_2PO_4, the saccharides such as mannitol have been used successfully to regulate the osmotic potential of the priming solution. Other products such as sodium polypropionate and synthetic sea water also have been used as an alternative osmolytes to salts and PEG in seed priming.

Various inorganic salts were first used as osmotica for seed treatments. In early work, Kotowski (1926) reported greater germination of pepper seeds, after they were soaked in a solution containing $NaNO_3$, $MnSo_4$ and $MgCl_2$. Ells (1963) using K_3PO_4 solution, improved the germination rate of the tomato at low temperatures (below 20^0C). However, this and other early reports attributed the advantage of the treatments to a nutritional influency on penetration into the embryo. High concentration in the embryo may affect germination adversely when inorganic salts are used as osmotica in the seed priming process.

Heydecker (1973) used different terms depending upon the methods adopted for the treatment such as

(i) Osmopriming - Soaking the seeds in osmotic solutions.
(ii) Solid matric priming - Use of solid matric as a carrier agent for osmotic solution.
(iii) Halopriming - Soaking the seeds in salt solutions.
(iv) Biopriming -Coating the seeds with biological agents like bacteria.

8. MECHANISM OF PRIMING

Seeds require water, oxygen and suitable temperature for germination. The water uptake in seeds follows a triphasic pattern with an initial rapid uptake phase known as imbibition (Phase 1) followed by lag period (Phase 2) and then a second increase in water uptake associated with seedling growth (Phase 3). Seeds are desiccation tolerant during phase 1 and 2 but become intolerant during phase 3. At each phase, water uptake is controlled by the availability of water to the seed. In seed priming regime, seed water potential is maintained at a level sufficient enough to initiate metabolic events in phase 2 of germination process, but which prevent radicle emergence. Germination response to priming are obtained approximately at a seed moisture content of 30% and it increased linearly over the range of 45 to 50%, the upper limit depending upon the species.

Factors influencing the efficacy of osmopriming are concentration of osmoticants, duration of treatment, temperature, light intensity, manner and extent of drying, type of seed and its response to osmoconditioning.

8.1. Solid matric priming (Matriconditioning)

For large scale priming, considerable volume of solution are required and aeration of seed during priming becomes difficult. Also the cost of chemicals, their removal from seeds and disposal are problematic. In order to overcome this, solid matric priming is advocated.

Solid matric priming consists of mixing seeds with an organic or inorganic carrier and water for a period of time. The moisture content of matric is brought to a level just below what is required for radicle protrusion. Seed water potential is regulated by the matric potential of the seed and during priming the water is largely held by the carrier. Seeds imbibe water from the carrier till the equilibrium is reached. This controlled hydration allows specific quantity of water directly to get in contact with the seeds inorder to achieve an elevated moisture content. The weight of water needed to increase a given seed mass to a desired moisture content can be calculated by:-

$$Wt.H_2O=Wt_1\ (Mcf-Mci)\ (100-Mcf^{-1})$$

Where $Wt.H_2O$ of water to be added to the initial weight of dry seeds ($wt._1$), Mc_1 the moisture content of the initial seed sample and Mcf, the final desired moisture.

8.2. Drum priming

The Drum priming method involves the hydration of seeds over a period of 24-48 hrs in a drum revolving at 1-2 cm/sec. Mixing of seed is very uniform and seed at the end of this period are plumb but surface dry. The degree of hydration is determined by a simple calibration test for each lot. A preset volume of water is injected during each cycle as regulated by a timer attached.

9. POST TREATMENT EFFECTS

Drying (dehydration) and subsequent storage of primed seeds are crucial component of this technology for its commercial application. For proper handling and storage, the seeds should be dried to acceptable moisture content and at the same time the beneficial effect of priming should also be maintained. Temperature and rate of dehydration following priming are critical factor affecting seed quality.

Rapid dehydration of imbibed pepper and carrot seeds resulted in delayed germination upon dehydration which completely eliminated the advantage gained by priming. Situation requiring drying seeds back to original water content before sowing, exposure to lower water potential would be required even if first drying stage occurred at higher water potential.

Preconditioning the seeds by first drying to - 4Mpa prior to drying at - 150 Mpa retained rapid subsequent germination rate.

After drying, the primed seeds are generally stored before sowing. The storage environment can have profound effect on primed seeds efficacy and longevity. The beneficial, neutral and deleterious effects of priming after storage has been reported. The RH, temperature and duration of storage are responsible for these conflicting reports.

10. BENEFITS

- Faster emergence and more uniform field stand in normal as well as in stress situations.
- Besides uniformity and synchrony, significant yield increase in many vegetable crops in particular, has been reported.
- Osmoconditioning would very effectively overcome the serious problem of soaking injury in many legumes.
- Osmoconditioned seeds can be rinsed and carefully dried for restorage for short periods.

11. LIMITATIONS

- Osmotic seed treatment is expensive, especially when osmotica such as PEG are used. Only low volume, high value seeds are suitable for this treatment.
- The bathing solution needs continuous aeration, otherwise the seed will suffer from the ill effects of anaerobic respiration.
- As the treatment is done over a long period, proper caution against microbial attack should be taken.
- Vigour of the seed at the time of treatment may determine the efficacy of treatment.

- Storability of the seeds are adversely affected by osmoconditioning because of advancement of germination reactions to a level not compatible with the drying back of the seed required for restorage.
- In certain materials, there is a possibility of induction of secondary dormancy, particularly under unfavourable temperature and light conditions.

12. CONCLUSION

Improvement of seed quality by physiological treatments is a simple, easy and inexpensive approach to enhance seed performance and agricultural productivity. Currently these techniques are being adopted mostly for the high value seeds. In order to exploit the potential of seed invigouration treatment with respect to other popularly cultivated crops species, further investigations with respect to low cost invigouration procedures and mechanization of these procedures are warranted.

REFERENCES

1. Ali, A.V., Souza Machado and A.S. Hamil (1990). Osmoconditioning of tomato and onion seed. *Scientia Hort., 43*: 213-224.
2. Bradford, K.J. (1986). Manipulation of seed water relations via osmotic priming to improve germination under stress conditions. *Hort. Sci., 21*: 1105-1112.
3. Cantliffe, D.J. (1981). Seed priming of lettuce for early and uniform emergence under conditions of environmental stress. *Acta. Hort., 122*: 29-38.
4. Ells, J.E. (1963). The influence of treating tomato seed with nutrient solutions on emergence rate and seedling growth. *Proc. Am. Sco. Hort. Sci., 83*: 684-687.
5. Heydecker, W. (1973). Germination of an idea : The priming of seeds. University of Nottingham, school of Agriculture Rep. 1973/74.
6. Kotowski, F. (1926). Chemical stimulants and germination of seeds. *Proc. Am. Sco. Hort. Sci., 23*: 173-176.

CONCEPTUAL AND PHYSIOLOGICAL DIFFERENCES BETWEEN SEED HARDENING AND OTHER SEED INVIGOURATION TREATMENTS

R.UMARANI[1] AND R.JERLIN[2]

1. INTRODUCTION

Seed invigouration implies an improvement in seed vigour by any post harvest treatment resulting in improved germinability, greater storability and better field performance than the corresponding untreated seed. An invigouration treatment should bring about a qualitative improvement in the seed which should persist after the treatment is winded up. True invigouration treatments are physiological treatments in the sense that all resultant effects would be the manifestations of improved physiological status of the invigourated seed.

Physiological seed invigouration treatments comprise (i) presowing, (ii) pre-storage and (iii) mid storage seed treatment.

S.No.	Presowing	Prestorage	Mid storage
1.	Seed fortification*	Seed halogenation	Hydration-Dehydration*
2.	Seed hardening*	Dry permeation	
3.	Seed pelleting		
4.	Seed priming*		

* Treatments involving soaking and drying of seeds.

The seed treatments *viz.*, seed fortification, seed hardening, seed priming, hydration and dehydration requires a same methodology of soaking and drying.

1 *Department of Seed Science and Technology, Tamil Nadu Agricultural University, Coimbatore- 641 003.*
2 *Agricultural Research Station, Bhavanisagar – 638 451.*

However, there are imperative but intricate differences in the methodologies and the ensuring mode of actions within the seed. All people who are engrossed in Seed Science and enthralled by seed mechanisms would appreciate the nuances in the methodology and modes of action of different yet similar seed treatments.

2. SEED FORTIFICATION

2.1. What is it?

It is process of enriching the seeds with bioactive chemicals for improving the germination and seedling vigour and to overcome dormancy (Khan, 1977). It also improves the seed storability (Basu, 1984).

2.2. Why it is done?

In order to improve the vigour of the seeds which can help to improve the seed germination, seedling vigour and storability. Specific chemicals can also help to overcome dormancy.

2.3. How it is done?

To the known volume of seeds, one-third volume of the nutrient solution is added and allowed to imbibe for short duration. The imbibed seeds should be dried under shade. If immediate sowing is planned, drying back to original moisture content is not important otherwise seeds should be dried back to original moisture content.

2.4. Mode of action

Seed fortification helps to improve germination and seed vigour due to the supplemental nutrients provided. The bioactive chemicals used for fortification may help to overcome the effects of chemical dormancy. Fortification of seeds with ferrous sulphate helps in increasing seed storability by bringing about significant reduction in lipid peroxidation.

2.5. Precautions

Care should be taken to prevent leakage of solutes from the seeds under treatment. The volume of solution used should be just enough so that all the seeds are imbibed. Excess of water if used may lead to reverse osmosis and cause loss of essential nutrients contained in the seed.

Solute leakage from the seeds can also be prevented by optimizing the duration of soaking.

3. SEED HARDENING

3.1. What is it?

It is the process of hydrating the seed to initiate the pre-germinative metabolism followed by dehydration which fixes the biochemical events.

3.2. Why it is done?

In order to impart resistance against stress conditions *viz.*, drought and cold, to the emerging seedlings.

3.3. How it is done?

The technique for "Presowing drought hardening" used by Henckel is basically as follows. Seeds or grains are allowed to take up a certain amount of water, and then they are kept moist at 10° - 25°C for several hours before drying in a steam of air. The best results are claimed for seed subjected to two - three cycles of wetting and drying, although for some seeds one cycle is sufficient. Different amounts of water are recommended for different species and cultivars of seed or grain.

3.4. Mode of action

The basis for presowing seed hardening is significant increase in hydrophilic property of protoplasmic colloids namely viscosity and elasticity; increased phosphorylation activity of their mitochondria. Reduction in solute leakage by regaining cell membrane integrity (Simon and Raja Harun, 1972), resumption of rate of protein and RNA synthesis characterized in the first period of imbibition and shortening of the time of DNA replication in the second hydration period. Simulation of long lived mRNA under moisture stress conditions; simultaneous protein and proline content increase after hardening treatments (Srivastava and Randhawa, 1983).

3.5. Precautions

The timing of the initial imbibition period is critical, because as germination and growth proceed the resistance to drying of the embryo decreases. On the other hand, the degree of "hardening" (i.e. the effect of treatment) included is claimed to be high when the embryo is in more advanced stage of germination, at the time of drying. The optimum duration of hardening should be a compromise between the two conflicting tendencies.

The temperature of soaking and drying cycles and the rate of drying are all important.

4. SEED PRIMING

4.1. What is it?

It is the process of controlled hydration of seeds to a level that permits pre-germinative metabolic activity to proceed, but that prevents actual emergence of the radicle.

4.2. Why it is done?

In order to increase the speed and uniformity of germination particularly under adverse conditions of temperature, moisture and salinity.

4.3. How it is done?

Seeds are soaked in variety of solutes, including solutions of various inorganic salts, sugars and polyethylene glycol (PEG) a chemically inert, high molecular weight compound that unlike the lower molecular weight compound probably does not penetrate the cell walls. The temperature suggested during priming is between 10°C - 15°C. The duration of priming varies with the crop from 1 day (sorghum) to 23 day (onion). The seeds are then rinsed and redried to original water content.

Heydecker *et al.* (1973) used different terms depends upon the methods adopted for priming.

(i) Osmopriming - Soaking the seeds in osmotic solutions

(ii) Halopriming - Soaking the seeds in salt solutions

(iii) Biopriming - Coating the seeds with biological agents like bacteria

(iv) Solid matric priming - This consists of mixing seeds with an organic or inorganic carrier and water for a particular period of time. The moisture content of the matric is brought to a level just below what is required for radical protrusion. Seed water potential is regulated by the matric potential of the seed and during priming the water is largely held by the carrier. Seeds can imbibe water from the carrier till the equilibrium is reached.

4.4. Mode of Action

Seeds in contact with an aqueous solution of this chemical begin to take up water in the normal way but cease to do so once they have reached equilibrium with the osmotic potential of the solution. This osmotic potential and therefore the seed water content can be adjusted to a level that permits the seeds - provided they are freely aerated to go through all the essential preparatory processes of germination but prevents cell elongation and in consequence, radicle emergence even after weeks of contact between seeds and osmotic solution. Moreover, during the waiting period at the osmotic barrier the 'slower' seeds tend to catch up with the 'faster' ones so that the subsequent germination is more uniform, especially at lower germination temperatures, than that of untreated seeds.

Favourable treatment conditions permit both the break down of food reserves and the synthesis of materials required for germination to occur, possibly to a greater extent than if the seeds had germinated immediately. This results in instantaneous growth, once the osmotic obstacle has been removed.

4.5. Precautions

(i) The treatment temperature, the concentration of the chemical (resulting in the osmotic potent) and the period of treatment are all important.

A temperature of 20°C appears to be too high for best effects on some temperate species.

There is little doubt that the best contribution of osmotic potential, temperature and duration of the treatment varies according to the species and possibly even the seed lot of any given cultivar, but as a general guide a treatment at –11 bars (288 g/kg H_2O) at 15°C for 14 day may be a good basis for trial.

If the seeds germinate during the treatment, the concentration of the solution should be increased at the next attempt.

(ii) The osmotic potential of the solution should be so adjusted to permit free aeration.

(iii) Seeds should be necessarily washed to remove the last traces of osmotic solutions.

5. HYDRATION - DEHYDRATION

5.1. What is it?

It is the process of soaking the low and medium vigour seeds in water with or without added chemicals usually for short durations to raise the seed moisture content to 25 - 30% and drying back the seeds to safe limits for dry storage.

5.2. Why it is done?

It is done to improve the vigour of medium and low vigour seeds by bringing about certain preventive, corrective and restorative action.

5.3. How it is done?

5.3.1. Types of H-D treatments

The wet treatments include soaking-drying, dipping-drying, spraying-drying, stepwise hydration-drying, moisture equilibration-drying, moisture equilibration soaking-drying, moisture conditioning-drying etc. The choice of the treatment depends upon the characteristics of seed and initial vigour status of the seeds.

5.4. Mode of Action

The hydration dehydration treatment may improve the vigour by controlling free radical reactions and consequent peroxidative damage to lipoprotein cell membranes.

$$R^{o} + R^{o} \xrightarrow[H_2O]{} 2\ R \text{ (Non radical product)}$$

Recombination of free radicals in the presence of a medium like water as depicted in the above equation helps to terminate the free radical induced lipid peroxidation (Basu, 1984).

5.5. Precautions

(i) The mid-term wet treatment should be given only to stored medium and low vigour seeds and not to freshly harvested seed.

(ii) The moisture content of the seed should be much less than that is necessary for germination.

(iii) The duration of hydration should be so adjusted to avoid unnecessary germination advancement which is incompatible with drying back so far as further storability is concerned.

(iv) Drying back to the original weight is essential as otherwise, re-storage of insufficiently dried seed would do more harm than good.

(v) Direct soaking of injury prone leguminous seeds should be avoided.

(vi) Lastly it needs to be clearly emphasized that the treatment would maintain vigour and viability in storage but would not make a seed germinable, which has already lost viability.

6. CONCLUSION

The treatments discussed above have to be prudently employed to defy drought, frost, salinity, low storability and low germinability. The distinguishing factors among those technologies with respect to their need, methods, mode of action and precautions discussed elaborately in the paper should help the scientists to have a clear and clever in sight into these technologies. Proper understanding of the treatmental methodologies is envisaged through this chapter.

REFERENCES

1. Basu, R.N. (1984). An appraisal of research on wet and dry physiological seed treatments and their applicability with special reference to tropical and sub-tropical countries. *Seed Sci.& Technol.*, 22 : 107-126.

2. Heydecker, W., J.Higgiris, and R.L.Gulliver. (1973). Accelerated germination by osmotic seed treatment. *Nature (London)*, *246*: 42-44.

3. Khan, A.A. (1977). The physiology and biochemistry of seed dormancy and germination. Elsevier / North-Holland Biomedical Press.

4. Simon, E.L. and Raja Harun. (1972). *J. Exp. Bot.*, *23*: 1076-1085.

5. Srivastava, K. and A.K.Randhawa. (1983). Eco physiological exploitation of triticale seeds with presowing treatments to develop hardiness against moisture stress. Acta agronomica academial scientiarum Hungaricae.

PART III

SEED PELLETING

GENESIS AND IMPROVEMENT OF THE CONCEPT OF SEED PELLETING

A. BHARATHI[1], P. NATESAN[1], M.BHASKARAN[2], SHERIN SUSAN JOHN[1] AND P.THANGAVEL[1]

1. INTRODUCTION

In Agriculture, Horticulture and Forestry to achieve a desired plant population and a high probability of successful establishment of each seed planted, precision planting is advocated. The most essential phases of precision planting is the singling of seeds for exact placement at a predetermined depth and spacing in the soil with a uniform coverage and at reasonable planting speed. Singling is difficult for small seeds which are low in density and irregular in shape. To overcome these difficulties, seed pelleting is advocated. Literally, it is not at all a new invention and is in vogue since long back.

Seed pelleting is the process of enclosing a seed into a small quantity of inert material just large enough to produce a globular unit of standard size to facilitate precision planting. The inert material creates natural water holding media and provides small amount of nutrients to young seedlings.

Seed pelleting is also a mechanism of applying needed materials in such a way that they affect the seed or soil at the seed-soil interface. Thus, seed pelleting provides an opportunity to package, effective quantities of materials such that they can influence the micro-environment of each seed. By not having to treat the remaining bulk of their soil, farmers may be able to save on the inputs required and the associated costs of applying them.

1 *Department of Seed Science and Technology, Tamil Nadu Agricultural University. Coimbatore – 641 003.*
2 *National Seed Research and Training Centre, Varanasi – 221 106.*

In India, practising of seed pelleting is limited; neither commercial nor government seed organizations has realized its importance. Whereas, in developed countries, it is very common and most of the seeds are sold with coating only. However, now attention has been invited on the importance of seed pelleting for aerial seeding in the afforestation of inaccessible areas in particular. For this situations, it is the cheapest, quick and quite useful method. This paper deals with the different aspects of seed pelleting *viz.,* genesis, and improvement in the concept of seed pelleting.

2. EVOLUTION OF SEED COATINGS (PELLETING)

Seed coating (including pelleting) was being practiced as early as 2000 years ago. The ancient Chinese, coated rice seeds in mud balls to anchor the seed in a flooded paddy field. The coating eliminated the problem of seed drifting when seeds were sown on the surface of the flooded paddy field. Many of our so called 'modern seed enhancements' (including pelleting) have their historical basis dating back hundreds to thousands of years ago . The principles and concepts have changed little with time. Improvements have been gained through better technology given by a better understanding of seed biology.

Seed coatings (pelleting) have evolved from those which protect the seed from fungal and insect attack to a diverse range of coatings, the objectives of which include the protection of rhizobia, supply of micro and macronutrients, protection from birds and rodents, supply of growth regulators, attraction of moisture, supply of oxygen, germination stimulation, germination delay, increase in seed weight or size, and the supply of selective herbicides or herbicide antidotes. Inspite of a considerable amount of research, reliable and effective seed coatings are not currently available for many crops. Even in the case of legume inoculant coatings, which have been intensively studied, there is still no universally accepted practice of inoculation, particularly for coatings applied well in advance of planting (as preinoculation).

Successful results with coatings (pelleting) in many crops, indicated that seed coatings have the potential to overcome some of the problems of plant establishment. On the other hand, by not specifying precisely how coatings have been prepared, researchers have made it virtually impossible for their work to be validated by others and hence, there are many reports of workers finding different results from supposedly similar coatings.

Because of the limitations mentioned above, the potential of seed coatings has not yet been explored thoroughly. When coating materials and techniques are identified more precisely, producing repeatable coatings will become relatively straight forward and an understanding of how each coating affects seeds under various conditions can be developed.

Factors that affect the performance of seed coatings (pelletings) can be grouped into seed, coating, soil and aerial environments and then further partitioned into physical, chemical and biological sectors (Fig.1).

Much of the research done to date has been conducted in such a way that general principle have been difficult to transfer from one area of seed coating research to another. Knott and Lorenz (1950) reviewed much of the early work on seed coatings, which largely concerned the development of relatively inert coatings that permitted the rounding and enlarging of small seeds (Vegetables and sugar beet seeds) sufficiently to facilitate precision mechanical planting. Seed coatings are reviewed mostly, with an emphasis on seed pelleting.

3. IMPROVEMENTS IN SEED PELLETING

There have been improvements in the pelleting process and formulations made by the seed industry. The pellets should not retard the germination rate or reduce the per cent stand in comparison with the non-pelleted control. The pellet may act as a barrier to oxygen diffusion, thus affecting germination. This problem has been overcomed by the addition of oxygen-liberating compounds into the pellet or modification of pellet breakdown after sowing. Alkaline earth, and other peroxidase-containing compounds have been used to release oxygen, and this has been commonly employed in pellets. Pellets have been developed, that fracture some plane during the early phases of water uptake, commercial development termed 'split-pellet'.

The pellet can also serve as a delivery system for the materials required at the time of sowing. In particular, plant protectants required at high loading rates can be applied during the coating. After the first layer of coating materials have been applied to the seed, the active ingredients, formulated as dry powders, can be applied in subsequent layers. Final coating applications with filler materials encapsulated in plant protectants, prevent worker exposure to the seed treatments. This application, termed 'pellet loading' avoids direct contact of the plant protectants with the seed surface. This physical separation avoids phytotoxicity caused by high concentrations of particular plant protectants. Pellet loading has been used successfully for onion seeds for application of fungicides and insect growth regulators. The pellet contains the fungicide Pro-Gro, a

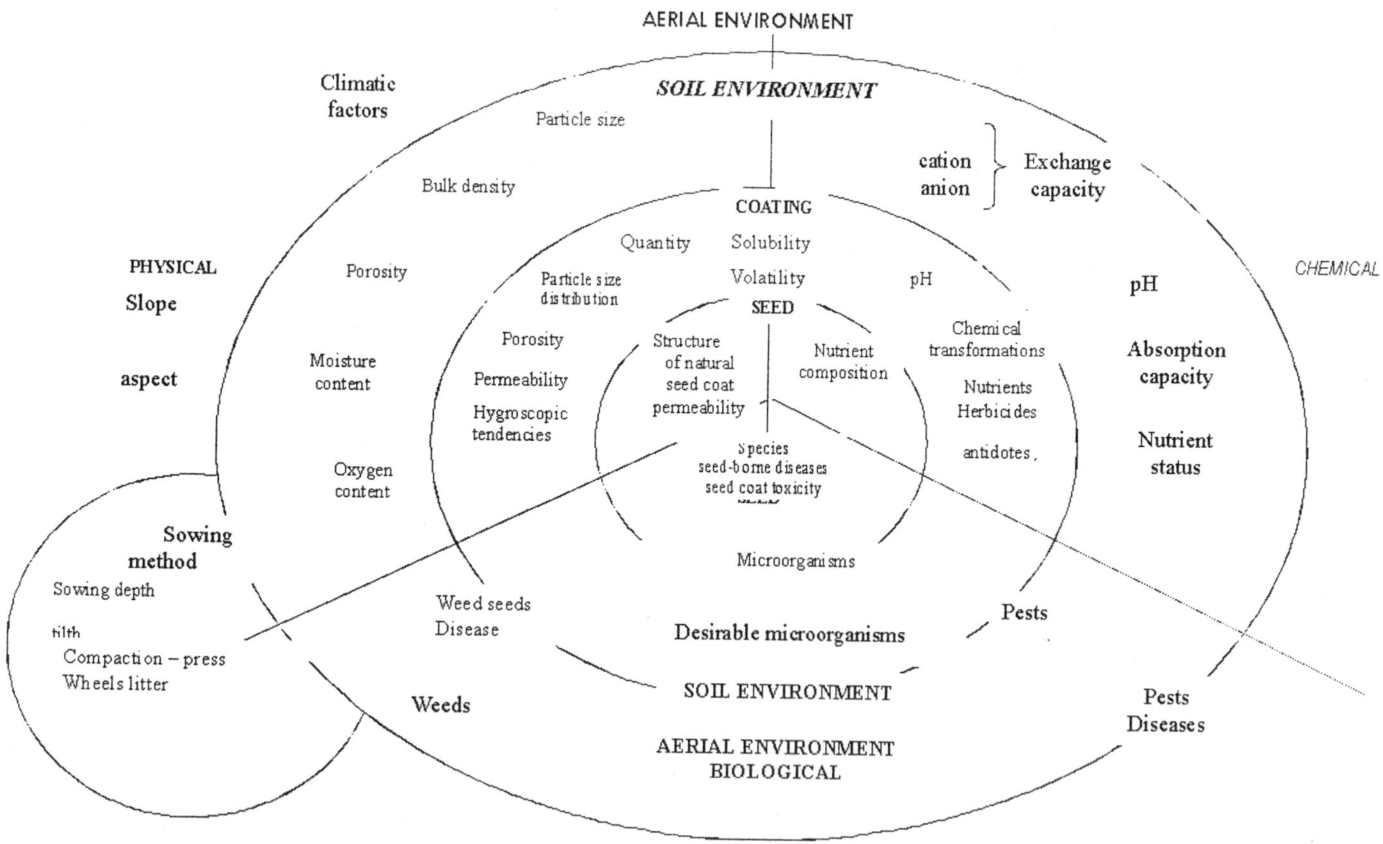

Figure. 1. Schematic diagram showing the relationships among factors affecting the performance of coated seeds

mixture of thiram and carboxin, at 20g a.i. kg seed^{-1} and the insect growth regulator, Trigard (cyromazine) at 50 g a.i. kg seed^{-1}.

The addition of hydrophobic compounds for sowing into wet soils (Taylor *et al.,* 1992), hydrophilic compounds in coating for sowing into dry soils and temperature sensitive coatings to regulate the onset of imbibition at a desired soil temperature are also improvements in seed pelleting.

These coatings improves the chances of successful germination and seedling establishment under field conditions. Improved pelleting treatments are generally restricted to the high value crop seeds for which the added cost of these treatments can be justified.

4. CONCLUSION

The seed pelletisation is very much useful for aerial seeding. Because the tiny seeds not quickly, come to the earth from the top. But the pelleted seeds come quickly so that it enhances the precision sowing. The application of fertilizers, fungicides, insecticides in seedling stage involves more cost and risk. But in this method we can easily apply over seeds, and also pelletisation increases the germinability and seedling vigour. Forest departments are following this useful techniques for aerial seeding.

REFERENCES

1. Taylor, A.G., P.S.Allen, M.A. Bennett, K.J. Bradford, J.S. Burris and M.K. Misra. (1992). Seed enhancements. *Seed Science Research, 8*:245-256.
2. James M. Scott. (1989). Seed coatings and treatments and their effects of plant establishment. *Advances in Agronomy,* 42:43-77.

SEED PELLETING - PRINCIPLES AND PRACTICES

V.KRISHNASAMY*

1. INTRODUCTION

Seed pelleting is the process of enclosing a seed with a small quantity of inert material just large enough to produce a globular unit of standard size to facilitate precision planting. The inert material creates natural water holding media and provides small amount of nutrients to young seedlings. Seed pelleting also serves as mechanism of applying needed materials in such a way that they affect the seed or soil at the seed soil interface. The main objective of seed pelleting is precision with added advantages of better establishment and increased productivity.

The advantages of seed pelleting are listed below by different researchers

- Increase in size
- Singling of seeds by prevention of clogging
- Precision planting
- Attraction of moisture
- Supply of growth regulators, nutrients
- Stimulation of germination
- Influence of micro-environment
- Saving of chemicals/fertilizers applied to soil
- Supply of oxygen
- Protects seed at aerial seedling by improving ballaistic ability

* *Centre for Plant Molecular Biology, Tamil Nadu Agricultural University,* Coimbatore – 641 003.

- Reduces seed rate
- Uniform field establishment
- Increase in yield
- Remedy for sowing at problematic soils
- Protection from birds, animals and insects.

2. THE PROCESS OF PELLETING

The three basic steps involved in pelleting are stated as stamping, coating and rolling. The materials needed for pelleting are seed, adhesive and filler materials. Initially the seeds are uniformly coated with adhesive in correct quantity . Then the filler materials are sprinkled on the coated seeds and are rolled on the filler material for effective and uniform coating.

3. SELECTION OF MATERIALS USED FOR PELLETING

3.1. Adhesive materials

The materials used for coating should be perfect as the physical integrity of coating is decided by the type of adhesive and this was highly influencial during handling, transport and planting operations of the pellets. The adhesives recommended by various authors are, gum arabic, methyl cellulose, gelatin, casein casemate salts, plastic rexins, polyvinyl acetate, methyl ethyl cellulose, polyurethane polyvinyl alcohol, polyvinyl acetate, poly electrolyte or dextran and poly ethylene oxide.

Some of the cheap and low cost binding materials are the rice gruel, maida gruel, sago gruel and starch (revive) gruel.

In selection of these adhesives, the concentration of the adhesive is important for development of required viscosity which can fix the material in efficient manner without breaking and dusting on shade drying.

The concentration recommended for methylcellulose is 3% W/V, methylethyl cellulose is 5% W/V, gum arabic is 45% and nitric coat is 4.3 % W/V solution in water. The low cost adhesives such as rice gruel, maida gruel etc. are used at the concentration of 5%, 10% depending upon filler material used for pelleting. Selection of adhesive is also based on selective purposes. Plastic resins, polyvinyl acetate and insoluble poly electrolyte complexes are used to bind pesticides to seeds; poly ethylene oxide is to prevent erosion of surface sown seeds, polyurethane to bind lime in a way that resists coat abrasion, blends of polyvinyl alcohol and polyvinyl acetate to bind vermiculite and poly electrolytes or dextran to aggregate soil around the seeds, thereby improving the aeration of sown seeds.

3.2. Charateristics of adhesive materials

1. The coating must be porous to allow movement of air to the seed.
2. The coating must weaken or breakdown when it comes in contact with soil moisture to prevent any physical impedance to seed germination.
3. Material used must not have any toxicity to the seeds.
4. It must be possible to apply the coating on commercial basis.

3.3. Filler materials

The materials used as filler materials for pelleting must be beneficial and harmless to both seed and the rhizosphere.

The most common materials used for protectants for *Rhizobia* include lime, gypsum, dolomite and rock phosphate. Other materials include clay minerals such as montmorillonite and vermiculite. Besides blood, peat, poultry manure, moss and mucilage are also used.

The size of particle selected for filler material vary with the type of filler material. The particle size is important for resistant coating on the seed material. The activated clay, lime and dolomite should pass through 300 mesh sieve. The dried blood, wood charcoal, milk powder and yeast extract should pass through 150 mesh sieve size. In case of dried blood, milk powder and yeast extract alone are used, they are used in combination with dolomite in different proportions *viz.,*

Dried blood	-	15:85 (W/W) blood	:	Dolomite
Milk powder	-	3:85 (W/W) milk powder	:	Dolomite
Yeast extract	-	1:99 (W/W) Yeast extract	:	Dolomite

In addition to the above explained materials, biofertilizers are also used as filler materials.

For low cost investment and environmental benefit, botanic leaf powder are used as filler materials. The common botanics used for pelleting are arappu, (*Albizia amara*), pungam (*Pongamia pinnata*), notchi (*Vitex negundo*), prosopis (*Prosopis juliflora*), neem (*Azadirachta indica*), moringa (*Moringa pterygosperma*) and tamarind (*Tamarindus indica*). These botanics are recommeded @ 200-300 g kg^{-1} of seed. The fineness of the powder should be in such a way that it passes through muslin cloth. These leaf powders contain an auxin like substances which regulates the growth of seedling in initial establishment.

3.4. The characters of filler material

- It should be non-toxic
- Friendly to both seed, adhesive and environment
- Easily soluble in water
- Easily available for commercial production
- Low cost

4. EQUIPMENT FOR PELLETING

Tamil Nadu Agricultural University has developed a pelleted machine (hand operated) where the drum rolls on a axis on rotation of the handle. The seeds are coated initially and then mixed with filler material and thus pelleted seeds are prepared.

4.1. For increased efficiency /cautions

1. Excess adhesive should not be added. It may lead to clogging of seeds.
2. The adhesive should be with required viscosity, otherwise the coating will not be perfect.
3. The filler material used should be a fine powder. Then only coating will be perfect and coating also will be retained for longer duration.
4. The coated seed should be spread in this layer of filler material otherwise more seed will be there in a single pellet, it will also bring the pellets to a uniform size.

5. TYPES OF PELLETING

Based on either-bio-active chemicals or filler material added for pelleting, seed pelleting is classied into different types. They are :

5.1. Inoculant pelleting

Different bio-fertilizers *viz. Rhizobia, Phosphobacteria, Asozpirillum* or *Azatobactor* are used as filler materials which are fixed to the seed with the help of an adhesive. This type of pelleting will help in improving the activity of micro organisms of rhizosphere and help in nitrogen fixation. The selection of biofertilizer vary with crop. VAM (Vesicular Arbuscular mycorrhizae) can also be used for pelleting the seeds.

5.2. Protective coating

The disease controlling, biocontrol agents like *Rhizobacteria bataticola* or *Bacillus* spp. or *streptomyces* spp. can also be used for pelleting the seed. Antibiotics are also used for control of diseases through pelleting.

Seed pelleting is also done with fungicides and pesticides and they are added to the adhesive and coated on the seeds. But caution should be taken in such a manner that the filler material should be compatible with the fungicides and pesticides used (eg.) for inoculant pelleting pesticides and fungicide inclusion will not be good always. Bird repellants like methiocarb, endirm or thiram, rodent repellants like mestranol and sesoreinal can also be used for pelleting to protect the seed from birds and rodents. Sometimes coating with nutrients, clays and lime itself is sufficient as the pest not recognise the seed as good.

5.3. Herbicides coating

Filler antidote or absorbent coatings can be used before sowing. Herbicide antidote like 1, 8 napthalic anhydride (NA) is found to be best for seed pelleting. The absorbent, activated carbon can also be used as filler material which is found good in protecting the seed from 2, 4-D and alachlor herbicide damage.

5.4. Nutrient coating

The nutrient coating with micro and macro nutrients enhances the germination and seedling growth, also make it less available to weed species and also avoids the wastage of nutrients. The micronutrients are required in less quantity than for soil or foliar application. The micronutrients used are $ZnSO_4$, $FeSO_4$, $CuSO_4$, KH_2PO_4, KCl, borax etc. The micronutrients are added at required quantity (for e.g, DAP 60g kg^{-1}, $ZnSO_4$ 300 mg kg^{-1}, borax 100 mg kg^{-1}) to the adhesive and are filled with the filler material.

5.5. Hydrophilic coating

Starch graft polymers (which can absorb upto 1000 times their own weight of water) and magnesium carbonate are capable of improving movement of air and water. The increase in germination is due to increase in the rate of imbibitions where the fine particles in the coating act as a "wick" or moisture attracting material or perhaps to improve seed soil contact.

5.6. Oxygen supplier coating

Seeds are coated with peroxides of zinc or calcium which aids in increased oxygen supply to the germinating seeds. The seeds which has got seed coats impermeable to oxygen will be highly benefited for enhancing germination.

6. SEED PELLETING IN FORESTRY

Aerial seedling is an useful seed treatment as this treatment increase the ballistic ability of the seed and acts as a cushion to the resting and thereby avoid injury to the dropped seeds. The seed pelleting is beneficial to large seeds also.

SELECTION OF ADHESIVES, FILLERS AND NUTRIENTS FOR EFFICIENT PELLETING AND THEIR FUNCTION

V.PALANISAMY[1], K.RAJA[2] AND P.SELVARAJU[1]

1. INTRODUCTION

The rationale for pre-plant seed treatments is to mobilize the seed's own resources and to augment them with external resources to maximize improvement in stand establishment and yield. Mobilization of these resources is particularly critical when using seeds of modern cultivars, which are selected for higher yields often under optional growing conditions. Physiological and non-physiological treatments are needed to remove several soil climatic and hydrological constraints under dryland ecological system. Physiological seed treatments that improve or enhance seed performance and that are based primarily on seed hydration include pre-soaking, wetting and drying, humidification, osmoconditioning, matric conditioning and pre-germination. Other physiological treatments that promote germination or remove blocks to germination require physical or chemical stimuli in addition to seed hydration. These include moist chilling, thermal shock, irradiation, aeration and hormone permeation.

There are several non-physiological seed treatments that can remove mechanical, soil and environmental constraints and that may indirectly improve seed germination and stand establishment. These include seed scarification, coating or pelleting treatment with bioactive chemicals and inoculation or coating of seeds with beneficial microbes.

Seed pelleting is a process that creates a complete nutrient environment in the immediate vicinity of the seedling. This nutrient

1 *Department of Crop Improvement, Anbil Dharmalingam Agricultural College and Research Institute, Tiruchirappalli-620 009*

2 *Regional Research Station, Virudhachalam – 606 001.*

environment can be created by adding a variety of substances, such as fertilizer and chemicals to the seed coat in such a way that they affect seed or soil at the seed / soil interface. The artificially created environment results in a combination of benefits that assist the seedling through its critical establishment phase. It may be defined as the application of solid materials to seed in sufficient quantity to make the pelleted seed substantially larger and/or heavier and approach a spherical or elliptical shape. While seed coating is a general term for the application of finely ground solid or liquids containing dissolved or suspended solids to form a more or less continuous layer covering the natural seed coat; includes pelleting and may other seed treatments. In seed pelleting treatments, the seed samples are coated with same inert material, the sticker, to which various chemical may be added.

2. BENEFITS OF SEED PELLETING

- Good seed / soil contact results in better moisture uptake by the seed, thus improving its chances of germination and establishment into a healthy seedling.
- Plant growth regulators or stimulants may promote rooting or hasten emergence of the seedling.
- Fungicides and insecticides are more effective when in direct contact with the seeds.
- Seed coating is an excellent application method for chemical seed treatments that are combined with the adhesive used in the seed coating process to bind solid coating materials to the seed.
- Small seeds become larger and heavier and this improves the scatter pattern in aerial seeding.
- Fungicides and insecticides are more effective when in direct contact with the seeds.
- Seeds may be protected against rodents by adding impalatable, repellent or poisonous substances.
- Pelleting can also be used to incorporate a brightly coloured substance with the seed, which renders it more visible against the soil and thus facilitates uniform sowing rates (Aldhous, 1972).
- Coating with an anti-desiccant material such as alginate can preserve the seed against drying out, until adequate rain has fallen to permit safe germination.
- Seed inoculation with root nodule bacteria (*Rhizobium*) to legume seeds is the most critical factor in its early establishment.

3. MATERIALS USED FOR SEED COATING / PELLETING

3.1. Adhesives / Stickers

The process of seed coating usually involves the use of adhesives (also known as glues, binders or stickers) to bind materials to the surface of the seeds; coating without them using, for example, water also will usually lead to fragile coats that are extremely prone to dusting and cracking and subsequently to the loss of the active ingredient. Such dusting and breakage can also lead to severe problems in handling the seed due to hazards to operators and to problem in mechanical planting. When fine particles are agitated, as in a coating drum, they tend to aggregate naturally even without adhesives (due to cohesive are chemical and electrostatic forces) and hence the need for an adhesive is for one that assists these natural forces of aggregation while allowing the packaging of particulars and 'densification' to continue while the particulars are tumbling. Thus, the adhesive required need not be extremely strong but must be an appropriate adhesive, one that has affinity for both the natural seed coat and for the coating material (Scott, 1989).

Miller and Bensin (1974) have shown that the degree of attraction or repulsion of moisture by the seed coating can have large effects on the germination of coated seeds. In terms of their binding ability and their propensity to cause undesirable agglomeration of seeds there are large differences between the performance of adhesives. Hirota (1972) examined a range of adhesives for binding diatomaceous earth to seed of *Vicia villosa* and found that a mixture of methyl cellulose and gum arabica performed the best.

The coating of seeds with activated carbon has been successfully achieved with gum arabica plus a platicizer (Sharples, 1981), methyl cellulose (Vogelsang, 1954) or polyvinyl acetate (Nagju, 1973).

Test of the performance of adhesives for adhering lime to grass seeds were conducted by Hathcock *et al.* (1984) who showed methyl cellulose to be the most effective; when the coated seeds were sieved for 1 min, methyl cellulose resulted in the best retention of lime, but this was still only 71% of the lime applied.

There are patents covering the use of mineral oil, plastic resins, polyvinyl acetate and insoluble polyelectrolyte complexes to bind pesticides to seeds, polyurethames to bind lime in a way that resists coat abrasian, blends of polyvinyl alcohol and polyvinyl acetate to bind vermiculite and polyelectrolytes or dextran to aggregate soil around the seed, thereby improving the aeration of the sown seeds.

3.2. Fillers or Coating Materials

The most common materials used as protectants for *Rhizobia* include lime, gypsum, dolomite or rock phosphate. Other materials include clay minerals such as montmorillonite and vermiculite, which have been used principally as carriers for chemicals or as pelleting materials. Several other materials that have been found to be useful for different purposes are bauxite, diatomaceous earth, pumice sand and talc.

Many organic materials have been used in coatings, usually as protectants or sources of nutrition of *Rhizobia* or the developing plant. These include blood, bone meal, peat poultry manure, moss and mucilage.

4. CHARACTERISTICS OF PELLETING MATERIALS

The chemical analysis, pH, purity and particle size distribution are very important in selecting the materials. The effect of particle size of solid materials in particular, is crucial to any coating or agglomeration process; such effects have been well researched in the iron ore pelletizing industry. Urich and Hon (1962), for example, reported that iron ore pellets made with smaller particles (73% < 15 μm) had a porosity one-third and a strength three times that of pellets made from larger particles (6% <15 μm). Not only is the mean particle size is important, but the particle size distribution also affects the quality of granulation. The addition of small size particles to a material containing relatively large particles will result in more rapid granulation, increased granulation strength and decreased porosity. Low porosity is undesirable because it may restrict the movement of air to the seed.

5. NUTRIENTS USED

Macro nutrients such as nitrogen, superphosphate, potash and other forms of urea, DAP, monosodium phosphate, ammonium phosphate etc., can be used. Micronutrients such as $ZnSO_4$, $MgSO_4$, $FeSO_4$ etc., can be used.

6. SEED PROTECTIVE COATINGS

Chemicals such as thiram, captan, metalaxyl, etridiazoles, arason, endrin and anthroquinone are among those commonly used. Thiram has been found effective against damping off. Several fungicides, including metalaxyl, etridiazole and captan incorporated into magnesium silicate gel controlled damping off in tomato seedlings by *Pythium* sp. (Giamonichale and Pill, 1984).

Some fungicides have been found to be incompatible with some insecticide seed treatments, however, the combinations, adversely affect germination and plant populations. Interactions have also been found between fungicides-insecticide seed treatments and soil-applied herbicides.

Seed coatings containing of insecticides or acaricides have been used widely on many plant species, offer in combination with fungicides. Examples of some of the pests controlled with seed treatments include leaf hoppers on rice, barley and soybean, onion fly on onions and shootfly in maize. Examples of the use of acaricide seed treatments include the control of spider mites in cotton and red-legged earth mites in pastures.

7. HERBICIDES ANTIDOTES COATINGS

The first commercially produced herbicide antidote was 1,8-naphthalic anhydride (NA) when it was used as a seed treatment to protect maize from S-ethyl dipropythiocarbonate (EPTC) damage. NA permits good weed control while maintaining selectivity (Hoffman, 1971).

8. HYDROPHILIC COATINGS

The use of hydrophilic polymers as coating materials was claimed in a patent on the use of water-soluble polyelectrolytes (with molecular weights greater than 10,000) to enhance germination by aggregating soil clay particles adjacent to the seed, thus improving the movement of air and water to the seeds. With crops, the use of hydrophilic polymer seed coatings on cowpea seeds resulted in a depression of emergence, whereas similar coatings on corn seeds produced better emergence (Baxter and Waters, 1986).

9. SOLID SUBSTANCES

Various grades of celite (diatomaceous silica) and Micro-cel (produced by hydrothermal reaction of diatomaceous silica, hydrated lime and water) and the finer grades of zonolite vermiculite (vermiculite is hydrated magnesium-aluminium-silicate; vermiculite concentrate is exploited to produce zonolite vermiculite) (Khan *et al.*, 1990). Other solid carriers, including other grades of vermiculite (Terra-Lite, vermiculite 3), expanded calcined clay and Growsorb (Callan *et al.*, 1990) have also been used to condition vegetable seeds.

10. SEED COLOURING DYE

A more suitable dye which is equally effective as colouring and which has no apparent adverse effect on subsequent germination is a waxoline dye such as Lithofar Red or the later Waxoline Red. It can also be used as a marker for special certified seed batches in storage (Aldhous, 1972; Wokeman, 1975).

11. SEED COATING WITH GROWTH REGULATORS

Seed coating and pelleting technology have been developed to suit different needs: for seed signalation; for adding bioactive chemicals, nutrients and beneficial microbes; and for sowing seeds under dry and wet conditions (Halmer, 1988). Several studies have examined the advantages of growth regulators permeation into seeds (via aqueous soak or via organic solvents, such as acetone and dichloromethane) with or without seed coating. In several instances, seed coatings have been shown to delay germination by interfering with oxygen and gas exchange; this effect can be alleviated to varying degrees by incorporating growth regulators into seed prior to coating them (Khan, 1977).

Certain types of coating could also have a promotive effect on seedling emergence. Khan (1977) compared the effects of seed coating in lettuce seeds by permeation of seeds with fungal toxins fusicoccin (0.5 mM) and cotylenin E (0.5 mM) via acetone; and the permeation of growth regulators followed by seed coating. Both the seed coating and the application of growth regulators improved the performance when compared to untreated raw seeds. Kouno (1989) obtained an improvement in germination of rice seed which accomplished by encapsulating the seed in a gel coating that included an air bubble.

12. PROPERTIES OF MATERIALS USED FOR SEED PELLETING / SEED COATING

12.1. Adhesives

- Must have affinity (combinability) for both seed coat and selected filler material.
- Should have the required degree of water solubility or insolubility for easy emergence.
- Should have the required strength and plasticity to present dusting and breakage.
- Should have the most appropriate viscosity for each application.

12.2. Coating / filler materials

- Must be porous to allow movement of air to the seed
- The coating must weaken or break down easily when it comes in contact with soil moisture to prevent any physical impedance

- Should be nontoxic to seeds
- Size of the particles should pass through 300 mesh sieve.

REFERENCES

1. Aldhous, J.R. (1972). Nursery Practice. Forestry Comm. Bull., No.43, London.
2. Baxter, L. and L.Waters. (1986). *J. Am. Soc. Horti. Sci., 111*: 31-34.
3. Callan, N.W., D.E.Mathre and J.B.Miller. (1990). Bio-priming seed treatment for biological control of *Pythium ultimum* pre-emergence damping-off in Sh2 Sweetcorn. *Plant Disease, 74*: 368-372.
4. Giammichele, L.A. and W.G.Pill. (1984). Protection of fluid-drilled tomato seedlings against damping-off by fungicide incorporation in a gel carrier. *Horti. Sci., 19*: 877-879.
5. Halmer, P. (1988). Technical and Commercial aspects of seed pelleting and film coating. pp.191-204. British Crop Protection Council, Thornton Health.
6. Hathcock, A.L., P.H.Dernoeden, J.J.Murray and D.Wehner. (1994). *Hort. Science, 19*: 442-443.
7. Hirota, H. (1972). *J. Jpn. Soc. Grasoil. Soc., 18*: 299-309.
8. Hoffman, O.L. (1971). U.S.Pat. No.3, 564, 768 (23 Feb.).
9. Khan, A.A. (1977). Preconditioning, germination performance of seeds. pp.283-316. In: A.A.Khan (ed.). The Physiology and Biochemistry of Seed Dormancy and Germination, Elsevier, Amsterdam.
10. Khan, A.A., H.Miura, J.Prusinski and S.Ilyas. (1990). Matric conditioning of seeds to improve seedling emergence. *Proc. National Symp. Stand Estab. Hort. Crops*, Minneapolis, Mn, :19-40.
11. Kouno, Y. (1989). Method of applying gel coating to plant seeds. U.S.Patent-4808430, February 28, 1989.
12. Miller, W.F. and R.F.Bensin. (1974). *N.Y.Food Life Sci. Q., 7*: 20-23.
13. Nagju, D. (1973). *Diss. Abstr. Int. B., 33* : 4079.
14. Scott, J.M. (1989). Seed Coatings and Treatments and their Effects on Plant Establishment. *Adv. Agron., 42* : 43-83.
15. Sharples, G.C. (1981). *Hort. Science, 16*: 661-662.
16. Urich, D.M. and T.M.Hon. (1962). In "Agglomeration" (A.A.Knepper, ed.) pp. 669-713, Wiley (Interscience), New York.
17. Vogelsang, P. (1954). Pat. No.2, 671, 985 (16 Mar.).
18. Wokeman, D.C. (1975). Colouring tree seeds before sowing and for long term storage using Waxoline dye. Res. Inf. Note No.6/75, For. Com. Res. and Dev. Division, Alice Holt, U.K.

SEED PELLETING RESEARCH AROUND THE WORLD IN RAINFED AND GARDENLAND ECOSYSTEMS

S.SUNDARESWARAN AND A.VIJAYAKUMAR

1. INTRODUCTION

Seeds vary greatly in size, shape and colour. In many crops, seeds are small and irregular in shape, posing difficulty in precision seeding and uniform plant spacing. In addition, seeds should be protected from a range of pests that attack germinating seeds or seedlings. Seed coating technologies can be employed in both the situations. Pelleting increases seed size and weight and alters seed shape for precision planters. Coated seed is less likely to be eaten by birds and rodents because of its larger size and the lime or phosphorus coating. The potential advantages of seed pelleting are precision planting and spacing of small, irregularly shaped seeds, reduced thinning costs, reduced thinning shock, more uniform seed microenvironment, inclusion of beneficial chemicals with the seed and use of fewer seeds. Seed coating has also been utilized to ameliorate environmental stresses such as drought and flooding. Recently there is a renewed interest in the use of coated seeds to aid in complete mechanization and to simplify greenhouse and home garden operations.

2. HISTORY OF SEED PELLETING

The idea of coating seeds is not new. Coated seed or pelletized seed was introduced commercially several years ago but has been around for much longer, the process can be traced back to the times of the pyramids in Egypt (Walsh *et al.*, 1998). In China, farmers coated the paddy seeds with mud balls while sowing in flooded fields, 2000 years ago. In India, farmers have been traditionally using the cow dung slurry for pelleting cotton seeds

---

Department of Seed Science and Technology, Tamil Nadu Agricultural University, Coimbatore-641 003.

to facilitate easy sowing. Burgesser (1950) reported that the first patent issued relating to seed coating was in 1868. Vogelsang (1948) was the first advocate of seed pelleting. He widely acclaimed the potential benefits of his seed pelleting process and formed a company to produce pelleted seeds commercially. Now, many companies are producing pelleted or coated seed products or developing seed coating techniques. A survey by Asgrow Seed Company revealed that in 1974 some 43% of the American and European vegetable growers used precision planting techniques with coated seed being a key factor in the success of these techniques (Anon, 1974).

To our knowledge, a thorough review of the literature on seed coating has not been published until 1988. Because, products have changed over the years and continued to change rapidly, much of the literature was of only historical interest, and Scott (1989) is the first person who made a thorough review on the effect of seed coating and pelleting treatments on plant establishment.

3. PELLETING PROCESS WORLDWIDE

The term "pelleted seeds" refers to the seeds that are coated, tableted, or taped. Working definitions have been developed in the seed industry for several seed coating technologies. Pelleting is defined as the deposition of a layer of inert materials that may obscure the original shape and size of the seed, resulting in a substantial weight increase and improved plantability; while film coating retains the shape and the general size of the raw seed with a minimal weight gain (Butler, 1993). The coating should result in a more or less continuous coating, which eliminates or minimizes product dust-off. Both the methods may contain polymers, pesticides, biological materials, colourants or dyes and other additives.

The three basic coating systems now practiced elsewhere are pellets, tablets and tapes. Burgesser (1950) differentiated between pellets and coated seeds as follows: The term 'coated seed' refers to a single seed coated with an inert material, primarily to increase its size to facilitate planting. Pelleted seeds are a mixture of seed and an inert material formed into pellets, usually by molding one or more seeds by some mechanical means. Pellets as defined by Burgesser are not being produced, however, the term "pelleted seed" is often used to describe coated seeds that result from the layering of a substance on the seed through a process of accretion.

Seed tablet, also called wafer, is much larger than a coated seed but still contains only a single seed. A tablet may consist of two halves of a tablet glued together or a single unit with the seed embedded in the middle.

The tablets are resulted from the process of compression of the material, usually vermiculite or charcoal.

The third process of coating involving the manufacture of seed tape that is a water-soluble plastic tape or a strip of paper with seeds located at specific intervals. Seeds are evenly spaced on the tape, which may then be folded over the seeds and bonded. When the tape is planted, the seeds are automatically spaced at the desired intervals.

The pelleting operation is generally performed on a batch basis and uses an inclined drum or coating pan (Scott, 1989; Ni, 1997). The speed varies with pan diameter and ranges from 10 to 35 rpm. Pelleting has been a labour-intensive operation that requires skilled operators and generally long working hours. The pelleting process has become automated, and computer-controlled coating equipment has been described recently (Scott *et al.*, 1997). In each case, seeds are coated with a combination of binder (adhesive) and filler (bulking agent). Commercial coating formulations usually are blends of binders and fillers, and are formulated as dry powders. The powder is shifted on to the seeds during the coating operation, and water is misted on. The water allows the formation of the pellet and activates the adhesive. Pelleting is a wet operation, so pellets must be dried at the completion of the coating process.

The pellet can also serve as a delivery system for other materials required at the time of sowing. In particular, plant protectants required at high loading rates can be applied during the coating. After applying, the first layer of coating materials to the seed, the active ingredients formulated as dry powders, can be applied in subsequent layers. Final coating applications with filler material encapsulate the plant protectants and prevent worker exposure to the seed treatments. This application, termed pellet loading, avoids direct contact of the plant protectants with the seed surface. This physical separation avoids phytotoxicity caused by high concentrations of particular plant protectants (Hill, 1997).

4. PELLETING MATERIALS USED ELSEWHERE

A number of materials have been used as binders and fillers (Scott, 1989; Ni, 1997). The binders (adhesives) used for pelleting include gum arabic, gelatin, starch, methylcellulose, polyvinyl alcohol, polyoxyethyleneglycol based waxes, and carboxymethyl cellulose. The commonly used fillers (particulate matter) for pelleting are calcium carbonate, limestone, gypsum, activated clay, dolomite, talc, vermiculite, diatomaceous earth, kaolin clay, bentonite, zeolite and peat. Different biofertilizers *viz., Rhizobium, Phosphobacteria, Azospirillum* and *Azatobactor* are used as filler materials, which are fixed to the seed with the help of an adhesive. For low cost investment and environmental benefit, botanical leaf powders are used as filler materials. The common botanicals used for pelleting are arappu (*Albizia amara*), pungam (*Pongamia pinnata*), notchi (*Vitex negundo*), prosopis (*Prosopis juliflora*), neem (*Azadirachta indica*), moringa (*Moringa pterygosperma*) and tamarind (*Tamarindus indica*). These leaf powders contain an auxin like substances, which regulates the growth of seedling and initial establishment.

Ahamed Raza (1997) while pelleting onion found both 5% and 10% maida gruel as best adhesive and the quantum of adhesive used did not influence the seed quality. Balaji (1990) reported that seed pelleting with biofertilizers using gum arabica @ 100 ml per kg of seeds for better establishment and elite seedling production in paddy, greengram, soybean, redgram, sunflower, groundnut and cotton. He also reported that the seed pelleting with *Acacia nilotica* leaf powder helped better at saturated water holding capacity and pelleting products acts as buffer to establish the best soil crop relationship thereby enhancing germination, growth and development of soybean and other field crops. Nargis (1995) and Viswanatha Reddy (1995) identified *Albizia amara* leaf powder as the best filler material as they supply the nutrients for better root growth. Jagathambal (1996) while working with sorghum identified pungam leaf powder as the best-suited filler material compared to arappu, prosopis, acacia, and calotropis leaf powder.

The particle size generally varies with the type of filler material. The particle size is important for resistant coating on the seed material. The activated clay, lime and dolomite should pass through 150-mesh sieve size. The botanicals are recommended @ 200-300 g per kg seed. The fineness of the powder should be in such a way that it passes through muslin cloth. The amount of filler used determines 'build-up' and is expressed as a ratio of filler to seed (w/w). Pellet build-up has been reported to be 2:1, 4:1 to 9:1, and 17:1 to 35:1 for sugar beet, onion and lettuce, respectively (Taylor *et al.*, 1997). Small seeded species such as *Nicotiana* and *Petunia* require much larger build-ups, 100:1 to 150:1 (Ni, 1997). The physical nature of the filler directly affects the pellet weight or density. The desired pellet density is determined by sowing requirements. Low-density pellets are used with vacuum-drum seeding equipment for greenhouse production, while higher density pellets are required for field sowings (Hill, 1997). The high-density pellets are used in high-speed sowing operations to prevent bouncing after pellets are dropped from the metering device to the soil, while an intermediate density pellet can be used for slower field planting operations.

5. ADVANCES IN SEED PELLETING RESEARCH ACROSS THE WORLD

Seed pelleting have been used especially on forage legumes all over the world for more than fifty years. Most people in the seed industry are familiar with Rhizo-Kote brand coating and the benefits that are possible to stand establishment. The chief components of these coatings are *Rhizobium* bacteria for nitrogen fixation, a fungicide to protect against *Pythium* and *Phytophthora* and finely ground limestone to provide an improved microenvironment for seed germination and physical protection. The next generation seed coatings are insecticides, biological protectants, plant nutrient additions, botanical leaf powders, polymers and various colourants. The important seed coating innovations that have been made recently and what are in the pipeline for the future are:

5.1. Fluid seeding

Commercial fluid fertilizer applicators have pioneered this use, which allows rapid seeding of up to 50 acres per hour. Coated seed is added directly to the liquid fertilizer tank just prior to application. If the seed is treated without coating, the Rhizobium inoculants and fungicde would be washed off, but the coating protects the Rhizobia and keeps the inoculants and fungicide on the seed long enough for delivery to the soil.

5.2. Inoculants pelleting with genetically engineered organisms

Seeds are pelleted with bio-fertilizers *viz., Rhizobia, Phosphobacteria, Azospirillum, Azatobactor* etc. This type of pelleting will help in improving the activity of microorganisms of rhizosphere and help in nitrogen fixation. *Vesicular Arbuscular Mycorrhizae* (VAM) can also be used for pelleting the seeds. Muruganantham (1996) compared inoculant and botanic pelleting in Cotton MCU 7 and revealed that performance of inoculant pelleting with *Azospirillum* and *Phosphobacteria* was similar to that of arappu and pungam leaf powder pelleting in enhancing the yield.

EPA has approved the first genetically engineered *Rhizobium* strain for seed treatment. This first generation product has demonstrated measurable increases in nitrogen fixation. The engineered strain's effectiveness is diminished, however, by competing Rhizobia frequently found in production soils. Hopefully the next generation products will overcome this limitation.

5.3. Protective coating with biocontrol agents

Foliar applied biocontrols such as *Bacillus thuringiensis* have been used for a number of years, but there is a little in the way of seed applied biocontrols. The disease controlling, bio-control agents like *Rhizobacteria bataticola, Trichoderma harzianum, Bacillus spp, Streptomyces spp* and *Pseudomonas spp* can also be used for pelleting the seed. These biocontrols have the unique advantage of being able to provide extended control as long as they are viable, unlike chemical control that are metabolized. Seed can provide storage protection and microenvironment for establishment of these organisms in field. Seed pelleting with different antagonists increase the germination rate and reduce the post emergence mortality in many crops. The *Trichoderma* seed pelleting reduced the seedling disease of cotton besides enhancing the germination rate (Alagarsamy *et al.*, 1987). Seed coating with *T. harzianum* recorded the lowest incidence of damping off (15%) and highest disease control (77%). The seedling stand and vigour index in nursery was also highest with *T. harzianum* (Hazarika *et al.*, 2000). Antibiotics are also used for control of diseases through pelleting.

5.4. Fertilizer protection coating

Seed is sometimes mixed with granular fertilizer and broadcasted in the field as one-step operation. The survival of Rhizobia decreases as the time of exposure to the fertilizer is increased. It has been demonstrated that the lime coating on seed can protect the *Rhizobia* in the coating for a sufficient amount of time for the seed to be planted. Lime pelleting has been successfully followed in Australia and New Zealand to overcome the injury caused by acid soil and fertilizer.

5.5. Herbicide protectant coating

Filler antidote or absorbent coatings can be used before sowing. Herbicide antidote like 1,8-napthalic anhydride is found to be the best for seed pelleting. The absorbent, activated carbon can also be used as filler material to protect the seed from 2,4-D and alachlor herbicide damage. Sorghum growers are familiar with coating seed with oxabetrinil (concep), a protectant that safeguards the seed from the subsequent application of metalachlor. Grass seed producers coat activated charcoal on the seed to provide herbicide protection. There may be other chemicals that will metabolize the herbicide or chemicals that can be added to the seed to provide physical protection from herbicides (Walsh *et al.*, 1998).

5.6. Pesticide coating

Pellet loading has been used successfully for onion seeds for application of fungicides and insecticides. The pellet containing the fungicide mixture of thiram and carboxin at 20 g per kg and the insecticide, trigard (cyromazine) at 50 g per kg was used by Taylor and Eckenrode (1993). Bird repellents like methiocarb, endrin and thiram, Rodent repellents like mestranol and sesoreinal can also be used for pelleting to protect the seed from birds and rodents. Sometimes coating with nutrients, clay and lime itself is sufficient, as the pest not recognise the seed as good.

5.7. Slow release chemicals coating

Seed coatings can carry as much as 3 to 5 kg of material per acre depending on the planting rate. This is an excellent way to deliver some nutrients and agricultural chemicals, but because of the rates applied and the proximity to the seed, these chemicals may be toxic to the germinating seed. Encapsulation and slow release chemistry is one way to modify these toxic effects and still maintain the seed as a delivery system (Walsh *et al.*, 1998).

5.8. Hydrophilic and hydrophobic coating

Baxler and Wates (1986) suggested the seed coating with hydrophilic polymers like hydrolyzed starch-graft polyarylonitrile (H-span) to maintain a high water potential around germinating seeds. Starch graft polymers (which can adsorb up to 1000 times their own weight of water) and magnesium carbonate are capable of improving movement of air and water. The increase in germination is due to increase in the rate of imbibition where the fine particles in the coating act as a "wick" or moisture attracting material or perhaps to improve seed soil contact. Coating with a hydrophobic polymer (Vinamul 3650) regulated the rate of water uptake, reduced imbibition damage and improved the emergence of soybean seeds. Seeds with ethyl cellulose also reduced imbibition damage and improved germination (Hwang and Sung, 1991).

5.9. Oxygen supplier coating

Though seed pelleting is highly beneficial for enhancing germination, in the seeds with seed coats impermeable to oxygen, pelleting may act as a barrier to oxygen diffusion, thus affecting germination (Sachs *et al.*, 1981). This problem was overcome by improvements in the pelleting process and formulations made by the seed industry, by the addition of oxygen-liberating compounds into the pellet or modification of pellet breakdown after sowing. Seeds coated with peroxides of zinc or calcium aids in increased oxygen supply to the germinating seeds, especially under submerged or anoxia condition. Alkaline earth and other peroxide containing compounds that release oxygen have been commonly employed in pellets (Langan *et al.*, 1986). Liver and Roberts (1984) reported that the polymer coatings can be employed to retard imbibition rates when seeds are sown in a wet soil and seed coatings with peroxide compounds provided oxygen to seeds under anoxic soil conditions. Pellets have been developed that can fracture during the early phases of water uptake, a commercial development termed 'split-pellet' by Incotec Company.

5.10. Pelleting with nutrients

Seed pelleting with micro and macronutrients enhances the germination and seedling growth and make it less available to weed species and avoid the wastage of nutrients. The micronutrients are required in less quantity than soil or foliar application. Silcock and Pruith (1981) reported that macro and micronutrients applied in seed coatings improved early plant growth. The micronutrients used are $ZnSO_4$, $FeSO_4$, $CuSO_4$, KH_2PO_4, KCl and Borax. The micronutrients are added at required quantity (DAP 60 g kg^{-1}, $ZnSO_4$ 300 mg kg^{-1}, Botax 100 mg kg^{-1}) to the adhesive and are filled with the filler material.

Cotton seeds pelleted with gypsum + DAP recorded higher germination, speed of germination, root length, shoot length and vigour index (Saraswathi, 1994). Anon (1999) reported that in both tomato and soybean, nutrient pelleting with $ZnSO_4$ @ 300 mg kg^{-1} using 5% maida gruel as adhesive and arappu leaf powder as filler material enhanced not only the field emergence but also the yield potential. Seed pelleting of ammonium molybdate 250 kg kg^{-1} and ferrous sulphate 500 mg kg^{-1} along with *Rhizobium* produced better results in terms of seed germination, plant height, leaf area index and yield in soybean (Ramesh and Thirumurugan, 2001). Coating of carrot seeds with gypsum + DAP + micronutrient mixture found to enhance the seedling vigour and crop performance than the uncoated seeds (Ramesh, 1996). Similarly. in Bellary onion, seeds pelleted with vermicompost, biofertilizer, DAP micronutrient mixture recorded increased plant height, number of leaves and bulb yield (Ahmed Raza, 1997). Narasimha Prasad (1994) noted increased seed yield and quality of the resulting seeds of soybean crop raised with seeds pelleted with biogas slurry at 500 g kg^{-1}.

5.11. Pelleting with botanicals

The common botanicals used for pelleting are arappu, pungam, notchi, prosopis, neem, moringa and tamarind. These leaf powders contain an auxin like substances, which regulates the growth of seedling and initial establishment. Ahmed Raza (1997) in onion found that the seeds coated with arappu leaf powder @ 500 g kg^{-1} increased the seed vigour and germinability. Jagathambal (1996) reported that pelleting of seeds with pungam leaf powder enhanced the yield in different locations and season. This pelleting has an additive effect while the hardened seeds are used. Punithavathi (1997) observed better germination, seedling vigour and seed yield when hardening with KCl (1% for 12 hours in 1:0.7 ratio) followed by pelleting with pungam leaf powder (60 g kg^{-1}). Nargis (1995) working with tomato seeds reported that the germination and vigour were enhanced by the leaf powder pelleting not only in fresh seeds but also in seeds artificially aged. In general, pelleting with all the leaf powder was found to be better in protecting the seeds from deterioration than the commonly recommended fungicidal (thiram) treatment. The performance of pungam leaf powder in pelleting is even better than vitamin E treatment, which is an anti-oxidant that controls deterioration rate of seeds.

5.12. Polymer film coating

Film coating is versatile as a coating system or as a component of coating system. Film coating is adapted from the pharmaceutical and confectionary industries for uniform application of materials to seeds. The film forming formulation consists of a mixture of polymer, plasticizer and colourants (Halmer, 1988; Robani, 1994) that are commercially available as ready-to-use liquids or as dry powders (Ni, 1997). Application of the film forming mixture results in uniform deposition of materials on each seed with little variation among seeds (Halmer, 1988). Like pelleting, the formed film may act as a physical barrier, which has been reported to reduce the leaching of inhibitors from the seed coverings and may restrict oxygen diffusion to the embryo (Duan and Burris, 1997).

Film-coated seeds have better flow characteristics in the planter due to reduced friction between seeds (Hill, 1997). Film coating and pelleting may be combined in two different processes. Film coating may be performed as the final step after pelleting to provide a dust-free pellet. An intermediate between film coating and pelleting is termed encrustation. This hybrid technique provides good build-up, enhances plantability and is less expensive than pelleting. Film coating provides an ideal method for the application of chemical and/or biological seed treatments (Taylor and Harman, 1990; Taylor *et al.*, 1994; McGee, 1995). Relatively high loading rates of plant protectants can be applied with film coating. However, a spatial separation between the plant protectants and seed surface is not obtained as described for pellet loading. A major impetus for using film coating is to reduce the exposure of workers to chemicals from treated seed.

5.13. Temperature controlled polymer coating

Seeds can be coated with a polymer that will prevent the seed from imbibing the water needed for germination until a desired temperature is reached. The set point at which the polymer becomes permeable to water can be readily adjusted by changing the polymer composition. The Landec Corporation's intellicot polymer coating will make it possible to prevent seeds from germinating too early in the field and will synchronize germination better. As long as the soil temperature is too cold (below 55°F) for healthy germination, the polymer maintains a dense crystalline structure around the seed. This forms a barrier between the seed and the moisture in the soil, restricting the seed's ability to absorb water. When the soil warms up (above 55°F), the polymer changes its physical structure becomes amorphous, making it permeable to water, thus allowing the seed to absorb the moisture it needs to germinate. Angamuthu (1991) reported that seeds pelleted with polymer film containing fungicide and growth regulator increased the germination and plant height in minor millets. Pelleting of sorghum seeds with hydropolymer (alcosorb) in combination with micronutrients and DAP registered maximum germination and vigour in terms of root length, shoot length and vigour index (Selvaraju, 1992).

5.14. Seed markers and colourants

Special colorants, dyes and other chemicals can be added to seed coatings to distinguish seeds carrying different genetic traits or to tell varieties apart. Colourants provide an aesthetic appeal to the seeds and increase the visibility of seeds after sowing. This could also help to prevent misapplication of agricultural chemicals or even be used in cases of liability claims. Wilbur-Ellis Company in USA first offered a seed colourant system in water-soluble packets in 1991. The entire packet dissolves in water to make the same amount of solution as one quart of liquid colorant. These dissolvable packets leave no residue, and eliminate triple-rinsing and container disposal rpoblems. The 'Color-coat' water-soluble packets that are formulated with a high-tech polymer coating material which hold pesticides and other ingredients uniformly to the seed were then made available for most types of seed.

6. SEED PELLETING RESEARCH IN FORESTRY

Seed pelleting is very much useful for aerial seeding in forestation programme. Because the tiny seeds does not drop quickly to the earth, but the pelleted seeds fall quickly so that it enhances the precision sowing. The application of fertilizers, fungicides, insecticides in seedling stage involves more cost and risk. However, in this method, we can easily apply over seeds. Polarization also improves the germinability and seedling vigour. Now the forest departments are following this useful technique for aerial seedling.

Seed pelleting work done by the Tamil Nadu Forest Department indicated that the seeds of *Acacias* and Tamarind should be scarified and the seeds of *Zizyphus jujupa* and *Sapindus emarginatus* should be depulped manually before pelleting. The pelleting is done in the power operated pelletiser in the sequence, seed + glue solution (Titebond and Fevicol @ 1:1 ratio) + thiram (2 g/kg) + filler (talc and dolomite @ 1:1 ratio) + micronutrient (Tracel @ 2 g/kg) + Aldrin (2 ml/kg) + colouring agent. The sequence of pelleting followed by the Karnataka Forest Department for aerial seeding is, seed + Emisan (1 g/kg) + gum arabica + sand, red earth and manure mixture @ 1:1:1 ratio. While the filler improves the ballistic property for aerial seedling, the pesticides coating acts as rodent/bird repellant.

7. DEVELOPMENTS IN SEED COATING MACHINERIES

Jeffs and Tuppen (1986) reviewed the seed-coating equipment developed for laboratory and commercial applications. In film coating technique, active material is dispensed or dissolved in a liquid adhesive and applied to seeds either with a fluidized bed treater or with pharmaceutical coating drum. This method permits application of multiple coatings with 1-10% increase in seed weight.

Gustafson, Inc, USA has begun marketing the Driacoater, state-of-the-art coating machinery for film coating and pelletizing the seeds. The Driacoater, designed and manufactured by Driam USA, utilizes the same technology found in other Driam coating equipment, which is widely used in the pharmaceutical and confectionary industries. The Driacoater is fully automated easy operation and is offered in a variety of commercial and lab model, with capacities ranging from 6 to 5000 pounds. It utilizes a multi-sided drum provide gentle and continuous mixing action and prevent seed from shifting the drum. The entire coating process is handled in a single, continuous operation, eliminating the need to move seed from machine to machine.

The Dumoulin IDA-1000 X has unique "double air flow system" that blows drying air from above, below and up. Hence, the coating is more uniform, the seeds are handled more gently and drying is faster. This system of 'fluidizing' the mass reduces damage from friction and impact of the product against the drum (Burris *et al.*, 1994). The shape of the pan is shallower so the weight of the product mass is less, also reducing cracking or chipping. The combination of these two design features provides a fast and uniform spread of the coating. Clean-up or change over is very simple with the insertion of a special cleaning pan in which the drum can revolve.

8. ORGANIZATIONS INVOLVED IN SEED COATING RESEARCH ACROSS THE WORLD

Apart from the research organizations and universities across the world, the private companies involved in seed coating research and development of seed coating machineries are, Aventis Crop Science Limited, Switzerland; Syngenta, Switzerland; GTC, Germany; Grow Tec Inc; Incotec, The Netherlands; The Landec Ag, Indiana; Cotton Incorporated USA; Cel Pril, Atlanta; Magnus Kahl Seeds Pvt. Ltd. Australia; Kisan Brothers Private Limited, India; Alberta Agriculture, USA; Becker Underwood; Satech Corporation, The Netherlands; Sub-Sahara International Inc.; Cimbria Heid, Denmark; Seed Processing Holland; Westrup, Denmark; Gustafson, USA; Seed Dynamics Inc, USA and Shanghai Zhanghwei Company, China.

REFERENCES

1. Ahamed Raza, M. (1997). Seed Technological studies on bellary onion **Allium cepa* L.) Cv. Agrifound dark red. M.Sc.(Ag.) Thesis, TNAU, Coimbatore.

2. Alagarsamy, G., S.Mohan and R.Jeyarajan. (1987). Effect of seed pelleting with antagonists in the management of seedling disease of cotton. *J. Biol. Control, 1(1)*: 66-67.

3. Angamuthu, K. (1991). Sturies on seed pelleting, physiological maturity, sowing quality and storage in minor millet. M.Sc.(Ag.) Thesis, TNAU, Coimbatore.

4. Anonymous. (1974). Study shows 43% using precision planting methods. *Seedsmens Digest, 25(12)* : 72, 78.

5. Anonymous. (1999). All India Co-ordinated National Seed Project (Crops), Annual Report (1998-99), IARI, New Delhi.

6. Balaji, D.S. (1990). Studies on the seed soil relationship in certain crops – paddy, greengram, soybean, redgram, sunflower, groundnut and cotton. M.Sc.(Ag.) Thesis, TNAU, Coimbatore.

7. Baxler, J. and L.Wates. (1986). Effect of a hydrophilic polymer seed coating on the imbibition, respiration and germination of sweet corn at four matric potentials. J. *Am. Soc. Hortic. Sci., 111*: 517.

8. Burgesser, F.W. (1950). The use of coated seeds in modern agriculture. *Proc. Amer. Soc. Sugar Beet. Tech., 6*: 79-84.

9. Burris, J.S., L.M.Prijic and Y.Chen. (1994). A small-scale laboratory fluidized bed seed-coating apparatus. In: Seed Treatment: Progress and Prospects, (Ed) T.Martin. British Crop Protection Council, Surrey, England, pp.419-423.

10. Butler, R. (1993). Coatings, films and treatments. *Seed World, October* : 19-24.

11. Duan, X. and J.S.Burris. (1997). Film coating impairs leaching of germination inhibitors in sugar beet seeds. *Crop Science, 37*: 515-520.

12. Halmer, P. (1988). Technical and commercial aspects of seed pelleting and film coating. In: Seed Treatment: Progress and Prospects, (Ed.) T.Martin. British Crop Protection Council, Surrey, England, pp.191-204.

13. Hazarika, D.K., S.Sarmah, T.Paramanick, K.Hazarika and A.K.Phookan. (2000). Biological management of tomato damping off caused by *Pythium aphanidermatum*. *Indian J. Plant Pathol., 18(1&2)*: 36-39.

14. Hill, H.J. (1997). New developments in seed technology. Proceedings of the Oregon Horticultural Society, Portland, Oregon, 88: 123-130.

15. Hwang, W.D. and F.J.M.Sung. (1991). Prevention of soaking injury in edible soybean seeds by ethyl cellulose coating. *Seed Sci.& Technol.,19*: 269-278.

16. Jagathambal, R. (1996). Presowing seed treatment to augment productivity of sorghum Cv. Co 26 under rainfed agriculture. Ph.D. Thesis, TNAU, Coimbatore.

17. Jeffs, K.A. and R.J.Tuppen. (1986). Application of pesticides to seeds and requirements for efficient treatment. In: Seed Treatment, (Ed.) K.A.Joffs, British Crop Protection Council, Surrey, England, p.17.

18. Langan, T.D., J.W.Pendleton and E.S.Oplinger. (1986). Peroxide coated seed emergence in water-saturated soil. *Agronomy Journal, 78*: 769-772.

19. Liver, J.P. and E.H.Roberts. (1984). Peroxides in seed coatings. *Outlook Agric., 13*: 147.

20. McGee, D.C. (1995). Advances in seed treatment technology. Proceedings of the Asia and Pacific Seed Association (APSA), New Delhi, pp.1-14.

21. Muruganantham, K. (1996). Influence of seed pelletization, viability, vigour and productivity in rice fallow cotton Cv. MCU 7. M.Sc.(Ag.) Thesis, TNAU, Coimbatore.

22. Narasinha Prasad, K. (1994). Studies on certain aspects of seed management in soybean (*Glycine max* Merill.). Ph.D. Thesis, TNAU, Coimbatore.

23. Nargis, S. (1995). Influence of pelleting, magnetic treatment and radiation on the performance of differentially aged seeds of Tomato, *Lycopersicon esculentum* Mill. Cv. PKM 1. M.Sc.(Ag.) Thesis, TNAU, Coimbatore.

24. Ni, B.R. (1997). Seed coating, film coating and pelleting. In: Seed Industry and Agricultural Development, Chinese Association of Agricultural Sciences, DOA, Ministry of Agriculture, Beijing, China Agriculture Press, pp.737-747.

25. Punithavathy, M. (1997). Seed hardening and pelleting studies in ragi Cv. CO 13. M.Sc.(Ag.) Thesis, TNAU, Coimbatore.

26. Ramesh, D. (1996). Studies on enhancement of seed germination and vigour in cotton Cv. Zino. M.Sc.(Ag.) Thesis, TNAU, Coimbatore.

27. Ramesh, K. and V.Thirumurugan. (2001). Effect of seed pelleting and foliar nutrition on growth of soybean. *Madras Agric. J., 88(7-9)*: 465-468.

28. Robani, H. (1994). Film-coating horticultural seed. *Hort. Technology, 4*: 104-105.

29. Sachs, M., D.J.Cantliffe and T.A.Nell. (1981). Germination of clay-coated sweet pepper seeds. *Journal of the American Society for Horticultural Sciences, 106*: 385-389.

30. Sarasvathy, G. (1994). Studies on seed pelleting in relation to sowing qualities of cotton Cv. LRA 5166 (*Gossypium hirsutum* L.). M.Sc.(Ag.) Thesis, TNAU, Coimbatore.

31. Scott, J.M. 1989. Seed coatings and treatments and their effects on plant establishment. *Advances in Agronomy, 42* : 43-83.

32. Scott, J.M., G.J.Blair and A.C.Andrews. (1997). The mechanics of coating seeds in a small rotating drum. *Seed Sci.& Technol., 25*: 281-292.

33. Selvaraju, K. (1992). Studies on certain aspects of seed management practicesfor seed production under moisture stress conditions in sorghum (*Sorghum bicolor* L. Moench). M.Sc.(Ag.) Thesis, TNAU, Coimbatore.

34. Silcock, R.G. and F.T.Pruith. (1981). Seed coating and localized application of phosphate for improving seedling growth of grasses in acid, sandy red earths. *Aust. J. Agric. Res., 33* : 785.

35. Taylor, A.G. (1997). Seed storage, germination and quality. In: The Physiology of Vegetabl;e Crops, (Ed.) H.c.Vien, CAB International, Wallingford, pp.1-36.

36. Taylor, A.G. and C.J.Eckenrode. (1993). Seed coating technologies to apply trigard for the control of onion maggot and to reduce pesticide application. pp.73-78. In: Efforts pertinent to the Integrated Pest Management at Cornell university. *NYS IPM Publication, 117* :.73-78.

37. Taylor, A.G. and G.E.harman. (1990). Concepts and technologies of selected treatments. *Annual Review of Phytopathology, 28*: 321-329.

38. Taylor, A.G., D.F.Grabe and D.H.Paine. (1997). Moisture content and water activity of pelleted and film-coated seeds. *Seed Sci.& Technol.,19*: 24-32.

39. Taylor, A.G., G.E.Herman and P.A.Nielson. (1994). Biological seed treatments using *Trichoderma harzianum* for horticultural crops. *Hort. Technology, 4*: 105-109.

40. Viswanatha Reddy, K. (1995). Studies on presowing seed treatment in differentially aged seeds of brinjal (*Solanum melongena* L.) Cv. PKM 1. M.Sc.(Ag.) Thesis, TNAU, Coimbatore.

41. Vogelsang, P. (1948). Seed pellets. What they are. *News Letter, Assoc. Offic. Seed Anal., 22*: 20-24.

42. Walsh, J., K.Rooney and J.Canestrino. (1998). Seed coating innovations. Proceedings of Alfalfa Symposium, 3-4 December, Reno Nevada, California.

CURRENT SCENARIO OF SEED PELLETING IN AGRICULTURAL CROPS

A.S.PONNUSWAMY

1. INTRODUCTION

The farmer uses insecticides, herbicides, fertilizers, fungicides and other additives such as bacteria to grow a better crop and achieve a better economic return on his investment. The way in which we could roll up all things that the farmers wishes to add to the seed to enhance his crop and come out with a product that would be large coated magic seeds (Ramsey, 1977). It is often times good to remember that as a seed industry, our primary concern is the establishment of crop by the most economical and efficient means possible.

In Agriculture, to achieve a desired plant population and successful establishment of each seed planted, several presowing seed treatments are advocated. One such presowing seed treatment is seed pelleting.

2. SEED PELLETING

Seed pelleting is the process of enclosing a seed into a small quantity of inert material just large enough to produce a globular unit of standard size to facilitate precision planting. The inert material creates natural water holding media and provides small amount of nutrients to young seedlings. Pellets should always be a stable unit but should then easily disintegrate to release the seed.

In India, seed pelleting is a limited factor. Neither commercial nor government seed organizations realized the importance of seed pelleting

Department of Seed Science and Technology, Tamil Nadu Agricultural University, Coimbatore – 641 003.

where as in developed countries, seed industries have evolved their own patented formulas and most of the seeds are sold with coated only.

The main objective / aim of seed pelleting is precision planting with added advantages of better establishment and increased productivity. The merits of seed pelleting are listed below by different scientists.

- Pelleting regulates the size of seeds for precision planting by machine / man.
- Reduces the amount of seeds required to plant and cut the work for thinning the crop.
- Singling of seeds and prevention of clogging.
- Attraction of moisture.
- Supply of plant growth regulators and micronutrients.
- Influence of micro-environment.
- Saving of chemicals / fertilizers applied to the soil.
- Ensure uniform field establishment.
- Increased yield.
- Remedy for sowing at problematic soils.
- Protection from birds, animals and insects.
- Pelleting provides uniform sized seeds which makes it easy to handle.

3. DISADVANTAGES OF PELLETING

There are some drawbacks in pelleting the seeds as listed below

- When the seed is pelleted, its cost and weights are more than bare seeds.
- If proper pelleting techniques / machine is not used, there would be missing of seed during pelleting or some times more than one seed may contain in a pellet.
- More moisture is required to germinate since pellet must be dissolved first.
- Delayed field emergence at low moisture level.

4. PROCESS OF PELLETING

The three basic steps involved in pelleting are stated as stamping, coating and rolling. The materials needed for pelleting are seed, adhesive and filler materials. The seed are uniformly coated with adhesive in correct quantity initially. Then the filler materials are sprinkled on the coated seeds and are rolled on the filler material for effective and uniform coating.

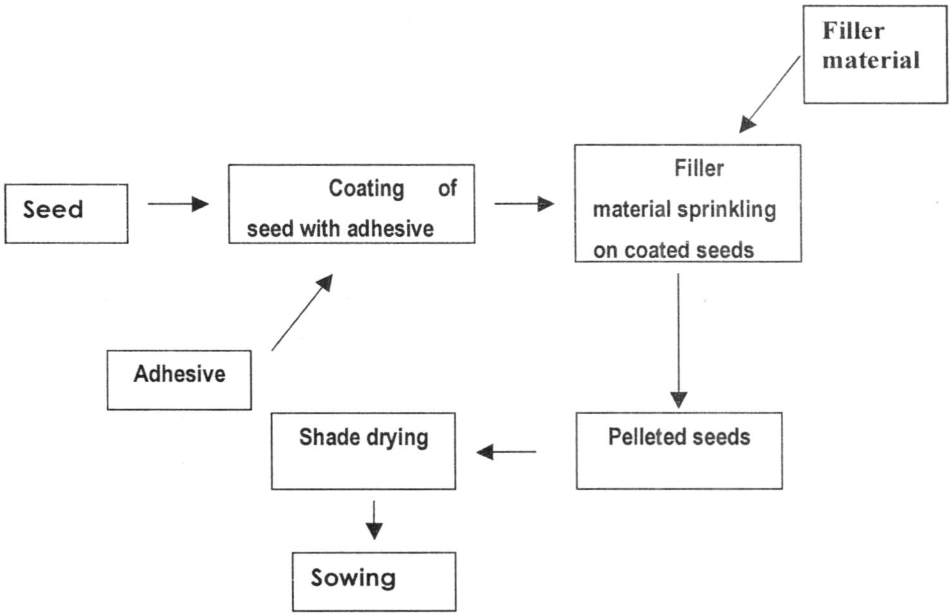

Table 1. Types of seed pelleting

Type	Materials used
(i) Inoculant pelleting	Biofertilizer, *viz., Rhizobia, Phosphobacteria, Azospirillum, Azatobactor*, VAM
(ii) Protective coating	Biocontrol agent like *Rhizobacteria bataticola, Bacillus sp, Streptomyces sp*, pesticides, fungicides.
(iii). Herbicide coating	Filler antidote or absorbent coating. Herbicide antidote like 1.8 napthalic anhydride (NA).
(iv) Nutrient coating	Coating with micro and macronutrients (eg.) $ZnSO_4$, $FeSO_4$, $CuSO_4$, KCl, Borax etc
(v) Hydrophillic coating	Starch graft polymers, magnesium carbonate
(vi) Oxygen supplier coating	Peroxides of zinc and calcium

5. PELLETING AND ITS EFFECT (OR INFLUENCE) ON VARIOUS AGRICULTURAL CROPS

As revealed elsewhere the pelleting of seeds is beneficial to agricultural crops for improving the initial establishment and in turn the final yield potential.

Soybean seed pelleted with *Acacia nilotica* leaf powder performed better at saturated water holding capacity with enhanced germination (Balaji, 1990). Angamuthu (1991) reported that seeds pelleted with polymer film containing fungicides and a growth regulator increased the germination, plant height and root weight etc. in minor millets. Jeyabal *et al.* (1992) found that seed coated with biodigested slurry 50 per cent, super phospate (2 per cent), *Bradyrhizobium* (2 per cent) and *Phosphobacteria* (2 per cent) increased the number of filled pods and yield by 29.6 and 37.2 per cent respectively over the uncoated seeds in soybean. Rice fallow greengram seeds pelleted with biogas fresh slurry at 10, 25 and 10 per cent with or without DAP 5 per cent and murate of potash (MOP) influenced favourably the grain yield. Slurry at 50 per cent with 5 per cent DAP gave increased grain yield over unpelleted seeds. Pelleting of seeds with hydropolymer (alcosorb) registered maximum germination and vigour interms of root and shoot length and vigour index (Selvaraju, 1992).

Sorghum seeds pelleted with four different fertilizers increased the germination per cent. In green house, KCl, single superphosphate and lime gave better results than untreated seeds (Magalhaes *et al.*, 1994). Saraswathi (1994) reported that the cotton seeds pelleted with gypsum + DAP recorded the highest germination, root length, shoot length and vigour index. Muruganantham (1996) found that the performance of inoculant pelleting with *Azospirillum* and *Phosphobacteria* was similar to that of arappu and pungam leaf powder pelleting in enhancing yield (Table 2).

Table 2. Effect of pelleting on germination and seedling vigour & yield of cotton cv. MCU 7

Treatment	**Germination (%)**	**Vigour Index**	**Seed cotton yield**	**Seed yield (kg 10 m^{-2})**
T_0 Control	78	2200	1.16	0.78
T_1 *Azospirillum*	81	2475	1.34	0.90
T_2 *Phosphobacteria*	80	2480	1.34	0.90
T_3 *AZO + Azospirillum + Phosphobacteria*	79	2338	1.32	0.89
T_4 Arappu	83	2523	1.37	0.92
T_5 Neem	78	2324	1.32	0.89
T_6 Pungam	81	2438	1.36	0.91
T_7 Karuvel	80	2384	1.31	0.88
T_8 Prosopis	77	2256	1.26	0.85
CD	2.1	104	0.65	0.24

Pelleting of sorghum seeds with pungam leaf powder enhances the yield in different locations and season. This pelleting has an additive effect while the hardened seeds are used for pelleting (Jagathambal, 1996) (Table 3).

Table 3. Influence of pelleting cum hardening in sorghum cv.CO 26

Treatment	Seed yield (Kg/ha)			
	Rabi' 94		Rabi '95	
	CBE	APK	CBE	APK
Control	810	1050	840	530
Seed hardening	860	1150	950	566
With 1% prosophis + 1% Pungam leaf extract				
Seed pelleted with pungam	980	1400	1050	610
Leaf powder				
T_2+T_3	1080	1550	1130	660

Blackgram seeds fortified with $ZnSO_4$ + $MnSO_4$ followed by pelleting with DAP registered higher field emergence per cent, drymatter production, pod production, number of seeds pod^{-1} and it also increased the seed yield to 24.7 per cent over control (Vijaya and Ponnuswamy, 1998) (Table 4 and 5). Palanisamy and Punithavathi (1998) revealed that ragi seeds hardened with KCl (1 per cent) followed by pelleting with pungam leaf powder recorded higher germination and increased seedling growth, drymatter production, vigour index and field emergence over control seeds

Rathinavel and Dharmalingam (1999) reported that the fuzzy and delinted seeds of cotton MCU 7 pelleted using 5 per cent maida as an adhesive and finely powdered arappu leaf powder + DAP + micronutrient + *Azospirillum* provided higher germination associated with better seedling growth.

Blackgram variety Vamban 1 seeds pelleted with biodigested slurry + Super phosphate + Sodium molybdate + *Rhizobium* + *phosphobacteria* and carbendazim increased the seed yield up to 17.0 per cent. However pelleting treatments did not influence the quality of resultant seeds (Sabir-Aharmed, 1999). Weixintian *et al.* (1999) found that the sesame seeds coated with 2 per cent carbendazim, calcium magnesium phosphate (or) gypsum, promoted earlier maturation and produced the highest yield.

Table 4. Effect of fortification and pelleting treatments on yield attributes and resultant seed quality in blackgram cv. CoBG 282/1.

Treatments	Treatment details	Number of seeds pod^{-1}	Seed yield (g plant^{-1})	Seed yield (kg plot^{-1})	100 seed weight (g)	Seed recovery (%)	Germination (%)
T_0	Control	5.7	4.8	1.09	3.60	90 (71.57)	94 (75.82)
T_1	Fortified with micronutrients	6.0	5.3	1.19	3.62	92 (73.57)	95 (77.08)
T_2	Pelleted with micronutrients	6.1	5.3	1.21	3.63	92 (73.57)	94 (75.82)
T_3	Pelleted with macronutrients	6.2	5.5	1.29	3.72	93 (74.66)	95 (77.08)
T_4	Fortified with micronutrients + pelleted with macronutrient	6.3	5.6	1.37	3.80	94 (77.08)	96 (78.46)
SED		0.09	0.07	0.10	0.009	0.31	-
CD (P = 0.05)		0.20	0.15	0.23	0.02	0.68	NS

(Figures in parentheses are transformed values)

Table 5. Cost benefit ratio of seed fortification and pelleting in black gram

Treatment (as detailed in 4a)	Cost of cultivation (Rs./ha)	Computed yield (kg/ha)	Total income (Rs.)	Increased yield over control (kg/ha)	% of increased yield over control	Increased income over control (Rs./ha)	% of Increased income over control	Cost benefit ratio
T_0	6425	1372	24696	-	-	-	-	1 : 3 .8
T_1	6476	1498	26964	126	9.18	2268	9.18	1 : 4 .1
T_2	6533	1523	27414	151	11.0	2718	11.0	1 : 4 .2
T_3	6534	1625	29250	253	18.0	4554	18.3	1 : 4 .4
T_4	6534	1725	31055	353	25.7	6359	25.6	1 : 4 .7

Rice seeds coated with rock phosphate (RP) @ 0.5 kg kg^{-1} of seed stimulate only rice growth on low phosphorus soil and increased root length and increased shoot dry weight by 400 – 870 per cent (Ros *et al.*, 2000). Rathinavel and Dharmalingam (1999) found that the cotton seeds pelleted with arappu leaf powder + micronutrients + *Azospirillum* increased yield and its components.

Gao *et al.* (2000) observed that the wheat seeds coated with more than 0.3 g riadimenol kg^{-1} of seed and 0.6 g of tebuconazole kg^{-1} of seed reduced the germination and seedling emergence, but promotes the number of roots,. Seeds pelleted with sulphur @ 2.5 g kg^{-1} of seed or gypsum @ 8 - 20 g kg^{-1} of seed reduced the germination than the control (Bhaskar *et al.*, 2000). Abdul *et al.* (2000) reported that the seeds coated with calper and interaction of calper coated rice seeds with the 32/24^{0}C temperature regime gave the best germination (93.23 per cent). Kavitha (2002) found that greengram seeds hardened with prosopis leaf extract (1 per cent) followed by pelleting with 40 g DAP + 100 mg $ZnSO_4$ + 100 mg $FeSO_4$ + 250 mg Ammonium molybdate kg^{-1} of seed significantly increased the germination, vigour index and yield (Table 6).

Table 6. Effect of hardening and seed pelleting on seed physiological parameters and yield of blackgram cv. Vamban 3.

Treatments	Germination (%)	DMP (g $seedling^{-10}$)	VI	Field emergence at 10 DAS	Survival and establishment at 30 DAS	Seed yield (kg ha^{-1})
Control	94 (75.94)	0.25	3444	86 (68.00)	66 (54.43)	703.14
$ZnSO_4$ hardening	95 (77.71)	0.27	4060	98 (72.06)	75 (59.83)	757.50
Prosopis hardening + DAP + $ZnSO_4$ + $FeSO_4$ + Ammonium molybdate pelleting	98 (81.39)	0.28	4444	92 (73.44)	80 (63.15)	789.37
CD (P=0.05)	1.75	0.010	169	1.97	1.36	26.69

REFERENCES

1. Abdul, A., M. Mohammad, M.A. Rabbanim and M. Siddique. (2000). Temperature regimes and calper coating effect on some local and exotic cultivars of rice. *Sarhad Journal of Agriculture,* 16 (4): 397- 401.
2. Angamuthu, K. (1991). Studies on seed pelleting, physiological maturity, sowing quality and storage in minor millet. M.Sc (Ag.) thesis, Tamil Nadu Agricultural University, Coimbatore.

3. Balaji, D.S. (1990). Studies on the seed soil relationship in certain crops paddy, greengram, soybean, redgram, sunflower, groundnut and cotton. M.Sc (Ag.) Thesis, Tamil Nadu Agricultural University, Coimbatore

4. Bhaskar, S., K.Shivashankar, G.Janardhan and C.Ramachandra, (2000). Gypsum and sulphur nutrition to groundnut in rice fields. *International Journal of Tropical Agriculture, 18 (4)* : 355 – 360.

5. Gao. R., C.Den., X.Goa and J.Li. (2000). Effect of seed coating treatments with triadimenal and tebuconnide on the growth and development of wheat seedlings. *Acta phytophylacica, 27(4)* : 359-363.

6. Jagathambal, R. (1996). Presowing seed treatment to augment productivity of sorghum cv. CO 26 under rainfed agriculture, Ph.D., Thesis, Tamil Nadu Agricultural University, Coimbatore.

7. Jeyabal, A., G. Kuppusamy and A.R. Lakshmanan. (1992). Effect of seed coating on yield attributes and yield of soybean (*Glycine max* L.). *J. Agron. & Crop Sci., 169*: 145-150.

8. Kavitha, S. 2002. Seed hardening and pelleting for maximizing the productivity of black gram (*Vigna mungo* (L.) Hepper) cv. Vamban 3 under rainfed conditions. M.Sc. (Ag.) Thesis, Tamil Nadu Agricultural University, Coimbatore.

9. Magalhaes, P.C., D.M.N.Ferreira, C.A.Vasconcelos, J.T. Azevedo and C.S.Borba. (1994). Effects of pelleting on germination and development of sorghum cultivars. *Revista Brasileira de sementes, 16(1)* : 20-25.

10. Muruganantham, K. (1996). Influence of seed pelletization, viability, vigour and productivity in rice fallow cotton. cv. MCU – 7. M.Sc (Ag.) Thesis, Tamil Nadu Agricultural University, Coimbatore.

11. Palanisamy, V. and N.Punithavathi. (1998). Effect of seed hardening and pelleting on seed quality in ragi. *Madras Agric. J., 85 (10-12)* : 583-285.

12. Ramsey, C. (1977). Coating – A New era for seed. *Seeds and Farms,*: 21-22.

13. Rathinavel, K and C.Dharmalingam. (1999). Effect of seed pelleting on elite seedling production in cotton cv. MCU 7 (*Gossypium hirsutum* L.) *Crop Res., 18(1)* : 137-141.

14. Ros, C. R.W.Bell and P.F.White (2000). Phosphorous seed coating and soaking for improving seedling growth of *Oryra sativa* (rice) cv. IR 66. *Seed Sci. & Technol.,, 28* : 391-401.

15. Sabir-Ahamed, A. (1999). Effect of seed pelleting on field performance of blackgram. *Legume Res., 22(2)* : 109-112.

16. Saraswathi, G .(1994). Studies on seed pelleting in relation to sowing qualities of cotton cv. LRA – 516. *Gossypium hirsutum* L. M.Sc (Ag.) Thesis, Tamil Nadu Agricultural University, Coimbatore.

17. Selvaraju, K.(1992). Studies on certain aspects of seed management practices for seed production under moisture stress condition in sorghum cv. CO 26 (***Sorghum bicolor*** (L). Moench). M.Sc (Ag.) Thesis, TNAU, Coimbtore.

18. Vijaya, J.and. A.S. Ponnuswamy. (1998). Studies on seed fortification and pelleting in blackgram. *Madras Agric. J., 85 (10-12)* : 549 - 551.

19. Wei Xintian, Cuilanen, Maweidong, Lu Feng yin; Xure Shugin and pang Tie. (1999). Formulation and utilization of water absorbing coating agents for sesame seed. *J. of Henan Agricultural Sciences,12* : 9-11.

CURRENT SCENARIO OF SEED PELLETING IN HORTICULTURAL AND FORESTRY CROPS

M. BHASKARAN[1], A. BHARATHI[2], P. NATESAN[2],
R. UMARANI[2] AND K. VANANGAMUDI[3]

1. INTRODUCTION

The seed pelletisation is very much useful for aerial seeding. Because the tiny seeds does not drop quickly to the earth, but the pelleted seeds fall quickly so that it enhances the precision sowing. The application of fertilizers, fungicides, insecticides in seedling stage involves more cost and risk. But, in this method we can easily apply over seeds. Also, pelletisation increase the germinability and seedling vigour. Now the Forest department are following this useful technique for aerial seeding.

In vegetables, it is the atmost concern to explore the possibility of overcoming any type of adverse condition that prevents germination and establishment. In this direction, seed pelleting is a rewarding exercise.

2. ROLE OF SEED PELLETING IN VEGETABLES AND FORESTRY

Seed pelleting work done at Department of Seed Science and Technology and Forest Genetics Division of Tamil Nadu Forest Department, Coimbatore indicates that seeds of *Acacias* and Tamarind should be given prescribed treatment and shade dried for 24 h before pelleting. Seeds of *Zizyphus jujuba* and *Sapindus emarginatus* should be depulped manually. The pelleting is done in the power operated pelletisers.

---

1 *National Seed Research and Training Centre, Varanasi – 221 106.*
2 *Department of Seed Science and Technology, Tamil Nadu Agricultural University, Coimbatore – 641 003.*
3 *Dean (Agriculture), Tamil Nadu Agricultural University, Coimbatore - 641 003.*

The sequence of pelleting: seed – glue solution (Titebond and Fevicol @ 1:1 ratio)+ thiram @ 2g kg^{-1} of seed) + thin coat of filler (Talc and dolomite @ 1:1 ratio) + micronutrient (Tracel @ 2g kg^{-1} of seed) + filler and glue materials + Aldrin @ 2ml kg^{-1} of seed + colouring agent. Pelleted seed are later dried in shade.

While the filler improves the ballistic property for aerial seedling, the pesticides coat acts as rodent/bird repellant. Lime improves the porosity as well as dissolves the coating materials immediately after receipt of the moisture. Pelleting is done in a ratio of 1:2 or 1:3 ratio (seed: coating) or 1:1.05 to 1:1.54 ratio (weight basis). For aerial seedling, Beaver aircraft of Agricultural wing or Vayudoot with 600kg capacity/sorty can be engaged. The sequence of pelletisation followed by the Karnataka Forest Department for aerial seedling is, seed + Emisan @ 1g kg^{-1} of seed + gum arabica + sand, red earth and manure mixture @ 1:1:1 ratio. Pelleting is done in rotary concrete mixture.

Pelleting studies on *Acacia nilotica* were done using gum *Acacia* (white) @ 100 ml/kg of seed as adhesive and activated charcoal powder @ 150 g kg^{-1} as filler, and various insecticides, fertilizers and biofertilizers as coating material. Biofertilizer coating enhanced germination, seedling growth, dry weight of seedlings and root, also number of nodules/plant. Germination ranged from 43% (*A. amara* leaf powder) to 79% (Mycorrhizae). Mycorrhizae also enhanced root growth upto 22.5 cm. GA pelleted seeds produced shortest root of 8.88 cm, but the shoot was longest (17.66 cm). *Phosphobacteria* maximized the number of nodules per plant (Vanangamudi, 1992).

Fresh seeds of neem (*Azadirachta indica*) possessing 25% moisture were pelleted with biofertilizers, such as VA-mycorrhizal fungi, *phosphobacteria*, and *Azospirillum* each @ 40-50 g kg^{-1} of seed, biocides namely furadon 3G, Dithane M45, Captan 75% WDP each @ 2-4 g kg^{-1} of seed and Allitin 50 EC (Dially disulfide, a synthetic compound of garlic oil) @ 10 ml kg^{-1} of seed, nutrients like urea, diammonium phosphate, super phosphate and potassium dihydrogen phosphate each @ 2-5 g kg^{-1} of seed and a hydrophilic substance namely magnesium carbonate @ 2 g kg^{-1} of seed. The unpelleted seeds served as the control. The pelleted seeds along with control seeds were germinated in the laboratory as well as in the nursery. The study indicated that seed-pelleting with bio-fertilizers significantly enhanced the germination value, germination percentage and vigour in terms of root and shoot length, dry-matter production and vigour over the control in both the experiments. VA-mycorrhizal fungi was the best, followed by

Phosphobacteria and *Azospirillum*. Allitin was also found to improve the viability and vigour significantly. However, biocides drastically reduced the germination and vigour over control. Pelleted seeds deteriorated faster (35%) in storage under ambient conditions compared to unpelleted seeds which recorded 85 per cent germination after 3 months of storage (Vanangamudi *et al.*, 1993).

Neem-seeds, dried to 30, 20 and 15% moisture content immediately after extraction, were pelleted with macro and micro-nutrients and fungicides. The pelleted seeds along with non-pelleted ones (15% moisture content), stored under ambient conditions were evaluated for germination on the 3rd, 10th, 20th and 30th day after pelleting. Irrespective of the treatments, the germination was above 85 per cent on the 3rd day of test. The moisture content of pelleted seeds when lowered to 11 to 12 per cent had taken 18 to 20 days for 50 per cent emergence and complete germination respectively, against 10 days and 21 days for non-pelleted seeds. In another experiment, the pelleted seeds recorded 46 per cent germination against 75 per cent of non-pelleted seeds at 30 per cent water-holding capacity in the field. At 85 per cent water-holding capacity, the germination was completely affected. The study indicated that 10-12 day-old seeds with 15 per cent moisture content may be pelleted with macro and micro-nutrients along with identified fungicides for the production of vigourous seedlings (Ponnuswamy and Karivaratharaju, 1993).

Mani (1997) reported that pelleting of seeds with macro + micronutrients + biofertilizer exhibited quality seedling attributes for elite seedling production in Acacias. Though thirty and forty per cent water holding capacity proved to be optimal for both pelleted and unpelleted seed germination, the pelleted seeds showed its superiority over unpelleted seed at higher water holding capacity (75%).

The tamarind seeds pelleted with *Azospirillum* and inoculated with VAM produced better seedlings than other biofertilizers and uninoculated control. The relative enhancement of seedling growth characteristics by VAM might be due to the accumulation of phosphorus.

3. VEGETABLES

Doneen and Mac Gilliway, (1948) reported that vegetable seeds sown in two different soil types at various soil moisture content exhibited delayed germination. (Although pelleted seeds have been widely tested, their performance under soil conditions has not been reliable). Millier and Sooter

(1967) found that the rate of emergence in pelleted seeds in soil was much lower than that of non-pelleted seeds; but the adverse effect of pelleting disappeared when the pellets were cracked, indicating the problem to be the micro-environment of the germinating seed.

Seed pelleting in onion increased the yield by sitmulating the plant growth process (Gaizhutes, 1967). Longden (1975) stated that pelleting was adopted to make seed round so as to facilitate precision sowing in vegetables besides accurate dosing of seeds with chemicals.

According to Moore (1975), the field emergence and efficient viability of plant weight at 20 days were equal in both pelleted and unpelleted seeds in lettuce, while Hovadik (1976) explicated delayed germination and reduced emergence of pelleted seeds in parsely, carrot, lettuce, cauliflower, radish, onion and cabbage.

Olesen and Elkins (1977) stated that soil acidity that inhibited rhizobial multiplication in the rhizophere of developing seedlings might be overcome by lime pelleting of seed.

Adverse effect of high soil moisture content was reduced when tomato seeds were pelleted (Tarakanov *et al.*, 1977).

Shapoval, (1978) reported that use of pelleted sugarbeet seeds increased germination, and there by decreased sowing rate; and registered nearly equidistant emergence which decreased labour cost on thinning besides, it increased the average root yields by 1.6 t/ha.

According to Eric and Roos (1979) pelleted seeds of carrot and onion can safely be stored for upto three years under low temperture (10^0C) and low RH (50%) condition.

Wieser (1982) observed earlier ripening of fruits by two days by using seeds pelleted with an organic solvent binder and late ripening of fruits by two days as compared with unpelleted seeds in tomato.

Seeds pelleted with N,P,K, Fe, Cu, Mn, Brand Zn gave 3-4 percent higher field germination over unpelleted seeds in tomato (Konstantinov, 1983). Similarly pelleted onion seeds registered 10-23% higher field emergence over non pelleted ones (Schmidit, 1985).

Clay pelleting seeds in sugarbeet restored and reduced seedling emergence from wet soil (Dunning *et al.*, 1985).

Seed germination in vegetables especially in gourds are affected by various soil types and conditions such as soil moisture content, soil temperature, salinity and alkalinity of the soil. Soil is quite heterogeneous, germination and seedling establishment may be altered by micro heterogenous nature of the soil.

Among the vegetables, cucurbitaceous seeds are more sensitive to soil moisture levels, hence field emergence in these crops being always problematic even with seeds of high germinability. Either a little, high (or) low soil moisture results in poor emergence necessitating resowing (Krishnaswamy, 1991).

Seed pelleting with arappu powder @ 500 g per kg of seed registered a spectacular improvement in seed quality attributes in terms of germination, speed of germination, root and shoot length, dry matter production and vigour index in the gourds namely ash gourd, bitter gourd and ribbed gourd and the efficacy of the treatment became more pronounced at 60 per cent water holding capacity in the two types of soil i.e., sandy loam and clay soil. The pelleting product namely arappu acting as a buffer/wick to establish better seed soil relationship, regulated the moisture supply for better emergence, growth and development (Renuga Devi, 1992).

REFERENCES

1. Biradar, B.B. (1988). Precision planting by seed pelleting. *Seeds and Farms*, May : 32-34.
2. Konstantinov, G. (1983). Growing direct sown tomato cultivar Druzh ba from pelleted seed. *Gradinariki Lzarska Nauka., 20 (4)* : 53-57.
3. Loneragan, J.F., Meyer, D., Few cett, R.G and Anderson, A.J. (1955). *J. Aust. Inst. Agric. Sci., 21*: 264-265.
4. Longden, P.C. (1975). Sugar beet seed pelleting. *Adas. Quart. Rev., 18*: 73-80.
5. Mani, G. (1997). Seed technological studies in *Acacia nilotica* and *Acacia leucophloea* M.Sc. (Ag.) Thesis. Department of Seed Science and Technology, Tamil Nadu Agricultural University, Coimbatore
6. Miller, F.W and Sooter, C. (1967). Improving emergence of pelleted vegetable seed. *Transactions of ASAE* : 658-66.
7. Olsen, F.J. and D.M. Elkins. (1977). *Agronomy J., 69* : 871-874.
8. Ponnuswamy, A.S. and T.V. Karivaratharaju. (1993). Neem Seed Pelleting. World Neem Conference. Abstract p 65 , 24th-28th Feb. Bangalore, India.
9. Renugadevi, J. (1992). Studies on seed treatment and storage in ashgourd (***Benincasa hispida*** Thumb.), bittergourd (***Momordica charantia*** Linn.) and ribbedgourd (***Luffa acutangula*** Roxb.). M.Sc. (Ag.) Thesis, Tamil Nadu Agricultural University, Coimbatore.

10. Scott, J.M. (1989). Seed coating and treatments and their effects on plant establishment. *Advances in Agronomy, 12* : 43-83.

11. Vanangamudi, K. (1992). Seed pelleting studies in *Acacia nilotica* Personal Communication.

12. Vanangamudi, K., S. Padmavathi, V. Manonmani and C. Surendran. (1993). Effect of pelleting with biofetilizers, biocides and nutrients on the viability and vigour of Neem Seed. World Neem Conference. Abstract p 64 : 24th-28th.

SEED PELLETING FOR AERIAL SEEDING

V.SIVAKUMAR[1], B.GURUDEV SINGH[1],
R.ANANDALAKSHMI[1] AND K.VANANGAMUDI[2]

1. INTRODUCTION

Species of pastures and agricultural crops like soybeans, wheat and rice in which the broadcasting is successful, many techniques were developed as a result of industrialization and mechanisation. For increasing the area under forest cover in short period of time and reducing the gap between the afforestation and deforestation, the technique of aerial seeding was adopted in inaccessible areas such as mountain slopes, hilltops, ravines, etc.

Direct sowing of seeds in the field is well known in forestry. In East and West Africa and Sudan, species like *Acacia nilotica, Acacia senegal, Cassia siamea, Azadirachta indica, Gmelina arborea* and *Anacardium occidentale* are established by direct sowing. Forests have also been established using this technique (Lyon and Thomson,1979). But this technique is not widely practiced in many countries due to various reasons *viz.*, the seeds are eaten by insects, birds and animals, the success rate was very low and the total operation cost was high when compared to the seedling survival. After the invention of chemicals to protect seeds from birds during 1953 and from insects and rodents during 1956 in USA, aerial seeding became popular after the technique of seed pelleting was adopted in the seeding programs.

2. SEED PELLETING

Seed pelleting is dressing the seeds with suitable inert materials, fungicides, insecticides, pesticides, fertilizers, plant growth regulators, etc.,

1 *Institute of Forest Genetics and Tree Breeding, Coimbatore, Tamil Nadu, India.*
2 *Dean (Agriculture), Tamil Nadu, Agricultural University, Coimbatore - 641 003.*

along with water retentive or hygroscopic polymers (Jalshakti) using binders. The main objective of pelleting is to use seeds for direct or aerial seeding. The qualities of the pelletized seed are far superior for most of the characters to ordinarily available seed (Ram prasad, 1988).

There are many advantages in using the seeds after pelleting. In case of small sized seeds, the pelleting technique helps to improve the seeding distribution and ballistic properties. The seeds are completely protected from fungus, insects, birds and animals. The nutritional requirements of the germinated seedlings are fulfilled. Unlike organic fertilizers, the bio-fertilizer inoculation given will be helpful for relatively longer period of time. The pelleting may help the seeds from desiccation, high soil pH, dry hot wind and sunlight. Species which do not germinate in open may be pelletized and used for aerial seeding.

Pelletized seeds of Bamboo, Sissoo and Subabul near Visakhapatnam during 1986 and Neem, Tamarind, Red Sanders, Subabul and *Shorea talura*, near Tirupathi during 1987 were used for aerial seeding (Shankaranarayan and Suresh kumar,1986).About 2.5 to 3 cm size pellets made of tank silt, Red-clay, Farm yard manure and other ingredients were used for seeding. But these seeding attempts were failed due to severe drought immediately after aerial seeding.

Before pelleting, the seeds are dried to standard moisture required for the species. The standard moisture content of Eucalyptus seeds was found to be 22-25 %. (Shankaranarayan and Suresh kumar,1986). Then the seeds are graded to separate chaff, empty and small seeds. Uniform sized seeds are separated and coated with a suitable inert material like Dolomite (passing through 300 mesh with pH 7.6), Calcium carbonate (with pH 8.2), activated char coal, talc or ash and biodegradable glues. The nutrients, fungicides, pesticides, etc., required are given one after another. Simultaneously after every coating, the seeds are dried to standard moisture.

Various adhesives like gum arabic and Methyl cellulose are used for pelleting of seeds (Lahiri, 1991). However, Methyl cellulose (3% W/V solution) and Methyl ethyl cellulose (5% W/V solution) are widely used when compared to gum arabic (upto 45% W/V solution) due to its ease of use and cost effectiveness. Nitro coat (4:3 W/V) is a water-soluble gum and nutrient coating used for nodule bacteria in legume seeds. For binding pesticides, plastic resins, Polyvinyl acetate and insoluble poly electrolyte complexes are used. To prevent erosion problem, Poly ethylene oxide is used with seeds which binds with the surface. Similarly to form soil aggregate around the seed, poly electrolytes or dextran are used (Ram prasad, 1988).

Various kinds of fungicidal and pesticidal formulations are available to protect seeds. Chemical formulations like Methicarb and Thiram against birds and Mestranol and Resorcinal against rodent are used. Suitable pesticide formulation and concentration should be selected after considering their influence on the local bird population. In Alagar koil area of Dindigul District, Tamil Nadu, aerial seeding carried out with seed pellets mixed with pesticides resulted in death of many peacocks during 1987-88.

In few species, addition of nutrients during pelleting improves the seedling emergence and establishment. The Phosphorus fertilizers are treated with the seeds without any damage to the seeds. But the Urea-nitrogen damages the radicle of the germinating seeds at higher concentration. Ammonium Sulphate, Mono or Di- Ammonium Phosphates are used in *Acacia nilotica* seeds (Shankaranarayan, and Suresh kumar,1986).

Encapsulation of seeds with alginate entrapped rhizobia beads was developed as a new seed-applied inoculum technique for utilizing legume-Rhizobium symbiosis. Besides lime, gypsum, dolomite or rock phosphate, dried blood (15 parts mixed with 85 parts of Dolomite), milk powder (3 parts mixed with 97 parts of Dolomite) and Yeast extract (1 part mixed with 99 parts of Dolomite) are used with rhizobia as protectants. Studies carried out in *Sesbania sesban* and *Desmodium tortuosum* with alginate beads trapped water and immobilized rhizobia ensured germination of seeds and effective nodulation of seedlings at cheap cost to be used in aerial sowing. Addition of a synthetic polymer, Jalshakti, in the beads enhanced the activity of rhizobia upto 30 weeks.

Studies carried out in *Acacia nilotica* and *Albizia amara* revealed that the inoculation of VAM improved the germination and seedling growth (Anon, 1981). Among VAM, *Phosphobacteria* and *Azospirillum,* VAM was found to be the best in *Azadirachta indica* seeds, which improved both seed germination and seedling growth (Karivaradaraaju *et al.*, 1999). Rapid reduction of viability of the pelletized seeds is also noticed in *Azadirachta indica,* therefore the seeds should be pelletized one or two days before going for aerial seeding.

3. AERIAL SEEDING DARTS

Aerial darts are nothing but specialized small containers filled with growing media used for aerial seeding. The aerial darts are 11.5 cm long and consists of a plastic nose cone (to aid penetration into the soil) that supports a paper cone containing 10 ml of growing medium. Hydrophilic gels are also incorporated into the growing media to increase water retention

(Biradar, 1988). This kind of aerial seeding container and helicopter delivery system have been developed to improve the effectiveness of direct seeding as a afforestation method (Jha *et al.,* 1994).

There are various kinds of medium used in aerial darts. Three major functions of a growing medium are to provide support, give adequate aeration and nutrients and the highest water retention. A medium is mostly composed of materials like soil, sand, moss peat, sawdust, bark, vermiculite, etc. Some of the growing media commercially available for using in aerial darts are given in Table 1.

Table 1. Description of media used in aerial darts.

Medium	Company	Description
Pro-mix A	Les Tourbieres Premier Ltee., Riviere du Loup, Que.	Sphagnum moss peat (60%), vermiculite (40%), starter nutrients, wetting agent
Ball Growing Mix I	Ball Superior Ltd., 1155 Birchview Dr., Mississauga, Ont.	Compost pine bark, vermiculite, perlite, starter nutrients, wetting agent
Ball Growing Mix II	Ball Superior Ltd., 1155 Birchview Dr., Mississauga, ont.	Same as Ball Growing Mix I but with sphagnum moss peat
Turface	International Minerals and Chemicals Corp., 421 E. Hawley, Mundalieln, IL.	Calcined clay
Gro-bark	Gro-bark (Ontario) Ltd., R.R.1, Box 190, Claremont, Ont.	Pine bark sawdust sieved through 6 mm screen

Some hydrophilic gels can also be incorporated into the growing media to increase the water retention (Vanangamudi, 1992 ; Vanangamudi *et al.,* 1993 ; Lennox, and Lumis, 1987). These gels are mostly starch-hydrolyzed polyacrylonitrile copolymers or acrylamide and acrylic acid salt copolymers. These chemicals can absorb several hundred times of their weights in water. They are biodegradable, nonionic and stable for 6-12 months in media. They can absorb and supply water to the plants with sufficient aeration. Different kinds of hydrophilic gels used for aerial darts in the market are given in Table 2.

Table 2. Description of hydrophilic gels used in aerial darts

Medium	Company	Description
Agrosoke	Chemical Discoveries (UK) Ltd., 4, Hurst Dale, Devisdale Rd., Altrincham, Cheshire, England	Copolymer of acrylamide, crystal, absorbs 40 times its weight in water
SGP 104K	Henkel Corp., General Mills Division, 4620 West 77th st., Minneapolis, MN	Hydrolyzed starch ployacrylonitrile graft copolymer powder, absorbs several hundred times its weight in water
Terra-Sorb	Industrial Services International Inc., P.O.Box 10834, Bradenton, FL.	Hydrolyzed starch polyacrylonitrite graft copolymer using potassium hydroxide; flake, absorbs several hundred times its weight in water
Aqua Lox	Southern Hydro Mulch Supply Inc., P.O.Box 59413, Dallas, TX	Hydrolyzed starch-graft copolymer product derived from corn starch and acrylonitrite; powder absorbs more than 400 times its weight in water

Different soilless growing media were studied for their physical characteristics like total porosity (TP), air space (AS), easily available water (EAW) and water buffering capacity (WBC) to be used in aerial darts (Table 3). Of all the media tested, only Pro-mix without Terra Sorb and Turface-peat-vermiculite with Terra-Sorb had the total porosity proposed by Deboodt and Verdonck (1972). Addition of sand to the bark vermiculite medium reduced the water holding capacity while Terra-Sorb increased the total porosity and water holding capacity.

The bark-vermiculite mixes, Turface-peat-vermiculite and Pro-mix were recommended as good media for aerial darts as they had adequate aeration and highest water retention. The hydrophilic gels were effective when added with surfactants like Aqua-GRO.

4. SEEDING OF PELLETED SEEDS

Aerial seeding is mostly carried out using fixed-wing aircraft or helicopters flying at 15-25 m height. Over 50 years helicopters were used for aerial seeding in forests of USA. The aircrafts are efficient in large stretch of lands, but height and swath width are difficult to control in rough terrain. The helicopters can get closer to ground, fly at slower speed, land

Table 3. Percent volume of total porosity (TP), air space (AS), easily available water (EAW) and water buffering capacity (WBC) of several growing media amended with the surfactant Aqua-GRO

Growing medium	TP[1]	AS[2]	EAW[3]	WBC[4]
B-V (1:1)	78.9^{ef}	28.1^{b}	20.3	3.9^{ab}
B-V-TS (1:1)	82.0^{cde}	27.0^{b}	4.9	7.7^{ab}
B-V (2:1)	79.5^{def}	32.2^{a}	4.5	4.1^{bc}
B-V-TS (2:1)	88.6^{b}	35.4^{a}	4.3	3.0^{bc}
B-V-S	61.8^{h}	25.4^{b}	5.5	2.8^{c}
B-V-S-TS	70.1^{g}	27.6^{b}	8.3	1.2^{c}
T-P-V	82.1^{cde}	23.8^{b}	13.7	2.6^{c}
T-P-V-TS	84.1^{bcd}	30.2^{b}	6.5	1.1^{c}
PM	85.2^{bc}	28.5^{b}	9.8	5.3^{a}
PM-TS	98.0^{a}	25.0^{b}	15.6	7.9^{a}
Ball I	77.1^{f}	30.8^{ab}	7.2	3.7^{bc}
Ball I –TS	80.3^{def}	31.2^{ab}	6.8	4.3^{bc}
Ball II	77.3^{f}	35.8^{a}	8.7	2.3^{c}
Ball II –TS	83.1^{cde}	37.6^{a}	8.1	3.3^{c}
Ideal substrate[5]	85.0	20-30	20-30	4-10

B-V, Gro-bark- vermiculite; B-V-S, Grobark-vermiculite-sand; T-P-V, Turface-peat-vermiculite; PM, Pro-mix; TS, Terra-Sorb

1 Significant media by gel interaction at 5% level
2 Media significant at 1% level
3 Media not significant
4 Theoretical values described by DeBoodt and Verdonck (1972)

on site and reload the materials quickly and discuss plans readily with ground staff. In addition to that, the helicopters were found to be more efficient in sites where there are more edges and corners of irregular tracts of land, on steep mountains and in small areas upto 0.5 ha. On the other hand, the fixed-wing aircraft are cheaper to operate when compared to helicopters.

In 1986, Beaver type aircraft was used for aerial seeding in Morena and Bhind district of Chambal ravines and able to cover 500 ha per day. In the same area, Air Force Dakotas distributed the seeds at a height of 300 m and as a result much of the seeds were blown away (Anon, 1973). Efficient seeding was carried out in sand dune of Motigarh and Sardarpura during 1982 using Dakota aircraft of the Agricultural Aviation Department, New Delhi (Flannery and Busscher, 1982). Temporary air-strip of 600 m long and 30 m wide were determined in the prevailing wind direction of northwest-southeast direction. About 400 ha area was covered in 4 hours by this aircraft. The seeding time was more when helicopters were used. Helicopters used in Sundarbans were able to cover only 20 ha in a day (Still, 1976). The availability of the aircraft or helicopter is also a deciding factor for success of aerial seeding program, because the time of seeding is equally important as that of other factors affecting success of aerial seeding. To make the aircraft available at right time, the aerial seeding operations should be planned well in advance, especially in developing countries like India.

The area to be seeded should be delineated on the ground and maps or aerial photos are given to the pilot. Flagmen are positioned to guide the aircraft to avoid overlapping runs. Large-scale mosaics made from aerial photographs enables pilots to identify the areas to be sown without the use of ground-markers (Deboodt and Verdonck, 1972). The uniformity and density of seed distribution may be checked in advance before going for actual operation.

Seed hoppers and distributors that are attached to aircraft are developed. Fixed wing air craft usually use gravity flow of seed and an unpowered venturi flume-type distributor. The seed dropped into the flume is spread and carried away through the slipstream as it exits from the flume. In helicopters, power-driven slinger-type applicators are used for spreading the seeds (Page, 1969).

REFERENCES

1. Murthy., A.V.R.G.K. (1988). Pelletization and aerial seeding in forestry. *My Forest, 24* (4) : 279-287.
2. Anon. (1973). Benefits of agricultural hydrogel concentrate 50G. Union Carbide Corp., New Tork. Tech. Bull. 270.
3. Anon. (1981). Sowing Forests from the Air, National Academy Press, Washington, D.C.
4. Biradar, B.B. (1988). Precision planting by seed pelleting. *Seeds and Farms*, : 32-43.
5. Deboodt M.and O.Verdonck. (1972). The physical properties of the substrates in horticulture. *Acta Hort., 26:* 37-44.
6. Flannery R.L.and W.J.Busscher. (1982). Use of a synthetic polymer in potting soils to improve water holding capacity. *Commun. Soil Sci. Plant Anal., 13* (2) : 103-111.

7. Jha, P.K., S. Nair, and C.R.Babu. (1994). Development of an inexpensive legume-*Rhizobium* inoculation technology which may be used in aerial seeding. *Journal of Basic Microbiology, 34*(4) : 231-243.

8. Karivaradaraaju., T.V., K.Vananagamudi, R.Umarani, A.Bharathi, C.Surendran and S.Balaji. (1999). A Treatise on Tree Seed Technology. Tamil Nadu Agricultural University, Coimbatore, Tamil Nadu.

9. Lahiri., A.K. (1991). Aerial Seeding in Mangrove Swamps. *Indian Forester, 117* (3) : 159-161.

10. Lennox, T.L.and G.P. Lumis. (1987). Evaluation of physical properties of several growing media for use in aerial seeding containers., *Canadian Journal of Forest Research, 17* (2): 165-173.

11. Lyon N.F and J.S.Thomson. (1979). Forest Research Note. Ministry of Natural Resources Ontario No 25, pp.4.

12. Page, A.I. (1969). The use of large-scale aerial photo mosaics for planning and control of aerial seeding, *N.Z. J. For., 14* (1) :96-97.

13. Ram prasad. (1988). Effectiveness of Aerial seeding in Reclamation of Chambal ravines in Madhya Pradesh. *Indian Forester, 114*(1) : 1-18.

14. Shankaranarayan, K.A. and Suresh kumar. (1986). Aerial seeding of Sand dunes I.Trends in Seed germination and Seedling distribution. *Journal of Tropical Forestry, 2* (1) : 3-20.

15. Still. S.M. (1976). Growth of 'Sunny Mandlay' chrysanthemums in hardwood-bark-amended media as affected by insolubilized poly (ethylene-oxide). *Hort Science, 11*: 483-484.

16. Vanangamudi, K. (1992). Seed pelleting studies in *Acacia nilotica*. Personal Communication.

17. Vanangamudi, K., S.Padmavathi, V.Manonmani and C.Surendran. (1993). Effect of pelleting with bio-fertilizers, biocides and nutrients on the viability and vigour of Neem seed. World Neem Conference. Abstract p 64: 24th –28th.

18. Wood, A.D.. (1984). Developing a seed-containing dart and aerial delivery system for forestry applications. *For. Chron., 60* (2) : 86-92.

NATIONAL AND INTERNATIONAL SCENARIO OF FIRMS AND MACHINERY IN PELLETING AND COATING

R. UMARANI[1], R. JERLIN[2] AND K. VANANGAMUDI[3]

1. INTRODUCTION

Seed pelleting has been accepted as an important seed enhancement methodology that improves seeding as well as seed performance. This is plausible from the fact that number of companies in National and International levels have ventured into the field of seed pelleting. The companies with their own strong research team have mastered the technology. Each company has developed species specific technology and highly sophisticated computerised machinery for rapid, uniform and efficient pelleting of seeds. Its heartening to visit the websites of different seed companies, where they have demonstrated their mastery over the pelleting technology which indeed is the basis for developing efficient machinery. By sincere and concerned efforts the companies have made adaptation within the software and some technical changes which turned, the tried and tested seed coating machines into automatic seed pelleting medicines.

Before going into the machine/models produced by different companies let us try to appreciate the sound technical knowledge of each company with regard to seed pelleting.

2. "INCOTEC"

Pelleting is the application of a proportioned mixture of powders around the seed.

Department of Seed Science and Technology, Tamil Nadu Agricultural University, Coimbatore – 641 003.
Agricultural Research Station, Bhavanisagar – 638 451
Dean (Agriculture), Tamil Nadu, Agricultural University, Coimbatore - 641 003.

Some species are difficult to plant mechanically because of their shape (lettuce) or size (begonia). Pelleting makes seed more uniform in shape, weight and size. This makes it easier to precision plant using various types of planters. Seeds are planted at correct distances and depths with less seed loss. Approved fungicides and insecticides can also be incorporated into the pellet to protect some species from pests. The desired amount of active chemical ingredient can be placed near the emerging seedling, where it will be used more efficiently.

3. SPLIT PILL

A process unique for its splitting character. The coating material allows a free and quick access of moisture and oxygen to the seed. The pellet does not melt, but splits open and creates optimum conditions for a quick development of the seedling.

4. SPLIT PILL LIGHT

A light weight pellet suitable for sowing under fluctuating moisture levels.

5. SPLIT PILL SPECIAL

The Split Pill process in combination with pre-germination precision seed. For celery, the Split Pill Special is available with a Thiram Soak treatment, the Split Pill Special Dis.

6. SPLIT PILL THERMOCURE

The Split Pill in combination with the Thermocure treatment. The Thermocure treatment will provide maximum germination at very high temperatures. (For lettuce only)

7. SPLITKOTE

A process developed for direct sowing in outdoors. Splitkote has the same splitting characteristics as split pill, but differs from split pill by a different composition of the coating material. Splitkote gives good results in the field under varying conditions.

8. SPLITKOTE SPECIAL

Splikote special contains primed seed. The germination under stress conditions, i.e. high-low temperature, is better, and another effect of the priming is a faster and better emergence.

9. "SUB-SAHARA INTERNATIONAL INC"

9.1. Machine Features

Sub-Sahara is a distributor of rotary coating equipment. This type of equipment offers our customers a tool to better achieve their coating and treating objectives. Rotary coating machinery offers many advantages over traditional seed treating equipment.

- Precise material application
- Separate delivery systems for coating material and pesticide
- Highly efficient usage of coatings.
- Simple to clean
- Electronic digital scale system
- Capable of film coating almost for any seed type
- Low maintenance requirements
- Capable of pelleting seed with powder filled coating

10. "SEED EQUIPMENT CORPORATION, INC"

Film coating is normally used to apply a thin uniform layer of polymer over seeds without significantly increasing seed size or weight. The plasticizer polymer forms a flexible film that adheres and protects the fungicide, preventing dusting off and loss of fungicide during handling. The film is readily water soluble so as not to impede germination. The application process also allows a more precise, uniform application of fungicides, insecticides, colors and other additives than can be accomplished by a typical seed slurry treatment. Film coating helps to smooth and round the seed surface, which improves flowability and machine planting.

11. BENEFITS OF FILM COATING

- Enables accurate and even doses of chemicals and reduces chemical wastage.
- Saves valuable time during sowing period as seed requires no further treatment.
- Easier to observe sowing accuracy as coated bright green seed is easily visible.
- Improves the flow of seed through the seed hopper.
- Reduces grower exposure to agricultural chemicals as sowing process becomes dust free.

- Provides an effective strategy against infection by *Botrytis alli*, particularly when combined with *Botrytis alli* free seed.
- Assists in establishing onion plants in soils where white rot (*Sclerotium Cepivorum*) is present.
- Reduces the effects of 'damping off'.
- No changes to sowing equipment are required, as is the case with pelleted seed.
- Enables the application of trace elements to the seed to supply nutrient to the young seedling.

12. SEPIRET

12.1. What is SEPIRET seed film coating?

A thin film of biodegradable polymers which encase the seed treatment.

What seeds can SEPIRET film coatings be used on?

Oilseed rape, brassicae, sugar beet, vegetables, peas, beans, maize, cereals, linseed etc.

What advantages are there from SEPIRET film coatings?

- Improves the adherence of the seed treatment to the seed
- Aids the distribution of low concentrations of active ingredients to the seed'
- Eliminates dust during handling
- Provides greater safety for the user
- Improves seed flow and spacing between the seeds sown.

13. CELPRIL

13.1. Protect and Enhance Your Genetics

CelPass protects your best genetics by locking chosen seed protectants firmly to your seed. CelPass Film Coating gives your high value genetics every opportunity to perform the way they were intended. CelPass applies uniformly making it the ideal delivery mechanism for the new generation of high performance seed protectants. Seeds are totally encapsulated, assuring that seed protectants are bonded to your seed for maximum performance in the field. When seed protectants perform.... your genetics perform as well. Seed experts are already looking at the seed as the future delivery system not only for genetics, but for seed protectants and biologicals as well.

14. MACHINE FEATURES

ETS rotary film coating process uses a spinner to automize the polymer and other liquid additives such as fungicides onto the seed. Rotation of the coating chamber floor actively mixes the seed so that all seed surfaces are evenly coated. This process is actually a batch process, but is so quick and well automated (batch cycle times of 15-45 seconds) that a virtual continuous flow of finished product results. Equipment clean up is quick and easy (about 10 min), and the minimal amount of rinse water can usually be added back to the polymer solution for reapplication.

The design of this system allows concurrent addition of powder filler materials during the coating process. Trials with several vegetable, legume, grass and field crop species have proven at least 800% weight build-up is readily possible, thus offering an alternative for the slower, costly pan coating procedure. The process allows great flexibility for adding blended powders/additives or layering them individually onto the seed.

Details of seed pelleting companies and machinaries

S.No	Companies/Firms	Machinaries	Features/products
1	**Incotec (Integrated Coating and Seed Technology)** Netherlands, Japan, Italy, Spain, Israel, Brazil	Rotary coater	Split pill Split pill light Split pill special Split pill Thermoceae Splitkote Split kote special
2.	**Sub-Sahara International Inc** 13923-Emerald forest ct, Sugar land, Tx - 77478	Rotary coater	
3.	**Satec concept**	Batch coater Concept	Wheat/h Corn/h 3/25 - 5t 3-5t 3/25 Twin - 10t 7t 3/50 - 10t 7t 3/50 Twin - 20t 14 t with following redrying mixing chamber electronic seed balance high precision liquid dosing • Lab coaters • Manual operation • Mid size capacities • Large scale Production

4.	**Cimbria** A-2000 stockerau vienna, 31 Austria Email–heid @ cimbria.Com	Centri coater	CC Lab CC 50 DUO CC 10 CC 20 CC 50	
5.	**MKS** Magnus Kahl Seeds pvt. Ltd.	In house equipment	Supplies coated seeds	
6.	**Seed Tech +** Leeschweitzer, Ph.D 9355, Hope well Rd Salem OR 97304 USA	ETS Rotary coaters	R 30 A Rotary coater 30 inch chamber upto 400 lb h^{-1} Automatic Two liquid pumps 5x5 ft floor space Many options	
7.	Seed Equipment Corporation Inc.	R 6 M Manual	Capacity	lb min^{-1}
		R12 M Manual	0.2 – 0.5	
		R 12 A Auto	1-12	10-140
		R 24 M Manual	1-12	
		R 24 A Auto	15-30	30-90
		R 30 M Manual	30-120	
		R 30 A Auto	30-120	80-400
		Dual R 30A Auto	30-120	160-800
			Machine Features Manual • Seed is loaded into machine by hand • Treatments and coating materials are added by hand (or by manually controlled applicators) • Air operated discharge gate activated by panel button Automatic • Seed is loaded into machine by automated digital scale system (included) • Programmable peristaltic pumps apply treatments, polymer, or other liquids (two pumps standard, additional pumps optional)	

	• Computerized PLC panel to program scale system, peristaltic pumps cycle time, seed discharge, and optional precision powder feeders
	Options • Additional peristaltic pumps to apply other liquid ingredients • Precision powder feeders to apply dry ingredients for size and/or weight gain • Custom modifications for your specific applications • Design complete turnkey system including material handling and bagging • Lease/purchase options available

15. CONCLUSION

The International scenario of firms and machinery is really heartening. It is evident that developed countries have taken up Seed Technology seriously unlike the developing countries. It is obvious that research has to be carried out in our country to develop indigenous machinery to pellet the seeds with required ingredients, effectively and efficiently. It is well proven by the success of the international community that, indigenizing this technology will revolutionise our agriculture more.so ,with respect to high value - low volume crop seeds.

FUTURE PERSPECTIVES FOR RESEARCH ON SEED PELLETING

T.JAYARAJ[1], A.SABIR AHAMED[1] AND R.UMARANI[2]

As the beneficial effect of pelleted seeds becoming more popular, the demand for pelleted seed is increasing and would increase many fold in near future. In developed countries, the demand for pelleted seeds have already increased and the number of companies produce pelleted seeds. Increased competition in the pelleted seed market has fostered the development of more effective pellets with greater capabilities and wider planting characteristics. Therefore, as future perspective, improvement through research on pelleting technology on following areas is required.

First of all, a clear understanding for distinction between seed pelleting and seed coating should be brought into the minds of researchers and seed producers. Usually, seed pellet contains added substances applied to the seed to modify flat or irregularly shaped seed into a rounded seed or smaller-lighter seed into a larger heavier seed. Whereas, seed coating is applied to the seed that does not obscure its shape. Examples for coating are fungicides, microbiological treatments and micronutrients etc.

There are two components in seed pellet-bulking (or coating) material and binder. If the pelleting material is too hard after drying, it may be difficult for radicle emergence through the pelleted material. Research should be taken up to standardise the pelleting components, which are compatible with the seeds based on the seed physiology of each species. So that, seed quality is maintained and the germination is not hindered.

1 *Soil and Water Management Research Institute, Thanjavur – 613 501.*

2 *Department of Seed Science and Technology, Tamil Nadu Agricultural University, Coimbatore-641 003.*

The second component, the binder, holds the coating material together. The binder concentration is critical, because too much binder will delay germination and too little binder will cause chipping and cracking of pellets. Therefore, a thorough research on binder quality, quantity and concentration required to be used for each species has to be standardized.

Some pellets may form around multiple seeds. In other cases, particularly when the seed lot is not cleaned properly, pellets may form around inert matter such as, sand or undesired plant material. In such case, it will become an issue of certification and will pose lot of confusion when conducting a purity test with pelleted seeds. Therefore, it must be ensured that each pellet contains a seed and for that, suitable technologies should be developed to avoid empty pellets and multiple seeded pellets.

To avoid the deleterious effect of higher concentration of binding material (which becomes necessary to prevent breaking of pellets or chipping away of pellets), possibility coating with polymer can be explored to overcome the disadvantages of higher binder concentration.

An integrative approach should be developed to include other interactive treatments like seed priming, seed fortification, hydration-dehydration and mid-storage correction along with seed pelleting.

Results of various seed pelleting experiments conducted worldwide differs considerably due to differences in pelleting process and products used for such experiments. Therefore, it is imperative to develop, easy, reliable, uniform and reproducible methods for adoption world wide.

Though, the research on seed pelleting started with the vegetable seeds, of late, the technology drifted away towards agricultural crop seeds. As the vegetable seed industry gaining grounds in Indian scenario, a lot of research base is required for seed pelleting for vegetable seeds.

For large scale adoption of this pelleting technology, it is necessary to introduce automation of entire pelleting process and suitable machinery should be developed to pellet a wide range of crop seeds with precision and uniformity in pelleting process. Sequencing of pelleting process has to be standardized mainly for commercialisation of pelleting technology among the seed industry.

The pelleting technology has already been proved successful in developed countries. Historically, the increased usage of pelleted seeds occurred since 1970's in western world.

A voluminous informations are available on seed pelleting technologies in the websites related to seeds, thanks to the development of information technology. The need of the moment is to browse and update first hand information available on the latest technology of seed pelleting process and developments in mechanisation process. Visit to various seed industries abroad will enrich knowledge base in this field of seed science.

Large database on storage of pelleted seeds has to be generated in order to understand the influence of pelleting material, and environmental interactions on physiology and biochemistry of pelleted seeds and their ultimate impact on shelf life of such seeds.

Development of pelleting technology is of no use, if a proper sowing or seed placement technology is not developed. Therefore, method of either manual or mechanical sowing of pelleted seeds also to be developed and prescribed when such technology is taken to farmers field.

The success of this technology may not be complete if it is not taken to the field at user level. Therefore, conducting on farm trials with pelleted seeds of local interest will come handy in transferring this proven technology to the field for the benefit of farmers and ultimately stepping towards achieving sustainability in agriculture.

PART IV

SEED COATING

SEED COATING IN AGRICULTURAL CROPS - PRINCIPLES AND PRACTICES

M. BHASKARAN[1], A. BHARATHI[2], P. NATESAN[2],
R. UMARANI[2] AND K. VANANGAMUDI[3]

1. INTRODUCTION

Seed is the pivot of Agriculture; careful farming envisages and enforces first and foremost attention on seed. "Care with the seed - Joy with harvest" is the popular adage which enlightens about the importance incumbent on the seed.

Seed quality is one of the most talked about subjects among farming communities. To provide higher quality seeds, many seed companies have developed new production technique called seed enhancements. The two goals of seed enhancement, like seed designing and seed functioning can be improved through seed film coating. In view of this, many seed industries have developed specialized seed coating, seed pellets and film coatings.

2. SEED COATINGS

A seed coating is the substance applied to the seed that does not obscure its shape. Seed coating examples are fungicides, microbiological treatments and micronutrients. The major benefit of a seed coating is that the seed enhancement material is placed directly on to the seed. Small amount of chemical are needed as compared to broad casting is one of the most economical approaches for improving seed performance.

1 National Seed Research and Training Centre, Varanasi – 221 106
2 Department of Seed Science and Technology, Tamil Nadu Agricultural University, Coimbatore – 641 003
3 Dean (Agriculture), Tamil Nadu Agricultural University, Coimbatoire - 641 003.

3. FILM COATINGS

Film coating technology is a sophisticated process of applying precise amount of active ingredients along with a liquid material, directly on to the seed surface without obscuring its shape and total seed weight may increase up to 10%. Since the coatings are very thin, multiple coatings of various materials are possible. Most film coating contains a colorant that helps in visual monitoring and placing accuracy during automatic seeding. The coating materials can be designed for specific seed moisture requirements. Fungicides and insecticides can also be added. Film coatings can help to smooth the rough surface of seed, allowing the seed to flow more evenly during automatic seeding. Also provides strong identity for treated seeds.

3.1 Advantages of film coating

Film coating is mainly followed in the agricultural crop seeds, by different seed companies. The advantages of this method over the conventional seed treatment can be enumerated as follows.

1. Enables accurate and even doses of chemicals and reduces chemical wastage.
2. It makes room for including all the required ingredients like inoculants, protectants, nutrients, hydrophilic substances, herbicides, oxygen suppliers etc.
3. By encasing the seed within a thin film of biodegradable polymer, the adherence of the seed treatment to the seed is improved.
4. It ensures dust free handling, making treated seed, both user and environment friendly.
5. Addition of colorant helps visual monitoring of placement accuracy.
6. Coatings also results in more even seedling rate due to the smooth flow of the seeds.
7. Provides resistance against mechanical damage in the seed drill.
8. Improves the appearance and quality of treated seeds.
9. The major benefits of seed film coating are related to safety and easy too handling.
10. Polymer coating acts as a temperature switch and protective coating by regulating the seed coating intake of water, until the soil has warmed to a predetermined temperature.

3.2 Influence of seed coating polymer on vigour and yield potential

Ruban *et al.* (1983) reported that cotton seed coated with Penghiuram and polymer initially inhibited the activity of catalase andperoxidase and rate of seedling growth, but stimulated them later on.

Arvind Kumar *et al.* (1991) reported that seed yield of rainfed mustard increased by 18.1 percent over control on inclusion of hydropolymer jalasakthi in seed pelleting.

Markrov (1987) observed that sorghum seeds pelleted with polymer film coating, fungicide and growth regulator and sown early under low temperature condition registered normal germination, accelerated crop maturity by 9-21 days and increased yields compared with untreated seeds sown on normal date. Seeds coated with polymer film coating fungicides and micronutrients registered good germination energy and germination percent in Lucerne crop. (Konstantinov and Kolarova, 1988).

Willams *et al.* (1990) observed the effects on germination and emergence of coating fuzzy cotton seed with the easiflo polymer and fungicide mixture. Joshi *et al.* (1998) reported that sorghum seeds treated with 10-50 of hydrophilic polymer kg^{-1} seeds, improved seed germination and emergence rate at the lowest available soil moisture level. The best treatment was 10g kg^{-1} seeds.

Kajal sengupta *et al.* (2001) reported that plant heights, number of branches per plant, drymatter accumulation, pods m^{-2}, pod length, seed yield and harvest index were maximum with 1,875 m seed plus polymer kg for the rainfed greengram. Add 1000 seed weight and seeds per pod were maximum with seed plus at 1.25 ml and 5 g kg^{-1} respectively.

Smith *et al.* (2000) reported that coating with 24 mg mg of polymer per seed regulated the rate of water uptake, reduced imbibition damage and improved percentage germination and seedling emergence. Coating with a hydrophobic polymer can reduce rates of water uptake, lower solute leakage and improves vital tetrazolium chloride staining and partially improves emergence of soybean seedlings.

3.3 Effect of film coating on field emergence of the seedlings

Angamuthu (1991) observed that seeds pelleted with polymer and arappu leaf powder registered higher field emergence potential, plant height, ear length and seed yield in minor millets. The emergence rate Index of seeds was increased for captan + certop (a polymer) treatment. Also increased final emergence and seedling height at early planting (Rivas *et al.* 1998).

4. ROLE OF POLYMERS IN PROMOTING THE EARLY PLANTING OF THE CROPS BY ACTING AS A TEMPERATURE SWITCH

Crowely (2001) reported that Intellicoat coatings can provide protection to hybrid corn seed under cold stress conditions. Coated hybrid seed can retain its viability for a normal stand establishment for 2 to 3 weeks at 50°C in near saturated soil. Tony Vyn (2001) reported a frost that seriously damaged crop stand from uncoated seed planted in early April had little effect on a stand that emerged later due to seed coating.

Rusu *et al.* (2001) reported that polymer coatings enable emergence to occur when the best temperature for maize germination was reached in the soil, thus could be used for earlier maize sowing and for phasing sowing over a longer period of time.

Tony Vyn (2000) reported that polymer coating delays soybean germination for about two weeks, a period that accommodates the wheat harvest and leaves enough time for soybean to produce a good yield before a killing frost.

4.1 Polymer film coating for improving the health potential of the seeds

Ester *et al.* (2001) reported that the use of fipronil and imidacloprid film coated seed resulted in sufficient protection against the leek moth at low population densities. Whipps *et al.* (2001) reported that the film coating with *Cercospora minitans* to the infected sunflower seed decreased the germination of condia. Arantes *et al.* (2000) reported that seed germination and seedling growth were best and fungal infection was lowest with encapsulation or coating with gypsum + bentonite or the polymer sepiret 6182 combined with fungicides carboxin + thiram.

Current activities of the Department in respect to seed coating.

4.2 Seed coating polymer for maize seed

4.2.1. Pink Polykote

The objective was to standardize a suitable dosage of seed coating polymer namely "Polykote", supplied from the Little's Balm Pharmaceuticals Ltd, Chennai. The seed of maize cv. CO4 were coated with pink polykote @ 1,2,3,4 and 5g kg^{-1} of seed and evaluated for germination and seedling vigour.

Polymer coating was carried out as detailed below

1. Dry seed dressing.
2. Wet seed dressing – ie. water dilutions of the polymer @ 5 ml, 10 ml and 20 ml kg^{-1} of seeds.
3. Fevicol seed dressing @ 10%, 15% and 20% concentrations.

Among the above different methods water dilution @ 5 ml kg^{-1} of seeds, dry dressing and fevicol seed dressing @ 10% concentration were standardized as the optimum method of seed dressing (Table 1).

Table 1: Effect of pink polykote on germination and seedling vigour of maize

Polykote Treatments	Germination (%)	Root Length (cm)	Shoot length (cm)	DMP (g $seedling^{-1}$)	Vigour index
T_0 (0g/kg of seed)	96.50	14.935	13.505	0.802	77.37
T_1 (1g/kg of seed)	97.25	17.795	13.800	0.959	93.29
T_2 (2g/kg of seed)	97.75	17.595	13.355	0.945	92.39
T_3 (3g/kg of seed)	99.25	20.658	16.775	1.067	105.86
T_4 (4g/kg of seed)	98.75	20.205	12.005	0.946	93.39
T_5 (5g/kg of seed)	98.00	18.305	14.825	1.032	101.14

If is found that compared to control, the pink polykote coating of maize seeds at 3 g kg^{-1} of seed was found to improve the germination and seedling vigour. An increase of 2.7% in germination, 27.7% in root length, 19.5% in shoot length, 24.8% in drymatter production and 26.9% in vigour index was noticed. An increase in dosage of pink polykote over 3g/kg of seed was detrimental to germination and seedling vigour. A decrease of 1.2% germination, 11.2% in root length, 11% in shoot length, 3.3% in DMP and 4.5% in vigour index was recorded. (Table 1.)

It is concluded that coating of maize seeds using pink polykote @ 3g kg^{-1} of seed was found to be beneficial and shows promoting effect on germination and seedling vigour. An increase in dosage of pink polykote beyond 3 g kg^{-1} was detrimental to germination and seedling vigour and showed inhibition effect.

4.2.2. *Green Polykote*

The maize seeds cv. CO1 were coated with green seed coating polymer @ 3 g kg^{-1} of seed using different methods such as dry dressing and water dilution and evaluated for germination (%) and seedling vigour.

Treatment

T_0 - Control

T_1 - Dry dressing @ 3 g kg^{-1} of seed

T_2 - Water dilution @ ml kg^{-1} of seed

Table 2: Effect of green polykote on germination and seedling of maize

Treatments	Germination (%)	Root Length (cm)	Shoot length (cm)	DMP (g $seedling^{-1}$)	Vigour Index (G% x DMP)
T_0	75.0	13.060	11.420	0.8022	60.205
T_1	86.5	12.978	12.457	0.931	80.602
T_2	80.0	16.185	13.455	0.979	78.346
SEd	2.392	1.218	NS	0.041	4.690
CD (P=0.05)	5.412	2.756		0.092	10.611

Germination percentage and vigour index were best with dry dressing of polymer with the seed (Table 2). Where as the root length, shoot length and DMP was found to increase with water dilution of the polymer and then coated into the seeds.

4.3 Effect of pelleting with polymer on changes in seed quality of minor millets

The effect of pelleting with polymer, arappu powder and urea alone or in combination with each other was studied on intact and sprouted seeds of minor millets like Ragi, Panivaragu, Kuthiraivalu, Tenai and Samai.

Treatments

T_0 - Control (intact seed)

T_1 - Pelleting with polymer

T_2 - Pelleting with polymer + Arappu powder (*Albizia amara*)

T3 - Pelleting with polymer + urea

T4 - Pelleting with polymer + Arappu powder + urea

T5 - Sprouted seed

T6 - Pelleting with polymer

T7 - Pelleting with polymer + Arappu powder

T8 - Pelleting with polymer + urea

T9 - Pelleting with polymer + Arappu powder + urea

Methods and preparation of pelleted seeds

T_0- Control (Intact seed)

T_1- The intact seeds were coated with polymer (a seed treating chemical) @ 1.5 ml per 100 g of seed. The seeds were agitated well for thorough mixing.

T_2- With the above polymer coated seed, arappu powder @ 1 g per 100 g of seed was added, stirred well, kept for overnight.

T_3- Urea @ 1.5 g per 100 g of seed added to the polymer coated seed and kept for overnight.

T_4- Arappu powder and urea were added with the polymer coated seed.

T_5- Sprouted seeds for this, the seeds were soaked in potable water for 20 to 24 hrs. at 1:1 (W/v) (just mergence of the embryonic axis). Shade dried to original moisture content and taken for sowing.

T_6- Sprouted seeds were treated with polymer as in T_1

T_7- Sprouted seeds were treated with arappu powder as in T_2

T_8- Sprouted seeds were treated with urea as in T_3.

T_9- Sprouted seeds were treated as in T_4.

The adhesive along with pellet material was poured to 100 g of seed and 1 g of arappu powder was added to the seed and mixed rapidly for 5 minutes. Until the seeds are evenly coated. Care was taken that seed should not get aggregated. Then the pelleted seed was shade dried and kept for overnight.

Seed pelleting treatment (T2) with polymer @ 1.5 ml 100 g-1 + arappu powder (1:100) (*Azbizia amara*), revealed its superiority over the other treatments by recording maximum values of germination (%) and vigour index for all the minor millets studied. Also intact seeds performed better than the sprouted seeds (Table 3).

Table 3: Effect of pelleting with polymer on changes in seed quality of minor millets

Treat-ments	Ragi		Panivaragu		Kuthiralvali		Tenai		Samai	
	Germination (%)	Vigour index	Germination (%)	Vigour index	Germination (%)	Vigour index	Germination (%)	Vigour index	Germination (%)	Vigour index
T_0	56.40	545.8	48.00	1007.0	65.18	907.4	56.85	854.3	50.3	891.7
T_1	57.33	590.5	63.12	1062.4	64.15	914.7	59.07	869.6	51.5	893.4
T_2	66.3	672.4	74.74	1190.46	72.00	1073.7	64.00	1021.7	59.7	908.6
T_3	59.25	636.7	68.87	1093.6	67.00	954.3	61.71	969.1	55.2	894.1
T_4	56.00	620.8	66.0	1037.6	65.40	925.6	60.50	876.7	55.7	900.0
T_5	58.00	624.1	65.5	977.4	64.70	861.6	60.78	958.00	54.8	887.1
T_6	60.66	603.6	64.12	1003.8	68.50	953.0	61.21	949.10	53.7	890.5
T_7	62.33	640.6	65.5	1030.0	68.81	968.3	60.92	939.80	54.0	887.8
T_8	61.66	617.4	64.62	1038.6	67.81	951.4	59.64	916.10	53.0	882.3
T_9	58.1	598.8	61.75	1043.6	66.90	952.0	93.30	931.00	53.2	884.8
SEd	0.625	24.224	1.119	24.545	0.849	15.457	0.710	15.742	0.690	3.549
CD (P=0.05)	1.239	48.448	2.223	48.771	1.690	30.759	1.416	41.390	1.373	7.063

CURRENT SCENARIO OF SEED COATING POLYMER

K.VANANGAMUDI[1], A.BHARATHI[2], P. NATESAN[2],
M. BHASKARAN[3] AND G.SASTHRI[4]

1. INTRODUCTION

A high quality seed with better germination, vigour and seed health is the pre-requisite for a good crop. However in the field, various diseases and pests may still attack seeds. To overcome this, Integrated Pest Management strategy has to be adopted. The farmer is very much interested in the best seed management practices, which are safe, environmentally sound and scientifically proven to overcome the adverse situations. A strong reduction in the amounts of pesticides being used can be achieved by applying crop protection agents directly to the seed instead of spraying them in the field and can be incorporated in a pellet or a film coat layer, where it will be most effective i.e. directly on the seed. The agriculture industry is now responding to this trend with the development of more and more molecules that must be applied to the seeds and which require high quality application technologies. This technology may also be combined with other seed enhancement and protection technologies.

2. SEED COATING

A seed coating is the substance applied to the seed that does not obscure its shape. Seed coating examples are fungicides, microbiological treatments and micronutrients. The major benefit of seed coating is that

1 *Dean (Agriculture), Tamil Nadu Agricultural University, Coimbatore - 641 003.*
2 *Department of Seed Science and Technology, Tamil Nadu Agricultural University, Coimbatore-641003.*
3 *National Seed Research and Training Centre, Varanasi – 221 106.*
4 *Department of Plant Breeding and Genetics, NPRC, Vamban – 622 303*

the seed enhancement material is placed directly on to the seed. Smaller amounts of chemicals are needed as compared to broadcasting or surface dressing onto the growing medium. Seed coating is one of the most economical approach for improving seed performance.

Film coating is normally used to apply a thin, uniform layer of polymer over seeds without significantly increasing seed size or weight. The plasticizer polymer forms a flexible film that adheres and protects the fungicide, preventing dusting off and loss of fungicide during handling. The film is readily water-soluble (hydrophilic) so as not to impede germination.

The more precise and uniform application of fungicides, insecticides, colours and other additives can be accomplished by a typical seed slurry treatment. Film coating helps to smooth and round the seed surface, which improves flowability and machine planting.

The major benefits of seed film coating are mainly related to safety and easy handling. The other advantages of this method over the conventional one are

- Improves plantability and emergence of seeds to benefit seed companies and farmers.
- Enables accurate and even dosage of chemicals and reduces chemical wastage.
- It makes room for including all the required ingredients like inoculants, protectants, nutrients, plant growth promotors, hydrophobic/hydrophilic substances, herbicides, oxygen suppliers etc.
- By encasing the seed within a thin film of biodegradable polymer, the adherence of the seed treatment to the seed is improved.
- It ensures dust free handling, making treated seed both user and environment friendly.
- Addition of colourant helps visual monitoring of placement accuracy.
- Polymer coating results in more even seeding rate due to the smooth flow of the seeds
- Polymer acts as a temperature switch and protective coating by regulating the water uptake and subsequent germination of seed.

Thereby the two goals of seed enhancement like seed designing and seed functioning can be improved through seed film coating.

Polymer and colourant products can be added to the standard treatment slurry. They are compatible with the fungicides and insecticides commonly used in seed treatment. The polymer products can be used at near full strength or can be diluted with 1 - 4 parts water plus seed treatment chemicals. They contain excellent surfactants and spreaders and have a very quick drying time.

In the seed coating equation, there are three parts i.e. the right products, the right equipment and the right application process. A sufficient amount of total liquid must be present to allow for adequately covering the seed surfaces.

3. INFLUENCE OF SEED COATING POLYMER ON VIGOUR AND YIELD POTENTIAL

Several scientists studied the effect of polymer coating on different crops. Cotton seed coated with Penthiuram and polymer initially inhibited the activity of catalase and peroxidase and rate of seedling growth, but stimulated them later on. Sorghum seeds pelleted with polymer film coating, fungicide and growth regulator and sown early under low temperature condition registered normal germination, accelerated crop maturity by 9-21 days and increased yields compared with untreated seeds sown on normal date.

Seed coated with polymer film coating, fungicides and micronutrients registered good germination energy and germination per cent in lucerne crop. Seed germination and seedling growth were best and fungal infection was lowest with encapsulation or coating with gypsum + bentonite or the polymer Sepiret 6182 combined with fungicides carboxin + thiram.

The effects on germination and emergence of coating fuzzy cotton seed with the Easiflo polymer and fungicide mixture obviously indicated the Easiflo polymer itself did not significantly affect germination and emergence and concluded that the use of Easiflo polymer to treat fuzzy cotton seed for planting should not negatively affect seed germination and emergence. Sorghum seeds treated with 10 - 50 g of hydrophilic polymer kg^{-1} of seed improved seed germination and emergence rate at the lowest available soil moisture level. The best treatment was 10 g kg^{-1} seeds. Plant height, number of branches per plant, drymatter accumulation, pods m^{-2}, pod length, seed yield and harvest index were maximum with seed plus 1.875 ml of polymer kg^{-1} for the rainfed green gram and 1000 seed weight and seeds per pod were maximum with seed plus polymer at 1.25ml and 5 g kg^{-1} respectively.

Soybean with polymer coatings A and C emerged about one and two weeks later than the uncoated seed, respectively. Plant stand measurements one week after sowing were not significantly different among the treatments.

The biodegradable polymer seed coating permits farmers to sow soybean between rows of standing wheat in late spring, when wheat is less vulnerable to trampling by machinery. The coating delays soybean germination for about two weeks period that accommodates the wheat harvest and leaves enough time for soybean to produce a good yield before a killing frost .

A hydrophobic polymer was applied to soybean (*Glycine max* (L.) Merril.) seeds to regulate imbibition, reduce imbibition damage and improve germination and seedling emergence. Studies were conducted with genotypes prone to imbibition damage due to either seed coat splitting or low seed vigour. The effect of polymer application on seedling emergence and growth was tested under normal (75% soil water holding capacity (SWHC)) or short-term flooded (100% SWHC for the first 4 days and 75% SWHC subsequently) soil conditions. Coating with 24 mg of polymer per seed (Vinamul 3650) regulated the rate of water uptake, reduced imbibition damage and improved percentage germination and seedling emergence. In short-term flooding soil conditions, low seedling emergence was observed regardless of seed ageing and was particularly evident in seeds that possessed a high proportion of split testas. There was little difference between uncoated and coated seeds in the time of 50% emergence and seedling growth. These results suggest that coating with a hydrophobic polymer can reduce rates of water uptake, lower solute leakage and improve vital tetrazolium chloride staining and partially improve germination or emergence of soybean seedlings.

Table 1. Effect of 24 h soaking in water on percentage of normal seedlings, electrical conductivity, percentage of cotyledons fully stained with tetrazolium chloride (TTC) of unaged or aged (2 days A.A.), uncoated or polymer coated Forrest and Douglas seeds.

Cultivar	Normal seedlings (%)		Electrical conductivity $\mu S\ g^{-1}\ cm^{-1}$		Cotyledons fully stained (%)	
Forrest	Unaged	Aged	Unaged	Aged	Unaged	Aged
Uncoated	48	28	24	36	68	23
Coated	74	56	15	21	88	54
Douglas	Unaged	Aged	Unaged	Aged	Unaged	Aged
Uncoated	33	18	41	47	25	6
Coated	71	40	23	29	66	29
s.e.m. (n = 4)	1.7	2.9	1.4	1.5	3.1	2.6

Result : *Seeds coated with 24 mg of polymer resulted in more normal seedling, reduce leachate conductivity and completely stained with tetrazolium chloride.*

Table 2. Effect of 24 h soaking in water on percentage seedling emergence of unaged and aged Forrest and Douglas, uncoated or polymer coated seeds, at normal soil conditions (75% soil water holding capacity; swhc), short-term flooding conditions (100% swhc for 4 days and then 75% SWHC), and continous flooding conditions (100% swhc).

Cultivar	Seedling emergence under normal soil conditions (%)		Seedling emergence under short -term flooding conditions (%)		Seedling emergence under continuous flooding conditions (%)	
Forrest	Unaged	Aged	Unaged	Aged	Unaged	Aged
Uncoated	82	65	35	9	7	6
Coated	88	85	60	50	8	9
Douglas	Unaged	Aged	Unaged	Aged	Unaged	Aged
Uncoated	43	18	15	0	2	1
Coated	68	40	39	29	6	3
s.e.m. (n = 4)	3	2	3	4	2	3

Result : Seeds coated with 24 mg of polymer resulted in high seedling emergence under all the above soil conditions.

Reduced inbibition rate could be achieved by coating the seeds with hydrophobic polymers. Coating of soybean seeds with ethyl cellulose and found that it reduced imbibition damage and improved germination.

4. EFFECT OF FILM COATING ON FIELD EMERGENCE OF THE SEEDLINGS

Seeds of *Phalaris aquatica* coated with monocalcium phosphate suffered little emergence injury whereas *Medicago sativa* were damaged during emergence. However coating seeds with a polymer prior to applying the monocalcium phosphate resulted in substantial protection against injury. Seeds pelleted with polymer and arappu leaf powder registered higher field emergence potential, plant height, ear length and seed yield in minor millets.

Faster field emergence of rice seeds coated with 50g aqueous emulsion of sodium alginate per litre and dipping in 10g calcium oxide per litre than non coated seeds under moisture stress conditions and also coated seeds absorbed more moisture when compared to non coated seeds.

Polymers applied alone had no significant influence on emergence rate of either high or low vigour seeds in comparison with untreated controls. The Emergence Rate Index of seeds was increased for captan + Certop (a polymer) treatment. Also increased final emergence and seedling height. Polymers in combination with captan were as effective as captan alone in increasing the rate of emergence and improving seedling height at early planting. However *Pythium* control, using a polymer seed coating was inconsistent when used alone or in combination with captan . Experiment results reported a frost that seriously damaged a corn stand from uncoated seed planted in early April had little effect on a stand that emerged later due to seed coating.

5. ROLE OF POLYMERS IN PROMOTING THE EARLY PLANTING OF THE CROPS BY ACTING AS A TEMPERATURE SWITCH

Intellicoat polymer coatings can provide protection to hybrid corn seed under cold stress conditions. The coating acts as a physical barrier and slows down water uptake of corn seeds at low temperatures to alleviate chilling damage to the seedling in early spring. Coated hybrid seed can retain its viability for a normal stand establishment for 2 to 3 weeks at 5^0C in near saturated soil, while uncoated seed shows significant stand loss under the same conditions.

Water uptake was significantly reduced in corn seeds coated with Intelimer polymer compared with uncoated seeds after 48 hours at 10° and 25^0C. Germination of coated seed was significantly improved than uncoated seeds. Germination of the polymer coated seeds was 95.9% at 25^0C, after treatment in low temperature (10^0 C), and 35.7% after higher temperature (45^0 C) treatment. Seedling dry weight was greatest at low temperature. The seedling establishment rate was enhanced by treatment with Daran 8600, Sepiret and Sacrust, which are the recommended polymers for direct sown rice. Complex coating preparations with polymers which delay emergence, enabled emergence to occur when the best temperature for maize germination was reached in the soil, thus they could be used for earlier maize sowing and for phasing sowing over a longer period of time.

6. POLYMER FILM COATING FOR IMPROVING THE HEALTH POTENTIAL OF THE SEEDS

Seeds film coated with Fipronil and Imidacloprid showed effective control of thrips on the seed bed for 12 weeks and 13 weeks after transplanting. Film coating the seeds with Fipronil, Diflubenzuron, Imidacloprid and Teflubenzuron gave acceptable control of the larvae of the onion fly, whereas coating with Benfuracarb and Methiocarb was only moderately effective. The use of Fipronil and Imidacloprid film coated seeds resulted in sufficient protection against the leek moth at low population densities.

Oospore germination was unaffected by the film coating with *Pythium oligandrums* on sugarbeet. The coating gave significant control of damping off. The film coating with *Cercospora minitans* onto infected sunflower seeds decreased the germination of conidia.

7. EFFECT OF POLYKOTE ON MAIZE (*Zea mays* L.)

The pink polykote (supplied from the Little's Balm Pharmaceuticals Ltd, Chennai) coating of maize cv. CO 1 seeds at 3g kg^{-1} of seed improved the germination and seedling vigour. An increase of 2.7% in germination, 27.7% in root length, 19.5% in shoot length, 24.8% in drymatter production and 26.9% in vigour index was noticed (Table 3).

Table 3. Effect of pink polykote as slurry dressing on germination and seedling vigour of maize

Polykote Treatments (g kg^{-1} of seed)	Germination (%)	Root length (cm)	Shoot length (cm)	DMP (g/seedlings)	Vigour index
0	96.50 (79.25)	14.935	13.505	0.802	77.37
1	97.25 (80.49)	17.795	13.800	0.959	93.29
2	97.75 (81.55)	17.595	13.355	0.945	92.39
3	99.25 (85.70)	20.658	16.775	1.067	105.86
4	98.75 (83.67)	20.205	12.005	0.946	93.39
5	98.00 (82.01)	18.305	14.825	1.032	101.14
CD (P=0.05)	0.990	2.148	2.373	0.054	5.537

(Figures in parantheses indicate arcsine values)

9. EFFECT OF PELLETING WITH POLYMER ON SEED QUALITY OF MINOR MILLETS

The effect of pelleting with polymer, arappu powder and urea alone or in combination with each other was studied on intact and sprouted seeds of minor millets like Ragi, Panivaragu, Kuthiraivali Tenai and Samai (Table 4).

Seed pelleting treatment with polymer @ 1.5ml/100g + arappu powder (1:100) (*Albizia amara*), revealed its superiority over the other treatments by recording maximum values of germination (%) and vigour index for all the minor millets studied. Also intact seeds performed better than the sprouted seeds.

Table 4. Effect of pelleting with polymer on seed quality of minor millets

Treatments	Ragi		Panivaragu		Kuthiraivali		Tenai		Samai	
	Germination (%)	Vigour index	Germination (%)	Vigour index	Germination (%)	Vigour index	Germination (%)	Vigour index	Germination (%)	Vigour index
Control (intact - seed)	56	546	48	1007	65	907	57	854	50	892
Pelleting with polymer	57	590	63	1062	64	915	59	870	52	893
Pelleting with polymer + Arappu powder (*Albizia amara*)	66	672	75	1190	72	1074	64	1021	60	909
Pelleting with polymer + Urea	59	637	69	1094	67	954	62	969	55	894
Pelleting with polymer + Arappu powder + Urea	56	621	66	1038	65	926	61	877	56	900
Sprouted seed	58	624	66	977	65	862	61	958	55	887
Pelleting with polymer	61	604	64.	1004	69	953	61	949	54	891
Pelleting with polymer + Arappu powder	62	641	66	1030	69	968	61	940	54	888
Pelleting with polymer + Urea	62	617	65	1039	68	951	60	916	53	882
Pelleting with polymer + Arappu powder + Urea	58.1	599	62	1044	67	952	93	931	.53	885
CD (P=0.05)	1.239	48.448	2.223	48.771	1.690	30.759	1.416	41.39	1.373	7.063

10. DETERMINATION OF SHELF LIFE POTENTIAL OF SEEDS COATED WITH PINK POLYKOTE UNDER AMBIENT STORAGE CONDITIONS

The seeds of maize cv. CO 1 coated with pink polykote along with control was stored for 4 months under ambient conditions of storage and assessed for its germination and vigour potential at bimonthly intervals to document the effect of polykote on the shelf life of seed.

The results revealed that pink polykote @ 3 g kg^{-1} of seed + fungicide and insecticide is found to be best in registering higher germination potential and vigour, over a period of 4 months storage. But within the containers there is no significant difference in the germination and vigour potential. However polythene bags found better than gunny bags in storage in terms of germination and vigour maintenance.

Table 5. Effect of polymer coating on germination of stored seeds

Storage period	Gunny bag			Polythene bag		
	Control	Polykote @ 3g	Polykote @ 3g + Thiram (2g) + Imidacloprid (1 ml)	Control	Polykote @ 3g	Polykote @ 3 g + Thiram (2g) + Imidacloprid (1 ml)
Initial	96.5 (79.24)	99.0 (85.38)	99.5 (86.74)	96.5 (79.45)	99.2 (85.38)	99.5 (86.74)
2 months	95.0 (77.12)	98.0 (82.05)	98.5 (83.03)	95.5 (77.77)	98.5 (83.03)	99.0 (85.38)
4 months	92.0 (73.59)	96.5 (79.24)	97.0 (80.12)	93.0 (74.69)	97.5 (80.94)	98.00 (82.05)
CD (P=0.05)	0.602					

11. POLYMERS ON AGRONOMIC CHARACTERISTICS OF CANOLA

Polymers increase the safety to producers, treaters and retailers by reducing dust during handling. The polymer results in a more even seeding rate, due to the smooth flow of seed.

To evaluate the impact of seed treatments on agronomic characteristics of canola such as yield, quality and contribution margins, the seed treatments *viz.* foundation (check) and foundation plus Polymer were given.

At Russell no significant differences in yield, oil content or economic return were observed. Most of the stresses on this crop were related to environmental conditions and weeds, rather than disease or insect pressure (Table 7).

Table 7. Effect of polymer on yield, contribution margin and oil

Seed treatment trial - (Foundation) Russell, MB			
Treatment	**Yield (bu ac^{-1})**	**Contribution Margin ($ ac^{-1})**	**Oil (%)**
Foundation	18.6	(11.05)	42.7
Foundation + Polymer	18.0	(15.69)	42.7
LSD	3.72		1.22
cv %	12.2		1.7

At Whitewood no significant differences in yield or oil content were found. Four days after emergence plant counts were 46% higher for the polymer coating. This advantage dropped to 5%, 23 days after emergence. Contribution margins reflect yield variation and the cost of the polymer coating (Table 8, 9)

Table 8. Effect of polymer on average number of emerged plants m^{-2}

Seed treatment trial - (Foundation) Whitewood, SK			
Treatment	**4 DAE**	**13 DAE**	**25 DAE**
Foundation	46	84	96
Foundation + Polymer	67	104	101

DAE = Days After Emergence

Table 9 : Effect of polymer on yield, contribution margin and seed quality

Seed treatment trial - (Foundation) Whitewood, SK			
Treatment	**Yield (bu ac^{-1})**	**Contribution Margin ($ ac^{-1})**	**Oil (%)**
Foundation	28.6	41.12	46.4
Foundation + Polymer	29.8	49.99	47.1
LSD	1.51		0.91
cv %	3.1		1.2

At Naicam no significant differences in yield or oil content were observed. Plant counts were 20% higher in the polymer coating upto 16 days after emergence. Contribution margins reflect yield variation and the cost of the polymer coating (Table 10, 11).

Table 10. Effect of polymer on average number of emerged plants/m²

Seed treatment trial - (Foundation) Naicam, SK		
Treatment	**4 DAE**	**16 DAE**
Foundation	19	30
Foundation + Polymer	23	36

Table 11. Effect of polymer on yield, contribution margin and seed quality

Seed treatment trial - (Foundation) Naicam, SK			
Treatment	**Yield (bu ac^{-1})**	**Contribution Margin ($ ac^{-1})**	**Oil (%)**
Foundation	22.7	3.96	44.1
Foundation + Polymer	26.4	31.58	44.2
LSD	9.35		0.51
cv %	16.0		0.5

At Innisfail no significant differences in yield or oil content between treatments were observed. There was no maturity difference between treatments (Table 12).

Table 12. Effect of polymer on yield, contribution margin and oil

Seed treatment trial - (Foundation) Innisfall, AB			
Treatment	**Yield (bu ac^{-1})**	**Contribution Margin ($ ac^{-1})**	**Oil (%)**
Foundation	30.7	37.25	45.4
Foundation + Polymer	32.0	46.99	46.0
LSD	11.43		0.93
cv %	15.5		1.2

At Andrew net yield and oil content were slightly lower with the polymer treatment (Table 13).

Table 13. Effect of polymer on yield, contribution margin and oil

Seed treatment trial - (Foundation) Andrew			
Treatment	**Yield (bu ac^{-1})**	**Contribution Margin ($ ac^{-1})**	**Oil (%)**
Foundation	40.5	187	41.5
Foundation + Polymer	39.4	179	40.7
LSD	0.81		1.83
cv %	1.2		2.7

Yields were limited by climatic conditions at Wanham site. There were no significant differences in yield or oil content between the treatments (Table 14, 15).

Table 14. Effect of polymer on average number of emerged plants/m^2

Seed treatment trial - (Foundation) Wanham, AB			
Treatment	**9 DAE**	**19 DAE**	**28 DAE**
Foundation	100	106	94
Foundation + Polymer	98	101	95

DAE = Days After Emergence

Table 15. Effect of polymer on yield, contribution margin and seed quality

Seed treatment trial - (Foundation) Wanham, AB			
Treatment	**Yield (bu ac^{-1})**	**Contribution Margin (\$ ac^{-1})**	**Oil (%)**
Foundation	17.1	12.44	42.5
Foundation + Polymer	17.7	18.82	42.1
LSD	3.20		0.89
cv %	11.2		1.2

There were no significant differences in yield or oil content between treatments at Rolla (Table 16, 17).

Table 16. Effect of polymer on average number of emerged plants/m^2

Seed treatment trial - (Foundation) Rolla, BC			
Treatment	**4 DAE**	**12 DAE**	**18 DAE**
Foundation	178	184	184
Foundation + Polymer	168	180	180

DAE = Days After Emergence

Table 17. Effect of polymer on yield, contribution margin and seed quality

Seed treatment trial - (Foundation) Rolla, BC			
Treatment	**Yield (bu ac^{-1})**	**Contribution Margin (\$ ac^{-1})**	**Oil (%)**
Foundation	38.0	93.37	41.7
Foundation + Polymer	39.0	99.12	41.3
LSD	2.18		1.06
cv %	3.4		1.5

12. CONCLUSION

1. Polymer coating regulates the rate of water uptake.
2. It reduces the imbibition damage and improved the germination, seedling growth and vigour potential.
3. Polymer in combination with fungicides and insecticides improve the health potential of the seeds.

4. Polymer coating provides protection to seed under cold stress condition.

13. FUTURE THRUST

1. Since very few studies are available on polymer coating in major crops, it can be extended to various agricultural, horticultural and forestry seeds.
2. The actual cause for germination improvement due to polymer has to be studied in depth.
3. It is necessary to evaluate the shelf life of polymer coated seeds.
4. It is essential to know the mechanism of reduced imbibitions damage.

SEED COLOURING - AN INNOVATIVE APPROACH FOR SEED ENHANCEMENT

A. BHARATHI[1], K. VANANGAMUDI[2], P. NATESAN[1],
K. RAJA[3] AND S. KAVITHA[3]

1. INTRODUCTION

Seed colouring, which is something unrealistic and uneconomical" was a belief a decade ago. But now it has provoked the interest among many seed traders and seed companies, as it improves its marketability, brand identity and helps the farmers for easy identification of the crop and varieties based on colour. The practice of providing an exogenous colour coating to seed is only of recent interest. But it is very much prevalent in developed countries since a decade. Modern seed technology provides a wide selection of enhancements that can be aimed at translating a variety's genetic potential into improved harvest yield and quality.

2. IMPORTANCE

Colouring of seeds along with pelleting, film coatings, coating with biological is adopted in private sector seed companies in United States, Canada and Europe as these enable the seeds to be sown in defined in patterns besides modifying germination. Seeds of many cultivated crops have naturally irregular proportions and surfaces, wide size ranges or sticky or very small. These features can make it hard to handle and efficiently sow the seed. Hence "pelleting", "coloring" and "coating" are a family of treatments that are used to make seeds sown easily by altering their shape, weight and texture besides modifying its germination.

1 *Department of Seed Science and Technology, Tamil Nadu Agricultural University, Coimbatore – 641003*
2 *Dean (Agriculture), Tamil Nadu Agricultural University, Coimbatore - 641 003.*
3 *Central Institute for Cotton Research, Coimbatore – 641 003*

Few elite seed companies in India came forward for seed coloring with improved technologies. They import both the colouring material and sophisticated machinery at a huge cost. In other cases, the colouring materials and process adopted are very primitive without any scientific basis. A wide information gap is exiting in our country in this promising area of interest.

The Claims made in favour of seed colouring are many such as:

- To prevent inadvertent mixing of high value seeds with other seed of the same crop.
- To enable brand identification
- To add desirable ingredient along with colouring to offer protection to seed
- To provides a smooth and even surface, enhancing market value and easy sowing/ planting.
- To seal cracks on the seed coat and to improve the physical appearance.
- To give the seed a distinct and attractive look, to reduce dust and promote environment safety as well a workers safety in the field.

3. DIFFERENCE BETWEEN SEED FILM COATING, COLOURING AND PELLETING TECHNOLOGIES

	Seed film coating	Seed colouring	Seed pelleting
Adhesive	Seed coating polymer is required polymer itself has plasticzer	Only natural /synthetic dye is required.	Adhesives , nutrients,filler materials , biofertilizers are required
Fungicide and insecticide	Can be combined	Can be combined	Can be combined
Shape of the seed	Not obscured	Not modified	Shape is modified as globular unit
Size of the seed	Not modified	Not modified	Modified due to addition of filler materials
Filler materials	Not required	Not required	Required. Which may be gypsum (or) botanical leaf powder.
Nutrients and biofertilizers	Usualy combined	Not combined	Can be combined during the process of pelleting
Process	Have coating Process	Have colouring process	Have stamping, coating and rolling process
Safety and handing	More safe	More safe	No safety

Resistance to mechanical injury	More resistance	No resistance	More resistance
Stability	Is more	it varies according to dye/pigment used	Less stable
Prevention of dusting of fungicides	Prevents dusting of pesticide and fungicide	Not possible	Does not prevent
Wastage of pesticide and fungicide	Completely prevented	Not possible	Wastage is more during handling
Improvement in the physical appearance of seed	Very much improved	Very much improved	Improved but having no preference
Handling	Is easy	Is easy	Results in pollination
Identification	Useful for brand identify and identity of treated seeds since colourant substance is present	Useful for brand identify	Not useful
Effect on seed germination and seedling vigour	Improved	Improved	Improved but not preferred by consumers
Storability	Is good	Is good	Pelleted seed can not be stored for more than 3 to 4 months with better shelf life
Applicability	Suitable for dry, garden and wet land situations	Suitable for dry and garden land situation	Suitable for only garden / wet land situation
Inter linking of technologies	Colouring is combined and has advantage of colouring and coating technologies		Seed colouring can be combined with pelleting technology

SEED COATING

Seed film Coating

Sophisticated process of applying precise amount of active ingredients along with of active ingredients along with a liquid m aterial directly on to the seed surface without obscuring its shape

Seed Colouring

Application or precise or amount of precise or amount of dyes or pigments directly on to the seed to improve its brand identify and marketability.

Seed Pelleting

Process of enclosing the seed with small quantity of ingredients along with file material are coated to produce a globular unit of standard size to facilitation precision planting

4. FILM COATING AND COLOURING

Application of a thin, durable and water permeable coating material over the seed. It may have maximum effect at the time of sowing this method developed in UK in eighties, the shape of the seed not changed but increase in seed weight by 1 – 2 %.

4.1 BENEFITS OF SEED FILM COATING COLOURING

- Seed enhancement material is placed directly on to the seed
- Smaller amount of chemicals are needed as compared to broadcasting or surface dressing on to the growing medium
- It improves the seed performance

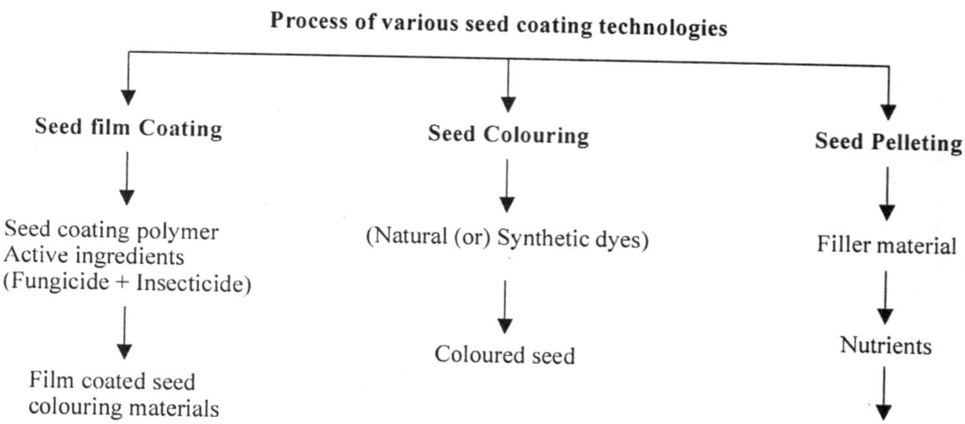

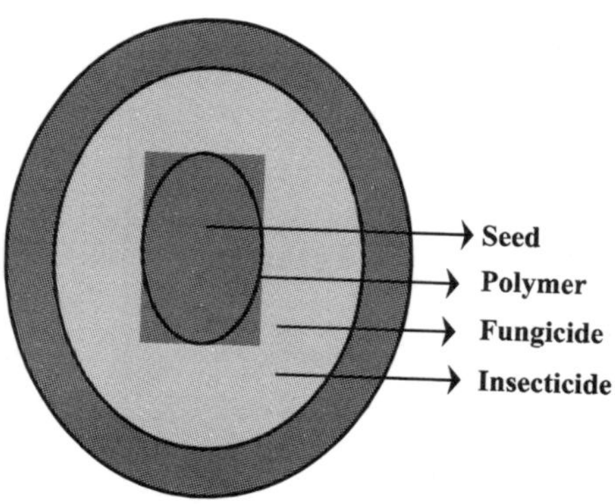

Plate 1. Diagrammatic representation of seed film coating / colouring technology

5. PARTICULARS OF PLANT PARTS USED FOR DYE EXTRACTION

	Botanical name	*Common name*	*Parts used*
1.	*Beta vulgaris*	Beetroot	Root
2.	*Syzygium cuminii*	Jamun	Fruit
3.	*Basella rubra*	Basella	Fruit
4.	*Opuntia* spp	Opuntia	Fruit
5.	*Curcuma longa*	Turmeric	Rhizome
6.	*Lawsonia inermis*	Henna	Leaf
7.	*Tagetes erecta*	Marigold	Flower
8.	*Hibiscus rosa sinensis*	Hibiscus	Flower
9.	*Bixa orellana*	Anatto	Seed

5.1 Synthetic chemical dyes

1. Jade green
2. Bromocresol green
3. Direct chenbaga
4. Turquoise blue
5. Sky blue
6. Congo red
7. Pink CM
8. Copper sulphate

5.2. Method of extraction and preparation of dyes

a. Beta vulgaris (Beet root)

The roots of *Beta vulgaris* can be cut into small pieces an fed into a pulper for grinding without adding any water, filter through a muslin cloth and use as such for colouring of seeds.

b. Syzygium cuminii (Jamun)

The dye can be extracted in two different ways as indicated below

1. Fully ripened fruits can be fed into a pulper for grinding without adding any water, filter through a muslin cloth and use as such.
2. From the fully ripened fruits the pulp can be separated from seeds and the pulp alone should be crushed and the dye can be extracted without adding water, filter through muslin cloth and use.

c. *Basella rubra* (Basella)

Fully ripened fruits can be collected and fed as such into a pulper for grinding without adding water, filter through muslin cloth and use for colouring of seeds.

d. *Opuntia* spp. (Opuntia)

Freshly ripened fruits can be used, the thorns are to be removed and then the fruits are cut into small pieces. The cut pieces are to be fed into a pulper for pulping without adding any water and filter through muslin cloth.

e. *Curcuma longa* (turmeric)

Dried turmeric rhizomes, devoid of fungal or insect infestation are to be used. Powder it with the help of an electrical mixed, sieve through a muslin cloth and then the powder can be mixed with water and use

f. *Lawsonia insermis* (Henna)

Fresh henna leaves (40g) can be boiled in one litre of water for about 10 minutes and filter the extract through a muslin cloth. The dark reddish brown extract can be obtained.

g. *Tagetes erecta* (Marigold)

Fresh petals from fully opened orange coloured flowers of African tall is to be used, boiled in water (at 1:2 W/V) for about 15 minutes and then filter through a muslin cloth. The filtrate obtained will be yellow in colour.

h. *Hibiscus rosasinensis* (Hibiscus)

Fresh petals from fully opened red coloured flowers are to be collected, macerate using minimum quantity of water (approximately 3-5 ml/15 flowers)and the extract is to be filtered through a muslin cloth and ca be used for seed colouring.

i. *Bixa orellana* (Anatto)

Seeds of *Bixa orellana* are to be soaked over night in an equal volume of water, then filter through a muslin cloth and the orange colour dye can be obtained.

5.3. Suitable natural dyes from botanicals and synthetic chemical dyes for seed colouring

The following natural dyes from botanicals and synthetic chemical dyes at different concentrations can be used.

5.3.1. *Natural dyes from botanicals*

- Fruit and seed extract of *Syzygium cumini* @120,160 and 200 ml per 100 g of seed
- Fruit pulp extract of *Syzygium cumini* @120,160 and 200 ml per 100 g of seed
- Leaf extract of *Lawsonia inermis* @120,160 and 200 ml per 100 g of seed
- Fruit and seed extract of *Basella rubra* @120,160 and 200 ml per 100 g of seed
- Rhizome extract of *Curcuma longa* 20,30 and 40 g per 100 g of seed
- Flower extract of *Tagetes erecta* @120,160 and 200 ml per 100 g of seed
- Root extract of *Beta vulgaris* @120,160 and 200 ml per 100 g of seed
- Flower extract of *Hibiscus rosasinensis* @120,160 and 200 ml per 100 g of seed
- Seed extract of *Bixa orellana* @120,160 and 200 ml per 100 g of seed
- Fruit extract of *Opuntia* spp. @120,160 and 200 ml per 100 g of seed

5.3.2 *Synthetic chemical dyes*

- Bromocreso green @ 0.5, 0.75 and 1.0%
- Sky blue @ 0.5, 0.75 and 1.0%
- Pink CM @ 0.5, 0.75 and 1.0%
- Jade green @ 0.5, 0.75 and 1.0%
- Direct chenbaga @ 0.5, 0.75 and 1.0%
- Congo red @ 0.5, 0.75 and 1.0%
- Turquoise blue @ 0.5, 0.75 and 1.0%
- Copper sulphate @ 0.5, 0.75 and 1.0%

6. SEED COATING MACHINE

Description of Seed film coating machine

Machine	:	Chemical Film Coating for all kinds of seeds which are more than 2 mm in diameter.
Application	:	Suitable for On-Line application for batch process.
Measuring	:	Motorized metering mechanism on timing bases depending on seed density Batch Quantity : 15 to 30 Kgs / batch per 35 seconds based on the seed variety.
Output	:	2.5 to 3.2 tons / hour max

Controls	:	PLC panel with self programming controls to change timings settings of seed qty, dosing qty & treatment time etc. during changeover.
Main Disc	:	Shall have a AC drive to have variable speed for the disc in order to treat any type of seeds like paddy, sunflower, bajra and jowar etc.

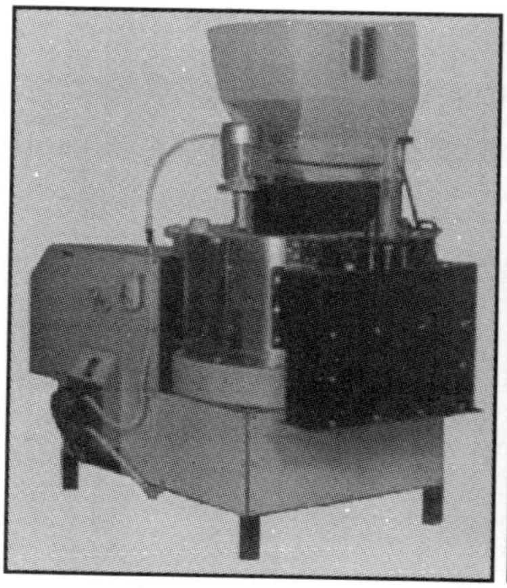

Plate 2. Seed film coating machine

Plate 3. Seed film coating machine with on line drier

Dosing Pump : Valve less non-contact Peristaltic Dosing Pump with Italian make geared mot or (maintenance free, life lubricated) for higher Accuracy can be operated in forward / reverse direction with any required speed by means of A/C Drive for achieving desired quantity. OPM panel for changing dosing quantity setting.

Mixing Tank : Completely made of SS-304 materials with independent drive and DOL starter for slurry preparation. Continuous stirring to avoid lumps and blockage of chemical.

7. RESEARCH FINDINGS ON SEED COLOURING

Vander Burg (1979) used water soluble fluorescent paint to colour seeds and observed that coloured seeds did not lose the paint very easily and stayed in good condition. Tonapi (1988) reported that among the 23 inorganic dyes used for colouring of sorghum seeds, methyl red, methyl orange, gentian violet and phenol red drastically affected the viability of seeds while rhodomine-D, malachite green fast green neutral red and methylene blue showed promoting effect on seed germination. Saraswathi (1994) opined that out of the 13 dyes used for colouring linted cotton seeds, methyl orange ranked first with highest germination of 85 per cent, whereas nigrosine reduced the germination drastically to 55 per cent.

Tonapi and Karivaratharaju (1994) reported that the studies on seed colouring involving four sorghum genotypes namely Co 24, Co 25, Co 26 and IH 35 brought out the deleterious effect of seed colouring dyes namely methyl red, methyl orange, genetian violet and phenol red, since they inhibited the seed germination and affected the vigour and viability. The dyes such as rhodomine-B, malachite green, fast green neutral red and methylene blue promoted seed germination, seedling growth and vigour potential. According to Vivekanadhan (1997) turmeric powder extracted from dried rhizomes was better for seed coloring in sorghum. He has also proved that textile organic dyes are harmful to the seeds and did not produce deep shades.

Basavaraj and Kurdikeri (2000) reported that amaranthus red, sunset yellow and methylene blue dye treated soybean seeds showed higher germination, vigour index and lesser EC values at 0.5 and 1.0 per cent concentration. Vishwaprasad and Kurdikeri (2000) reported that amaranthus red, sunset yellow, methylene blue and potassium permanganate either at 0.5 and 1.0 per cent concentration could be used for colouring maize seeds without any harmful effect on field performance. Seeds coloured with amaranthus red, sunset yellow and methylene blue recorded maximum germination and minimum EC values upto 10 months of storage. Whereas 10 per cent of potassium permanganate has shown deleterious effect on seed quality was reported by Vishwaprasad *et al.* (2001).

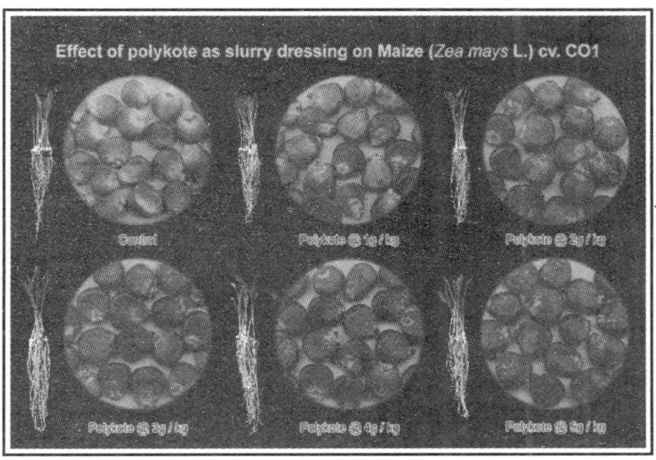

Plate 4. Effect of polykote as slurry dressing in maize

The pink polykote used @ 3g / kg of maize seed improved the germination and seedling vigour. Pink polykote @ 3g/kg of seed + fungicide (Thiram 2g) + Insecticide (Imidacloprid 1ml) maintained higher germination potential and vigour and also stored till nine months. From the Plate 4 showed that among the different concentration of polykote treated maize seeds, three g in 5 ml of water treated seeds observed higher root and shoot ratio and also produced healthy seedling.

8. DIFFERENT METHODS OF POLYKOTE IN MAIZE

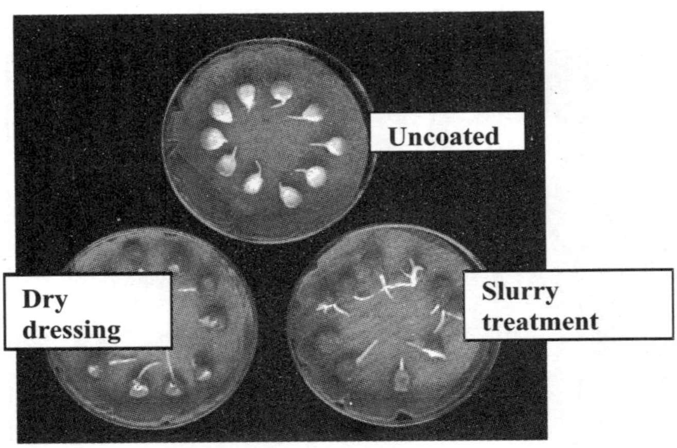

Plate 5. Seed coating polykote on maize germination: Petriplate method

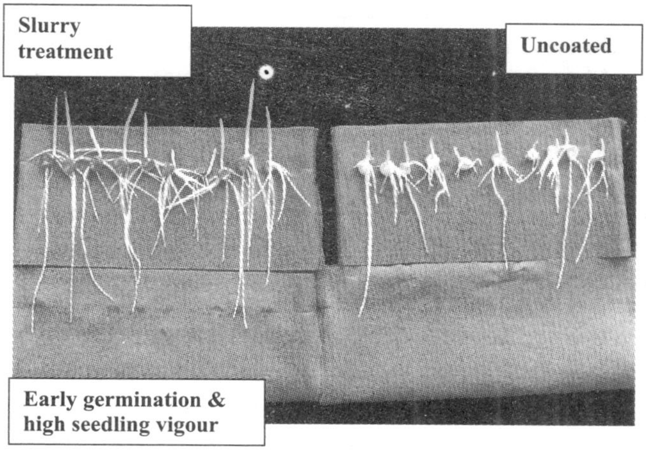

Plate 6. Seed coating polykote on Maize germination: Inclined plate method

Plate 7. Seed coating polykote on Maize germination: Petriplate method

9. PROPERTIES OF NATURAL (ORGANIC) DYES USED

a. Curcuma longa (Turmeric)

Pillai and Desai (1975) reported that turmeric as powder, significantly decreased the root - knot incidence in tomato when applied as soil amendment. The antimicrobial properties of essential oil of *Curcuma longa* against plant pathogenic bacteria and fungi was reported by Banerjee and Nigam (1978)

Venkataraman *et al.* (1978) studied the anti-fungal activity of certain plants of the family Zingiberaceae and reported the marked anti-fungal activity of the rhizome extracts of *Curcuma aromatica, Curcuma longa* and *Zingiber officinale*. Ghulam Jilani and Helen (1983) reported that turmeric powder was effective against *Sitophilus granarius* and *Rhyzopertha dominica*. Turmeric rhizome powder at 5 per cent level afforded good protection against *Callosobruchus chinensis* and *Trogoderma granarium* in stored grains (Chander and Ahmed, 1983).

Pranata (1984) reported that the extract of turmeric was 100-1000 per cent more toxic than other insecticides of plant origin for the control of the bruchid (*Callosobruchus maculatur*) on stored gram, but adult females were more tolerant than adult males. *Curcuma longa* when used as seed treatment was found to control the damping - off disease significantly as it had fungitoxic properties (Kishoe *et al.*, 1988).

According to Ghulam Jilani and Saxena (1990) turmeric oil and sweet flag oil were significantly more repellent during the first two weeks for *Rhizopertha dominica*. Rao (1991) reported that rhizome powder of *Curcuma longa* at 1 per cent level controlled the grain mycoflora of sorghum during storage.

Chander *et al.* (1992) reported that the progeny of treated insects was suppressed by 60 per cent at doses of 2 per cent turmeric powder, when used against the infestation by *Tribolium castaneum* for milled rice. Leela - Srinivas *et al.* (1992) reported that the water soluble peptide, turmerin from turmeric was an efficient anti-oxidant / DNA protectant / anti-mutagen which are heat stable. Aqueous extract of turmeric contains and anti-oxidant protein (TAP) which inhibits lipid peroxidation both in human as well as in plant system.

Mohiuddin *et al.* (1993) reported that oil of *Curcuma longa* had a repellency of 40. 1-60 per cent against *Tribolium castaneum* and *Rhyzoperthat dominica* for a period of eight weeks. Prakash *et al.* (1993) reported the efficacy of turmeric rhizome powder along with six other products against *Sitophilus oryzae*. Hexane extracts of turmeric rhizomes were inhibitory to growth of *Alternaria alternata, Aspergillus flavus, A. fumigatus and A. wentii*, causing spoilage of stored grains was reported by Kazmi *et al.* (1993)

Shivanna *et al.* (1994) studied the efficacy of turmeric powder as a pre storage treatment against *Callosobruchus chinensis* on red gram. Turmeric powder was the fourth best after sweet flag, custard apple and black pepper powder. Turmeric, an aromatic rhizome of East Indian Zingiberaceous plant,

Curcuma longa, has many beneficial properties, which include wound healing, anti-microbial and anti-insecticidal properties (Anon, 1997). Datar (1999) reported that rhizome extract of *Curcuma longa* was effective in controlling the mycelial growth o *Macrophomina phaseolina*.

Curcuma longa when used along with Thiophanate methyl was most effective against *Penicillium digitatum* where percentage control efficacy was increased above 60 (Khilare and Gangawane, 1997). Turmeric at 20 % concentration was found to reduce Alternaria fruit rot of pomegranate following garlic extract (Jitendra singh and Majumder, 2001).

b. Lawsonia inermis (Henna)

Leaf extract of *Lawsonia inermis* showed anti-bacterial and anti-fungal activity (Ray and Majumdar, 1976). Ahmed and Agnihotri (1977) reported the anti-fungal activity of extracts from *Lawsonia inermis* against *Alternaria brassicae, Collectotrichum papayae* and *Helminthosporium* spp., *in vitro*.

Charya *et al.* (1979) stated that the extracts of *Lawsonia inermis* completely inhibited the germination of *Drechslera rostrata* and *Curvularia lunata*. Babu and Reddy (1986) reported that the extract of *Lawsonia inermis* was effective in checking fruit rot of lemon. A gradual decrease in growth of *Helminthosproium oryzae* on increasing concentrations of the leaf extract of *Lawsonia inermis* in paddy was observed by Natarajan and Lalithakumari (1978).

Methanolic extracts of leaves, stem and buds of *Lawsonia inermis* gave 67-100 per cent larval mortality of *Meloidogyne javanica* at a dose of 4 mg / ml or at 1:5 dilution (kumari *et al.*, 1987). Leaf extract of *Lawsonia inermis* was active in vitro against a wide variety of microorganisms (Tacedo and Fatope, 1988).

Ghewaned (1989) reported the efficacy of aqueous leaf extracts of *Lawsonia inermis* against controlling the foliar diseases for groundnut. El-Sayed and Badawey (1991) reported the efficacy of powdered *Lawsonia inermis* on the inhibition of mycelial growth and delayed sporulation of *Aspergillus parasiticus* during seven day incubation and inhibition of growth was further intensified with higher concentration of henna. Korayem and Osman (1992) observed suppressive effect on root knot nematode by henna (*Lawsonia inermis*) when tomato andhenna were grown together.

c. *Tagetes erecta* (Marigold)

Flower extracts of *Tagetes lucida* caused greatest nematode mortality and inhibition in juvenile-hatching (Siddiqui and Alam, 1988). Asha Shivpuri (1997) reported that ethanol extracts of *Tagetus erecta* showed fungitoxic properties against pathogenic fungi like *Alternaria* sp., *Colleetotrichum* sp., *Fusarium* sp. and *Rhizoctonia* sp.

10. SUITABLE NATURAL DYES FROM BOTANICALS AND SYNTHETIC CHEMICAL DYES FOR SEED COLOURING

10.1 Natural dyes form botanicals

The colours produced by the different plant dyes can be identified using the munsen colour chart, the active principle for colouring and pH of the dye solutions are furnished below.

Botanical dyes	Colour appearance	Active principle	pH after seed treatment
Henna	Sand stone	Not known	4.30
Beetroot	Teak	Betanin	6.20
Hibiscus	Sand stone	Syanidin diglycoside	5.49
Basella	Leaf brown	Not known	5.20
Anatto	Dawn glow	Bixin	4.35
Jamun (without seed)	Brandy	Jamboline	6.75
Jamun (with seed)	Sand stone	Jamboline	6.75
Marigold	Golden brown	á-terthienyl	4.35
Turmeric	Lemon yellow	Curcumin	2.88
Opuntia	Red	Not known	3.00

10.2. Tomato seed colouring

10.2.1. Germination (Table 1)

There was no significant difference in germination due to seed coloring with natural dyes from botanicals (treatments), concentrations and their interactions.

No significant difference on germination percentage due to the use of various natural dyes from botanicals for seed colouring

Seed colouring with natural dyes - *Curcuma longa* rhizome powder @ 20 g / 100 g imp[roved the aesthetic value of the seed

10.2.2. Speed of germination

Seed colouring with natural dyes from botanicals showed significant difference for speed of germination. Seed colouring with turmeric rhizome

Table 1: Effect of natural dyes from botanicals on germination (%) of tomato seeds cv. PKM1

Concentration (C) ml/ 100 g	Botanical dyes (B)											
	Uncoloured	Jamun (with seed)	Jamun (without seed)	Henna	Basella	Turmeric*	Mari gold	Beet root	Hibiscus	Anatto	Opuntia	Mean
120	92 (73.7)	92 (73.7)	92 (73.7)	92 (73.6)	94 (75.8)	94 (75.8)	95 (77.1)	92 (73.7)	92 (73.6)	92 (73.6)	92 (73.7)	93 (74.4)
160	92 (73.7)	92 (73.6)	93 (74.7)	93 (74.7)	95 (77.1)	95 (77.1)	94 (76.0)	92 (73.6)	93 (74.7)	92 (73.7)	92 (73.7)	93 (74.8)
200	92 (73.7)	93 (74.7)	93 (74.7)	93 (74.7)	94 (76.0)	94 (76.0)	94 (75.8)	92 (73.6)	94 (75.8)	93 (74.7)	92 (73.7)	93 (74.9)
Mean	92 (73.7)	92 (74.0)	93 (74.4)	93 (74.3)	94 (76.3)	94 (76.3)	94 (76.3)	92 (73.6)	93 (74.7)	92 (74.0)	93 (74.4)	
				B	C	B x C			*C_1 - 20g / 100g of seed			
		SEd		1.19	0.62	2.07			C_2 - 30g / 100g of seed			
		CD (P=0.05)		NS	NS	NS			C_3 - 40g / 100g of seed			

powder recorded the highest speed of germination of 21.6 compared to uncoloured seeds (21.3). There was no significant difference among different concentrations, but their interaction effect was significant. Seed colouring with turmeric rhizome powder @ 20 and 40 g / 100 g of seed was found to be superior.

10.2.3. Shoot length

The shoot length differed significantly among seed colouring treatments with natural dyes from botanicals. Seed colouring using turmeric rhizome powder registered the highest shoot length of 6.2 cm compared to uncoloured seeds (5.9 cm). Concentrations were not significant, but their interaction was significant. Colouring seeds with trumeric rhizome powder @ 20 and 40 g / 100 of seed showed superiority over other concentration.

10.2.4. Root length

The treatments differed significantly for root length. Seed colouring with turmeric rhizome powder recorded the longest root of 12.4 cm compared to uncoloured seeds (12.1cm). The difference due to concentrations was not significant. However, interaction between treatment and concentration was significant. Seed colouring with turmeric rhizome powder @ 20 and 30 g / 100 g of seed was found to be the best.

10.2.5. Drymatter production

Drymatter production did not differ significantly due to seed colouring with natural dyes form botanicals, concentrations and their interactions.

10.2.6. Vigour index

Vigour index showed significant difference due to seed colouring with natural dyes from botanicals. Seed colouring with turmeric rhizome powder was found to record the maximum value of 1766 compared to the uncolored seeds (1656). There was no significant difference between concentrations and it's interaction.

10.2.7. Electrical conductivity

There was no significant difference in electrical conductivity due to seed colouring with natural dyes form botanicals, concentrations and their interactions.

10.2.8. Pathogen infection

There was no pathogen infection in uncoloured seeds as well the seeds coloured with natural dyes from botanicals at various concentrations.

Seed colouring with natural dyes from botanicals brought out the usefulness of *Curcuma longa* (turmeric) rhizome powder @ 20 g /100 g by minimizing the physiological deterioration of tomato seeds during storage. However, the other plant extracts of *Beta vulgaris, Syzygium cuminii, Basella rubra, Opuntia* spp., *Lawsonia inermis, Tagetes erecta, Hibiscus rosasinensis* and *Bixa orellana* did not affect the vigour and viability of tomato seeds during storage (Ramya,2003)

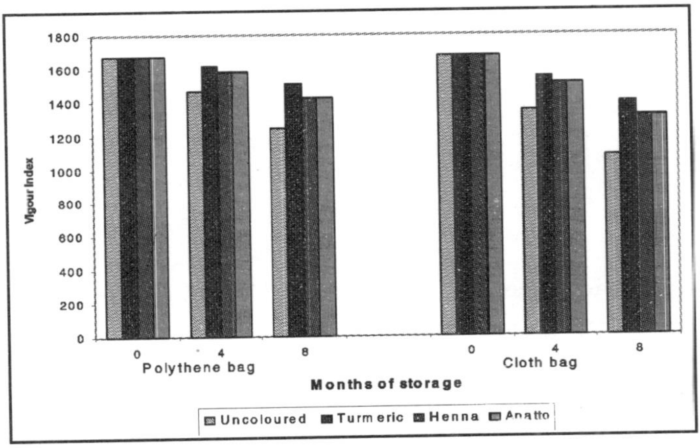

Figure 1: Effect of natural dyes from biotechnicals, containers and periods of storage on vigour index in tomato cv. PKM 1

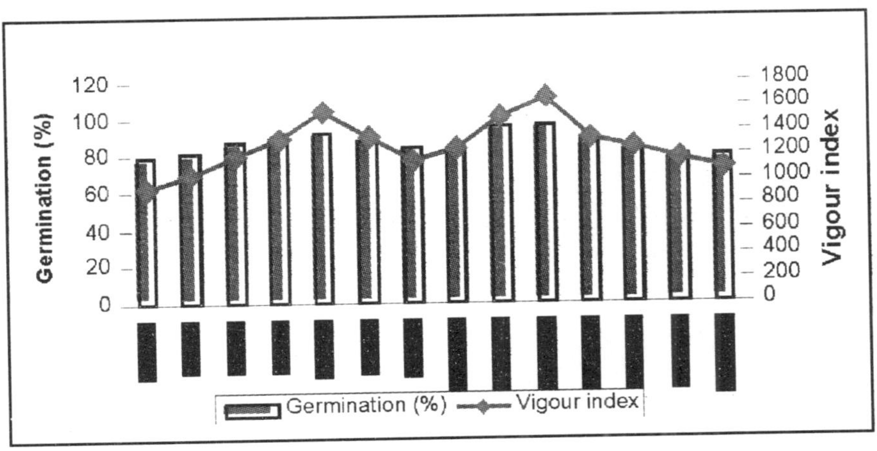

Figure 2. W. Red polykote @ 6g diluted in 20 ml of water/ kg of seed

10.3. CHEMICAL DYES

Tomato seeds when coloured with different synthetic chemical dyes makes the seed very attractive. Unlike the plant dyes, the seed colouring with synthetic chemical dyes did not result in lumping off seeds. The dyes when applied to the seeds were uniform and smooth. Seed coloring with different concentrations of copper sulphate produced abnormal seedlings and hence this treatment was not carried out for further studies. The colours produced by the different synthetic chemical dyes were identified using the munsen colour chart and pH of the dyes solutions were also identified and furnished below.

Synthetic chemical dyes	*Colour appearance after seed*	*pH Treatment*
Turquoise blue	Azure blue	7.87
Jade green	Deep green	9.93
Sky blue	Phiroza	9.38

Uncoated and coloured

Film coated and coloured

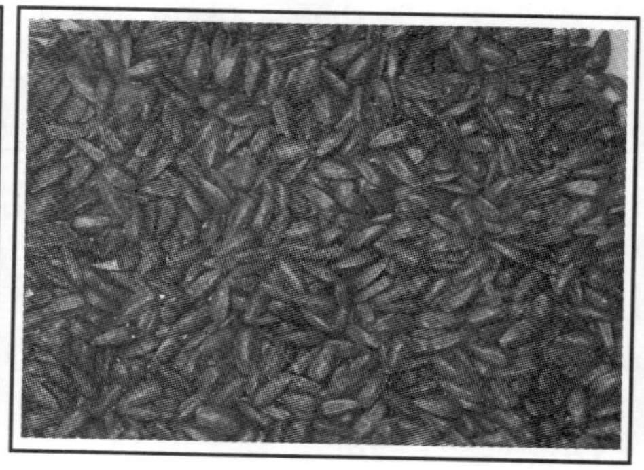

Plate 8. Black polykote @ 4g diluted in 5 ml of water/ kg of seed of sunflower seed

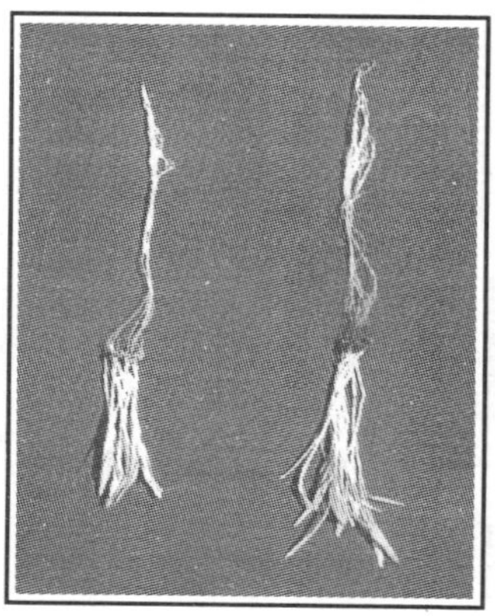

Plate 9. Black polykote @ 4g diluted in 5 ml of water/ kg of seed Sorghum

10.3.1. *Germination*

There was no significant difference in germination due to seed colouring with synthetic chemical dyes, concentration and its interactions.

10.3.2. *Speed of germination*

There was a significant difference among concentrations, but seed colouring using synthetic chemical dyes and it's interactions were non significant. Among the concentrations, 0.50 per cent was found to record the maximum value of 21.3 compared to 0.75 and 1.00 per cent.

10.3.3. *Shoot length*

Seed colouring treatments, concentrations and it's interactions were highly significant. Among the synthetic chemical dyes, direct chenbaga and turquoise blue recorded lower shoot length of 5.6 cm, compared to uncloured seed, which was on par with rest of the treatment. With respect to concentrations, 0.5 cm and 1.0 per cent recorded the shortest shoot of 5.7cm. Colouring seeds with bromocresol green, pink CM and congo red 0.50 per cent concentration registered the longest shoot when compared to other concentrations and treatments.

10.3.4. *Root length*

Seed colouring with synthetic chemical dyes, concentrations and it's interactions were highly significant.

Among the chemical dyes, direct chenbaga and turquoise blue recorded lower root length of 11.8cm compared to uncloured seeds (12.cm), which was on par with rest of the treatments. With respect to concentrations 0.50 per cent was found to record the longest root of 12.1cm and 0.75 per cent on par with 1.00 per cent recorded 11.9cm. Seed colouring with bromocresol green, pink CM and congo red @0.50 per cent concentrations was found to be superior.

10.3.5. *Drymatter production*

Seed colouring treatment and concentrations were significant but their interactions were not significant. Five out seven synthetic chemical dyes , namely pink CM, congo red, sky blue , bromocresol green and jade green @0.50 % were found highly suitable for seed colouring, whereas turquoise blue and direct chenbaga @ 0.50 % affected the vigour and viability of tomato seeds during storage (Ramya, 2003)

Among the treatments, seed colouring with direct chenbaga and turquoise blue recorded lower drymatter of 8.6 and 8.7 sl^{-5} compared to uncoloured seeds (8.9mg sl^{-5}), which was on par with rest of the treatments. Between the concentrations, 0.50 per cent registered higher drymatter of 8.9mg sl^{-5} and 1.00 per cent registered a lower drymatter of 8.7 mg sl^{-5}

10.3.6. Vigour index

Seed colouring with synthetic chemical dyes and concentrations were significant but their interactions were not significant.

Seed colouring with direct chenbaga and turquoise blue recorded the lowest vigour index value of 1600 and 103 respectively, compared to uncoloured seeds (1656), which was on par with rest of the chemical dyes. With regard to concentrations, 0.50 per cent recorded the highest vigour index value and 1.00 per cent recorded the lowest vigour index value.

10.3.7. Electrical conductivity

There was no significant difference in seed leachate electrical conductivity due to seed colouring with synthetic chemical dyes, concentrations and it's interaction.

10.3.8. Pathogen infection

There was no pathogen infection in uncoloured seeds as well as the seeds coloured with synthetic chemical dyes at various concentrations.

Bromocresol green	Bus green	10.15
Direct chenbaga	Golden brown	9.36
Congo red	Red	10.25
Pink CM	Rose	9.99

11. ADVANTAGES OF SEED FILM COATING AND COLOURING

- Safety and easy handling
- It improves plantability and emergence of seeds to benefit seed companies and farmers.
- It enables accurate and even dosage of chemicals and reduces chemical wastage.
- It makes room for including all the required ingredients like inoculants, protectants, nutrients, plant growth promotors, hydrophobic hydrophillic substances, herbicides, oxygen suppliers etc.
- By encasing the seed within a thin film of biodegradable polymer, the adherence of the seed treatment to the seed is improved.

- It ensures dust free handling, making treated seed both user and environment friendly.
- Addition of colourant helps visual monitoring of placement accuracy.
- Polymer coating results in more even seeding rate due to the smooth flow of the seeds
- Polymer acts as a temperature switch and protective coating by regulating the water uptake and subsequent germination of seed.

REFERENCES

1. Ahmed , S.R. and J.P. Agnihotri. (1977) . Antifungal activity of some plant extracts. *Ind. J.Mycology and Plant Pathol., 7(2)*:180 – 181.
2. Anonymous .(1977).Hort . Stat. (1977). Directorate of horticulture and Plantation Crops, Chennai.
3. Anonymous.(1977) . TURMERIC – the golden spice; The Hindu dt. 27.2.1977,p.28
4. Asha shivpuri, O.P.Sharma and S.L.Jhamaria.(1977). Fungitoxic properties of plant extracts against pathogenic fungi.*J.Mycol.Pl.Pathol.,27(1)*:29 – 31
5. Banerjee, A.and S.S.Nigam. (1978). Antimicrobial efficacy of the essential oil of *Curcuma longa. Ind.J.Med.Reds., 68(11)*: 864-866.
6. Basavaraj, G.and M.B.Kurdikeri.(2000). Effect of seed colouring on seed storability in soybean. *Seed.Res.,28(1)*: 39 –41.
7. Cander,H.and S.M. Ahmed. (1983). Potential of some plant products as grain protestants against insect infestation, *Bull. Grain Technol.*, 21(3): 179-188.
8. Chander, H., S.G. Kulkarni and S.K.Berry. (1992). Studies on turmeric and mustard oil as protectants against infestation on red flour beetle (*Tribolium castaneoum)* in stored milled rice. *J. insect Sci., 5(2)*: 200-222.
9. Charya, M.A.S., M S.. Reddy, B.P. Kumar and S.P. Reddy. (1979). Laboratory evaluation of some medicinal plant extracts against two pathogenic fungi, *New Botanist, 6*:171-174.
10. Dasgupta, P.R. and H.M. Austenson. (1973). Analysis of interrelationships among seedling vigour, field emergence and yield in wheat. *Agron.J., 65*: 417-422
11. Datar, V.V. (1999).Buieffucact of plant extracts against *Macrophonina phaseolina* (Tassi) Goid. The incident of charcoal-rot of sorghum. *J.Mycol. Pl.pathol., 29(2)*: 251-253
12. Edwards, R.L. and F.J. Sundstrom. (1987). After ripening and harvesting effects on tabasco pepper seed germination performance. *Hort. Science, 22(3)*: 473-475.
13. El-Sayed, A.M.A.A. and A.Badawey. (1991). Inhibitory effect of henna (lawsonia inermis) leaves and carrot root on aflatoxin production by *Aspergillus parasiticus, Mycotoxin Res.,7(2)*:61-68.

14. Gary Whiteaker. (2003). Vegetable seed quality: A local and international perspective. Website. http://www.wwha.wsu.edu/;proceeding whiteakar. doc.

15. Ghewande, M.P. (1989). Management of foliar diseases of groundnut (*Arachis hypogea*) using plant extracts. *Ind.J.Agric.Sci.*59(2): 133-134.

16. Ghulam Jilani and C.F.Su Helen. (1983). Laboratory study on several plant materials as insect replellants for protection of cereal grains. *J.Econ. Entomol., 76(1)*: 154-157.

17. Ghulam Jilani and R.C. Saxena. (1990). Repellent and feeding deterrent effects of turmeric oil, sweet flag oil, neem oil and a neem based insecticide against lesser gram borer.G.*Econ. Entomol., 83(2)*: 629-63

18. Hadas, A.and D.Russo. (1974). Water uptake by seeds as affected by water stress, capillary conductivity and seed soil water contact. I. Experimental study. *Agron. J.,66*: 643-647

19. Jitendra Singh and V.L.Majumdar. (2001). Efficacy of plant extracts against *Alternaria alternata* – the incident of fruit rot of pomegranate *Puncia granatum* L.) *J.Mycol. Pl.Pathol., 31 (3)*: 346-349

20. Kanawade,N.J., Ingle and H.G.More. (1987). Effect of packages on quality of certain vegetable seeds during storage. *J. Agrl. Engg., 20 (2)*: 25-29

21. Karivaratharaju, T.V., V.Palanisamy and K.Kumaresan. (1985). A note on the effect of fruit weight on seed quality in brinjal cv. MDU 1. *South Indian Horticulture, 33(3)*: 209-210.

22. Karivaratharaju, T.V., V.Palanisamy and K.Kumaresan. (1987). Effect of seed treatment and containers on the storability of brinjal. *Seed Res., 15(2):* 169-171.

23. Kazmi, A.R., L.Niaz and G.Jilani. (1993). Evaluation of some plant extracts for antifungal properties. *Pakistan J.Phytopathol., 5 (1-2)*: 93-97.

24. Khilare, V.C. and L.V.Gangawane. (1997). Application of medicinal plant extracts in the management of thiophanate Methyl resistant *Penicillium digitatum* causing Green mold of Mosambi. G. Mycol. *Pl.Pathol., 27 (2)*: 134-137

25. Korayem, A.M and H.A. Osman. (1992). Nematicidal potential of the henna plant (*Lawsonia inermis*) against the root knot nematode (*Meloidogyne incognita). Anzeiger-fur-schadling Skunda, Pflanzenschutz, Umweltschutz., 65 (1)*: 14-16

26. Kumari, R., K.K. Verma, K.S. Bhindra, and D.S. Bhatti. (1987). Screening of aerial parts of Datura, Ipomea, Tagetes and Lawsonia for their nematicidal activity on *Meloidogyne javanica. Agrl. Sci. Digest., 7(4)*:213-216.

27. Leela-Srinivas, V.K. Shalini and M.Shylaja. (1992). Turmerin : a water soluble antioxidant peptide from turmeric (*Curcuma longa*). *Archives of Biochemistry and Biophysics, 299 (2)*: 617-623

28. Maguire, J.D. (1962). Speed of germination aid in selection and evaluation of seedling emergence and vigour. *Crop Sci., 2* : 176-177.

29. Mead, J.F. (1976). In free radicals in biology. Pryor, W.A. (ed.) ongen, New York. Academic Press: 51-58

30. Mohiuddin, S., R.A. Qureshi, Z.Ahmed, S.S. Qureshi, K.Jamil, K.N. Jyothi and A.L.Prasunna. (1993). Laboratory evaluation of some vegetable oils as protectants of stored products. *Pak . J.Scientific Ind. Res., 36(9)*: 377-379.

31. Natarajan, M.R. and D.Lalithakumari. (1987). Antifungal activity of the leaf extract of *Lawsonia inermis* on *Drechslera oryzae. Ind . Phytopath., 40(3)*: 300-395.

32. Pillai, S.N. and M.V. Desai. (1975).Effect of turmeric on root knot and stunt nematodes. *Ind.J.Mycol.Pl.Patho., 5(1)* :20.

33. Prakash, A.Rao.J.and Gupta , S.P. (1993). Evalution of botanical pesticides as grain protectans against Rice weevil *Sitophilus oryzae*. Conference on botanical pesticides in integrated pest management. Indian soc. *Tobacco Sci., Rajamundry,* : 360 - 365.

34. Pranata , R.I., (1984). Possibility of using turmeric (*Curcuma longa*) for controlling storage insects. *Biotrop , Newslr., 45(3).*

35. Rao, V.K. (1991). Effect of certain indigenous plant materials on grain mycoflora of sorhum during storage . *Ind.J.Mycol.Pl.Pathol., 21(3)*:268 - 269.

36. Ramya , C. (2003). Studies on seed colouring, coating fruit maturity and fruit size variation on seed quality in tomato (*Lycopersicon esculentum* mill.) cv PKM1.M.Sc.(Ag) Thesis , Dept of Seed Science and Technology , Tamil Nadu Agricultural University, Coimbatore- 61 003.

37. Ray,P.G. and S.K. Majumdar. (1976). Antimicrobial activity of some Indian plants. *Econ. Bot., 30(4)*: 317 - 320.

38. Saraswathi , G. (1994). Studies on seed pelleting in relation to sowing qualities of cotton cv. LRA 5166.M.Sc.(Ag) Thesis, Tamil Nadu Agricultural University, Coimbutore -3.

39. Shivanna, S., S. Lingappa and B.V.Patil .(1994). Effectiveness of selected plant materials as protectants against pulse beetle, during storage of redgram. *Karnataka J. of Agric. Sci., 7(3)*: 285 - 290.

40. Siddiqui, M.A. and M.M.Alam. (1998). Toxicity of different plant parts of Tagetes lucida to plant parasitic nematodes. *Indian J.Nematol., 18(2)*: 181 - 185.

41. Tacedo, Y. and M.O.fatope. (1998). New phenolic glucossides from *Lawsonia inermis*. *J.Natural products.,51(4)*:725 - 729.

42. Tonapi A.V. and T.V. Karivatharaju. (1994). Studies on seed colouring in sorghum genotypes (*Sorghum bicolor* (L.,) Moench.). *Seed Tech News, 24(4)*:116.

43. Tonapi, A.V., (1998). Studies on the effect of harvesting conditions on vigour potential viability and storability of seeds in sorghum genotypes (*Sorghum biclor* (L) Moench.), Ph.D.Thesis, Tamil Nadu Agricultural University, Coimbatore - 3.

44. Vander Burg, W.J.(1979). Coloring seeds with fluorescent paint. *Seed Sci.& Technol.*, 7:565.

45. Venkatraman, S.T.R. Ramanujam and V.S.Venkatasubbu. (1998). Antifungal activity of certain plants belonging to the family Zingiberaceae. *Madras Argic J.* , *41(2)*: 92 – 94.

46. Vishwaprasad , M. and M.B. Kurdikeri. (2000). Influence of seed colouring on field performance in maize .*Seed Res.*, 28(1): 96.

47. Vishwaprasad, M. and M.B. Kurdikeri and Shekargouda. (2001). Storage potential of coloured seeds of maize. *Seed Res.*, *29*(1): 38 – 42.

48. Vivekanandan, R.(1997). New approaches in coluring of crop seeds. M.Sc. (Ag.) Thesis, Tamil Nadu Agricultural University, Coimbatore – 3.

49. Website *http:www.celpril.com/celpass1.htm.*

STUDIES ON SEED COLOURING IN CEREALS, PULSES, OIL SEEDS AND VEGETABLE SEEDS

VILAS A. TONAPI[1], P. HARINATH BABU[2] AND N.A. ANSARI[2]

1. INTRODUCTION

For grains produced as seeds the quality of the harvested produce is of considerable importance. However, mere discoloration of the seed does not mean complete loss of seed germinability beyond certification limit. Moderate discoloration infact is associated with only a slight decline in vigour potential.

Various types of dyes have been used successfully for colouring seeds including acid dyes, basic dyes, direct dyes and pigments. These dyes have been approved by Department of Agriculture in Canada and USA *viz.*, Rhodomine-B, Tetrazine, F.D. and C. Blue. Methylene Blue, Methyl Violet 2-B, D and C Red. D and C. Violet. D and C Green and Pigment red (Dupont de Nemours and Company).

But, till date there are very few and isolated studies in India to establish colour standards to pave the way for colouring the seeds by incorporating the provisions in the Seed Act of 1966. Since the literature has indicated absence of any relevant literature in developing colour standards in India except the work of Tonapi (1988) and Tonapi and Karivaratharaju (1994) in sorghum, Vivekanandan (1997) and Basavaraj *et al.* (1999) in soybean. In other countries too, colour standards are propriety of individual private companies. To be precise, processors colour seeds because it is required by law to avoid accidental use of treated seeds as food or feed. Some companies

1 *National Research Centre for Sorghum, Rajendranagar, Hyderabad - 500 030*

2 *Department of Plant Breeding and Genetics & Seed Science and Technology, Acharya N G Ranga Agricultural University, Rajendranagar, Hyderabad - 500 030*

colour seeds with a specific colour as a trademark just to identify their seeds. In this paper we discuss developing colour standards for selected cereals, pulses, oils seeds and vegetables and their resultant implications.

2. MATERIALS AND METHODS

The investigations, on seed colouring in were conducted with Paddy (Cv. MTU 7029, BPT 5204), Sorghum (cv. 27B, CSH 16), Maize (cv. Saranath), Bajra (cv. ICMH- 451), Redgram (cv. LBG- 30), Blackgram (cv. LBG-17), Bengalgram (cv. Annegeri), Castor (PCS-4), Sunflower (Morden), Safflower (cv. Bheema), Soybean (cv. JS 335) and Tomato (cv. Pusa Ruby) encompassing 25 dyes namely Rhodomine-B, Cotton blue, Fuchsine, Neutral Red, Gentian Violet, Methylene blue, Crystal violet, Congored, Fast green, Bromocresol Purple, Phenol red, Nigrosine, Erichro black T, Ammonium purpureate (mureoxide), Bromocresol green, Malachite green, Methyl red, Methyl orange, Tital yellow, Indigo carmine along with commercially available natural dyes in the market namely Kumkum, Yellow, Pink, Blue and Brick red (Table 1) to develop and recommend colour standards after assessing their effect on seed quality.

Table 1. Various dyes used in colouring the crop seeds

S. No.	Dye	Chemical composition
1	Indigo caramine	$C_{16}HgN_2O_2(So_3Na)_2$
2	Tital yellow	Dehydrothio-P-toluidine
3	Methyl orange	$Me_2NC_6H_4N : NC_6H_4SO_2Na$
4	Methyl Red	$Me_9NC_6H_4N : NC_6H_4COOH$
5	Nigrosine	$C_{38}H_{27}N_3$
6	Erichro Black-T	$C_{20}H_{12}N_3O_7Sna$
7	Ammonium purpureate	(Mureoxide) $(NH_4)_4P_2O_7$
8	Boromocresol green	$C_{21}H_{14}Br_4O_5S$
9	Bromocresol purple	$C_{19}H_{10}Br_2Cl_2O_2S$
10	Crystal violet	$C_{25}H_{30}ClN_3$
11	Malachite green	$Me_2NC_6H_4C_6H_5C : NC_6H_4Me_2Cl$
12	Congo red	$C_{32}H_{22}O_6N_6S_2Na_2$
13	Phenol Red	$C_{19}H_{14}O_5S$
14	Cotton Blue	$C_{32}H_{25}N_3O_9S_3Na_2$
15	Gentian violet	$C_{25}H_{30}CIN_3$
16	Fuchsine	$C_{20}H_{17}N_3Na_2O_9S_3$
17	Methylene Blue	$C_{19}H_{10}Br_2C_{12}O_25$

18	Rhodamine-B	$C_{28}H_{31}ClN_2O_3$
19	Neutral red	$Me_2NC_6H_3N : NC_6H_2MeNH_2HCl$
20	Fast green	$C_{37}H_{34}N_2Na_2O_{10}S_3$
		Natural dyes
21	Natural dye – Kum kum	-
22	Natural dye – Brick Red	-
23	Natural dye – Blue	-
24	Natural dye – Yellow	-
25	Natural dye – Pink	-
26	Control	-

The dye solutions were prepared at 0.75% concentration by dissolving 0.25 g of dye in 16.5 ml water and 15.0 ml ethylene glycol (Tonapi, 1988 ; Tonapi and Karivaratharaju, 1994). The individual dye colour was imparted on to 10 g seeds by vigourous shaking for 3-5 minutes in an Erlenmeyer flask till colour was imparted uniformly. The effect of seed colouring dyes on seed germination, root and shoot length of seedlings, dry weight of seedlings, speed of germination, field emergence and seed germination in exhaustion test was recorded following International Seed Testing Association Standards (Anon, 1985). Seed vigour was assessed through vigour index calculated as the product of root length and seed germination and expressed as absolute value (Abdul Baki and Anderson, 1973).

Electrical conductivity of seed leachate (Presley, 1958), a-amylase activity (Simpson and Naylor, 1962) and Dehydrogenase activity (Kittock and Law, 1968) were evaluated to assess the effect of dyes on biochemical composition of seeds. All the tests were conducted in four replications consisting 400 seeds. Fisher's method of analysis of variance was applied for data analysis. Critical differences were calculated at P= 0.05.

3. RESULTS AND DISCUSSION

The various dyes had significant promoting as well as inhibitory impact on mean seed germination (95-21%), vigour index (3349 - 271), field emergence potential (94 - 17.6%) and speed of germination (38.0 - 0.33) and their performance under stress tests across crops and cultivars under investigation (Fig. 1).

The influence of dyes on membrane integrity, enzyme activities and performance under stress also had significant influence. The results indicated that electrical conductivity of seed leachates was in the range of 128 to 431 ì mhos cm^{-1} indicating that seed colouring did influence the leakage of solutes seed steep water through their impact on membrane permeability. The μ-amylase activity, which is the precursor of germination and vigour potential of the seeds, was also affected (7.0 - 5.16 mm.) In addition, the Dehydrogenase activity indicating the degree of aliveness of seed as influenced by seed colouring also declined (0.4 to 0.32). The seed germination under stress in exhaustion test showed a significant decline (81 - 20.6%).

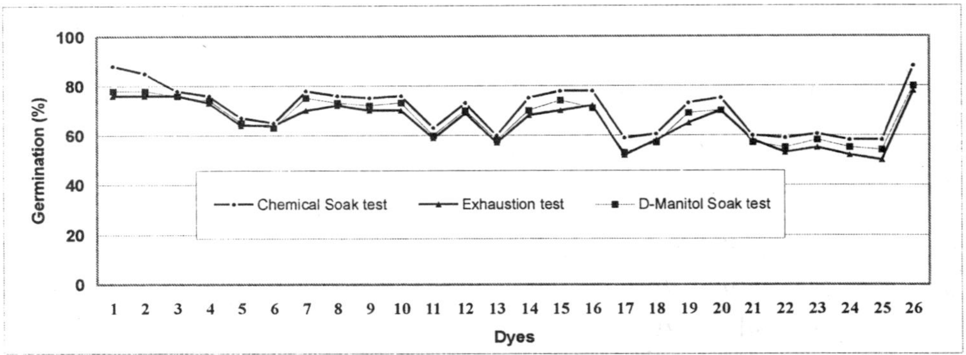

Fig. 1. Mean seed germination (%) under stress across crops and genotypes as influenced by seed colouring

(1) Rhodomine-B (2) Cotton blue (3) Fuchsine (4) Neutral Red (5) Gentian Violet (6) Methylene blue (7) Crystal violet (8) Congo red (9) Fast green (10) Bromocresol Purple (11) Phenol red (12) Nigrosine (13) Erichro black T (14) Ammonium purpureate (mureoxide) (15) Bromocresol green (16) Malachite green (17) Methyl red (18) Methyl orange (19) Tital yellow (20) Indigo carmine (21) Kumkum (22) Yellow (23) Pink (24) Blue (25) Brick red (26) Control

The promotional as well as detrimental influence of various dyes on seed quality parameters clearly indicated that all the dyes are basically chemical formulations, which upon colouring would permeate into the seed membrane upon seed hydration and might interfere with the germination process in both the directions, to inhibit or promote the same. These influences as observed in present investigations are in line with earlier findings of Tonapi (1988).

The detailed analysis of the results obtained in various crops helped us to categorize all the 25 dyes utilized in this investigations into 3 categories as (1) Best (promoting total seed quality components), (2) mid range (not very deleterious, but certainly declining seed germination and seed quality) and (3) most deleterious (deleterious to all seed quality components) (Table 2).

Table 2. Categorization of seed dyes based on their favourable and deleterious impact on seed quality

S. No.	Crop	Best Dyes	Mid range (Not very deleterious but significantly reducing seed quality)	Most deleterious
1	Sorghum (27B and CSH 16)	Rhodamine-B Crystal violet, Fast green. Fuchsine, Neutral red, Mureoxide, Congo red, Phenol red, Gential violet.	Cotton blue, Methylene blue, bromocresol purple, Nigrosine. Erichro black-T, Bromocresol green, Malachite green, Methyl red, Methyl orange, Titan yellow, Indigo caramine and Natural dye Yellow.	Natural dye Brick red, Pink, Blue and Kum kum.
2	Paddy (BPT 5204 and MTU 7029)	Fuchsine, Gentian violet, Neutral red, Bromocresol purple, Rhodamine-B, Fast green and Nigrosine.	Rhodamine, Cotton Blue, Methylene blue, Crystal violet, Congo red, Phenol red, Erichro black-T, Mureoxide, Bromocresol green, Malachite green, Methyl red, Methyl orange, Tital yellow, Indigo caramine and Natural dye Brick red.	Natural dye Kumkum, Pink, Blue and Yellow.
3	Maize	Rhodamine-B, Crystal violet, Fuchsine, Fast green	Cotton blue, Neutral red, Gentian violet, Methylene blue, Congo red, Bromocresol purple, Phenol red, Nigrosine, Erichro black-T, Mureoxide, Bromocresol green, Malachite green, Methyl red, Methyl orange, Titan	Natural dyes Kumkum and Pink.
4	Bajra	Rhodamine-B, Fuchsine	Cotton blue, Neutral red, Gentian violet, Methylene blue, Crystal violet, Congo red, Fast green, Bromocresol purple, Phenol red, Nigrosine, Erichro black-T, Mureoxide Bromocresol green, Malachite green, Methyl red, Methyl orange, Titan yellow, Indigo caramine, Natural dyes kumkum, Pink and Blue	Natural dyes Brick red and Yellow.
5	Redgram	Rhodamine -B. Fuchsine. Tital yellow, Methylene blue and Mureoxide.	Cotton blue, Neutral red, Gentian violet, Crystal violet,Congo red, Fast green Bromocresol purple,Phenol red, Nigrosine, Erichro black-T,Bromocresol green, Malachite green, Methyl red, Methyl orange, Indigo caramine and Natural dye Brick red.	Natural dyes Kumkum. Blue, Yellow and Pink.

6	Black gram	Rhodamine-B. Natural dye Yellow. Phenol red, Bromocresol purple and Fuchsine.	Cotton blue. Natural red. Gentian violet. Methylene blue, Crystal violet, Congo red, Fast green, Bromocresol green, Nigrosine, Erichro black-T, Mureoxide, Malachite green, Methyl red, Methyl orange, Tital yellow, Indigo caramine, Natural dyes Kumkum and Blue.	Natural dyes Pink and Brick red.
7	Bengal gram	Rhodamine-B. Titan yellow. Crystal violet and Methyl orange.	Cotton blue. Fuchsine. Neutral red. Gential violet. Methylene blue. Congo red. Fast green. Bromocresol purple. Phenol red. Nigrosine. Erichro black-T. Mureoxide. Bromocresol green. Malachite green, Methyl red. Indigo caramine. Natural dyes. Yellow and	Natural dyes Pink. Blue and Kumkum.
8	Castor	Rhodaine-B. Erichro black-T. Nigrosine. Cotton blue. Mureoxide and Neutral red.	Fuchsine. Gentian violet, Methylene blue. Crystal violet. Congo red. Fast green. Bromocresol purple. Phenol red. Bromocresol green. Malachite green. Methyl red. Methyl orange. Tital yellow. Indigo caramine. Natural dyes Kumkum. Yellow and Brick red.	Natural dyes Pink and Blue.
9	Sunflower	Rhodamine-B, Congo red, Bromocresol purple. Nigrosine. Phenol red. Cotton blue.	Fuchsine, Neutral red, Gentian violet, Methylene blue, Crystal violet, fast green, Erichro black-T, Mureoxide. Bromocresol green, Malachite green, Methyl red, Methyl orange, Tital yellow, Indigo caramine, Natural dyes Kumkum and Blue.	Natural dyes Pink. Yellow and Brick red.
10	Safflower	Rhodamine-B. Fuchsine. Neutral red.	Cotton blue. Gential violet. Methylene Blue. Crystal violet. Congo red. Fast green. Bromocresol purple. Phenol red. Nigrosine. Erichro black-T, Mureoxide. Bromcresol green, Malachite green, Methyl red, Methyl orange, Titan yellow, Indigo caramine, Natural dyes Yellow, Pink and Brick red.	Natural dyes Blue and Kumkum.

11	Soybean	Rhodamine-B. Fast green. Malachite green.	Cotton blue. Fuchsine. Neutral red. Gentian violet. Methylene blue. Crystal violet. Congo red. Bromocresol purple. Phenol red. Nigrosine. Erchro black-T. Mureoxide. Bromocresol green, Methylred, Methyl orange. Titan yellow. Indigo caramine. Natural dyes	Natural dyes. Yellow and Pink.
12	Tomato	Rhodamine-B. Fast green. Nigrosine, Methyl red.	Cotton blue, Fuchsine, Neutral red. GentIan violet, Methylene blue.Crystal violet, Congo red, Bromocresol purple. Phenol red, Erichro black-T. Mureoxide, Bromocresol green, Malachite green, Methyl orange. Titan yellow. Indigo caramine. Natural dyes. Yellow. Pink and Blue.	Natural dyes Kumkum and Brick red.

Hence, Rhodamine-B, Neutral red, Crystal violet. Fast green, Fuchsine, Mureoxide, Congo red, Phenol red and Gentian violet are the best suited dyes for seed colouring, since these dyes enhanced seedling characters, dry weight of seedlings, vigour index, field emergence potential, and speed of germination including their influences in preserving the enzyme activity were on par with control and promoting the performance of seed germination and under stress as indicated in exhaustion test. The most deleterious dyes were natural dyes Brick red, Pink, Blue and Kumkum, other dyes were in mid range.

Based on shine, luster, deep uniform and vibrant colouration imparted on to the seed in groundnut, a survey to know the choice of seed growers, processors and seed industry was conducted involving a total sample of 100. The most favoured seed dyes, in order of preference, by the seed industry on the basis of visual colour index and hues imparted are Rhodamine-B, Congo red, Phenol red, Fast green and Gentian violet (Table 3).

Since seed colouring is a technological incorporation of colour on to the seeds to improve their marketability and enable seed companies to have their brand identity, the process of colour standard development involved the decisions regarding the quantum of dye: water: chemical that can impart

Table 3. Best dyes and colours preferred by seed growers, processors and industry men as imparted to various crop seeds

S. No.	Crop	Best dyes as per investigation	Best dyes as per the visual observation (In the order of priority)
1	Sorghum (27B and CSH 16)	Rhodamine-B, Crystal violet, Fast green, Fuchsine, Neutral red, Mureooxide, Congo red, Plhenol red. Gentian violet.	Rhodamine-B, Congo red, Plhenol red. Fast green, Gentian violet.
2	Paddy (BPT 5204 and MTU 7029)	Fuchsine, Gentian violet, Neutral red, Bromocresol purple, Rhodamine-B, Fast green, Nigrosine	Rhodamine-B, Fast green, Fuchsine.
3	Maize	Rhodaine-B, Fuchsine, Fast green, crystal violet.	Rhodamine-B, Fast green.
4	Bajra	Rhodamine-B, Fuchsine	Rhodamine-B
5	Redgram	Rhodamine-B, Fuchsine, Titan yellow, Methylene blue, Mureoxide.	Rhodamine-B, Fuchsine, Titan Yellow
6	Black gram	Rhodamine-B, Natural dye Yellow, Phenol rel, Bromocresol purple, Fuchsine	Rhodamine-B, Fuchsine, Phenol red.
7	Bengal gram	Rhodamine-B, Crystal violet, Titan yellow, Methyl orange.	Rhodamine-B, Crystal violet, Titan yellow
8	Castor	Rhodamine-B, Erichro black, Nigrosine, Cotton blue, Mureoxide, Neutral red.	Rhodamine-B, Erichro black.
9	Sunflower	Rhodamine-B, Congo red, Bromocresol purple, Phenol red, Nigrosine, Cotton blue.	Rhodamine-B, Cotton blue.
10	Safflower	Rhodamine-B, Fuchsine, Neutral red	Rhodamine-B, Fuchsine, Neutral red.
11	Soybean	Rhodamine-B, Fast green, Malachite green.	Rhodamine-B, Fast green, Malachite green.
12	Tomato	Rhodamine-B, Fast green, Nigrosine, Methyl red.	Rhodamine-B, Fast green.

uniform luster on to the seed surface. In this regard, ethylene glycol was chosen in view of its ability to impart a kind of glossy and shiny luster on to the seeds. The trials on uniform colouration enabled us to arrive at a dye concentration of 0.75% prepared by dissolving 0.25 g dye powder in 16.5 ml water and 15 ml ethylene glycol. To prepare one liter solution of each

Table 4. Economics of seed colouring based on color standards developed for seed colouring for crop plants.

S. No	Name of the dye	Average quantity of dye solution (ml) needed to color of one kg seed	Average quantity of dye solution (ml) needed to color of one quintal	Quantity of dye powder required to prepare requisite solution	Cost of dye powder per gm (Rs.)	Total cost of dye for colouring one quintal seed (Rs.)	Quantity of ethylene glycol required to prepare requisite solution	Cost of ethylene glycol for colouring one quintal seed (Rs.)	Total cost to color one quintal seed
1	Rhodamine-B	8.735	873	6.92	2.80	19.40	416	87.30	106.70
2	Cotton Blue	13.235	1323	10.49	6.00	63.00	630	132.30	195.30
3	Fuchsine	8.958	895	7.10	7.20	51.10	426	89.40	140.50
4	Neutral Red	9.580	958	7.60	12.80	97.30	456	95.70	193.00
5	Gentian violet	8.546	854	6.77	6.00	40.60	406	85.20	125.80
6	Methylene Blue	10.069	1006	7.98	3.10	24.90	479	100.50	125.40
7	Crystal violet	7.993	799	6.34	4.40	27.90	380	79.80	233.10
8	Congo red	10.624	1062	8.42	1.80	15.20	505	106.00	121.20
9	Fast green	8.949	894	7.09	2.40	17.00	425	89.20	106.20
10	Bromocresol purple	11.374	1137	9.02	16.00	144.30	541	113.60	257.90
11	Phenol red	11.127	1112	8.82	8.40	74.00	529	111.00	185.00
12	Nigrosine	14.149	1414	11.22	3.00	93.60	673	141.30	174.90

13	Erichro black-T	12.967	1296	10.28	6.20	64.10	617	129.50	193.60
14	Mureoxide	14.255	1425	11.30	23.80	269.00	678	142.30	411.30
15	Bromocresol green	12.201	1220	9.68	67.20	650.00	580	121.80	771.80
16	Malachite green	9.371	937	7.43	2.80	20.80	446	93.60	114.40
17	Methyl Red	18.316	1831	14.53	4.10	60.40	871	182.90	243.30
18	Methyl orange	18.404	1840	14.60	2.50	37.40	876	183.90	221.30
19	Titan yellow	13.591	1359	10.78	23.00	248.00	647	135.80	383.80
20	Indigo caramine	15.783	1578	12.52	6.00	75.00	751	157.70	232.70
21	Natural dye Kumkum	36.478	3647	28.94	0.20	5.80	1736	364.50	370.30
22	Natural dye Yellow	32.201	3220	25.55	0.20	5.10	1533	321.90	327.00
23	Natural dye Pink	36.958	3695	29.32	0.20	5.80	1759	369.30	375.10
24	Natural dye Blue	34.966	3496	27.74	0.20	5.54	1664	349.40	354.90
25	Nat. dye Brick red	29.558	2955	23.45	0.20	4.70	1407	295.40	300.10

dye, we need 8 g dye powder, 550 ml water and 450 ml ethylene glycol. Excepting all dyes, natural dyes namely pink, blue, kumkum, yellow and brick red (though cheapest) could not impart uniform colour on to the seeds, but were required in quantities three times as that of remaining dyes and were highly toxic, as they inhibited seed germination and related quality indices. The average cost of the dye to colour one-quintal seed for different crops is given in Table 4.

In conclusion, through this study on colour standards for various crops, we would like to propose for making seed colouring mandatory by incorporating the same in the provisions of Indian seed act 1966 or draft seeds act 2000 by notifying the colour standards for cereal seed treatment as in USA, Canada and Europe. We propose that this provision can also be incorporated under the regulations of pest control act after suitable modifications with the text that " Where the physical properties of the control product are such that the presence of the control product may not be recognized when used and is likely to expose a person or domestic animal to severe health risk, the control product shall therefore be denaturated by means of colour, odour or such other means as the central seed committee may approve to provide a signal or warning as to its presence". Where the seed is packaged, the package should bear a label with words "the seed is treated and coloured with" followed by the name of the control measure product and the seed colouring dye, including the common name or chemical name of its active ingredient together with appropriate precautionary symbol and signal or warning words as the seed testing Committee approves. If the treated-coloured seed is sold and shipped in bulk, the shipping documents should bear information setting forth the common name or chemical name of active ingredient of both chemical and the seed colouring dye with a sub note that "seed colouring dye used is not injurious to seed". This will, in hence, may enable seed industry to adopt individual colours as their trademark including propriety colouring of parental lines to identify their seeds and seed colouring may substantially aid in preventing accidental usage of treated seeds as food or feed, or may help in upgrading the visual quality of blonded (discolored and rain soaked) and blended seeds (but, still maintaining seed germination at 75%) and above certification standards under emergent situations where there is a scarcity of seeds.

REFERENCES

1. Anonymous. (1985). International rules for seed testing. *Seed Sci.& Technol.,13(2)*: 307–355.

2. Abdul Baki and J.D. Anderson. (1973). Vigour determination in soybean seed by multiple criteria. *Crop Science, 13*: 630–633.

3. Basavaraj , G and M. B. Kurdikeri. (1999). Effect of seed colouring on field performance in soybean. *Seed Res., 27(2):* 207.

4. Kittock, D.L. and A.G Law. (1968). Relationship of seedling vigour to respiration nad tetrazolium chloride reduction by germinating wheat seeds. *Agronomy J., 60*: 286-288.

5. Presley J. J. (1958). Relation of protoplast permeability in cotton seed viability and predisposition to seedling disease. *Plant Disease Reporter, 42(7);* 852

6. Simpson J.D and J. M. Naylor. (1962). Dormancy studies in seeds of *Avena fatua* : a relationship between maltose, amylases and gibberellins. *Canadian Journal of Botany.,*: 50.

7. Vivekanandan, R. (1997). New approaches in colouring of crop seeds. M Sc(Ag) thesis submitted to Tamil Nadu Agricultural University, Coimbatore.

8. Tonapi, V.A. (1988). Studies on the effect of harvesting conditions on seed yield, quality and storability of sorghum genotypes (*Sorghum bicolor* (L.) Moench). Ph.D Thesis submitted to Tamil Nadu Agricultural University, Coimbatore- 641003, Tamil Nadu, India.

9. Tonapi, V.A and T. V. Karivaratharaju. (1994). Studies on seed colouring in sorghum genotypes (*Sorghum bicolor* (L.) Moench). *Seed Technology News, 24(4):* 116.

SEED COLOURING IN HORTICULTURAL CROPS

P.NATESAN[1], A.BHARATHI[1], H.RAMYA[1],
M.BHASKARAN[2] AND S. NANTHAKUMAR[3]

1. INTRODUCTION

Seed colouring ! – It's something unrealistic and uneconomical" It's a saying a decade ago. But now it has provoked the interest among many seed traders and seed companies. Seed colouring is mainly done to improve its marketability, brand identity and to enable the farmers for easy identification of the varieties based on colour.

The practice of providing an exogenous colour coating to seeds is only of recent interest. But it is very much prevalent in developed countries since a decade. Colouring of seeds along with pelleting, film coating, coating with biologicals is adopted in private sector seed companies in United States, Canada and Europe.

Several different types of dyes have been used successfully for colouring seeds, including acid dyes, basic dyes, direct dyes and pigments. Natural dyes have several advantages on health, safety and ecological point of view. In addition, most of them have medicinal and antimicrobial properties.

Earlier, natural dyes were used for cosmetics and colouring fabrics. Later on, synthetic dyes were discovered and came into use very widely in textile industry almost replacing the natural dyes.

It is believed that earliest dyes were quite possibly discovered by accident and might have been obtained colours from available fruits and nuts used for food and later from leaves, stems, flowers and roots of shrubs

1 *Department of Seed Science and Technology*, Tamil Nadu Agricultural University, Coimbatore – 641003
2 *National Seed Research and Training Centre, Varanasi – 221 106.*
3 *Department of Horticulture, Adhiparasakthi Agricultural College, G. B Nagar, Kalavai – 632 506*

or bark of trees. Earlier records show nearly one million natural sources of dyes of which 500 plant species have been identified as useful.

Processors do this colouring of seeds because it is required by law to avoid accidental use of treated seeds as food or feed. But some companies do this for brand identity, in such case it is necessary to prove that they are non toxic with respect to seed generation, vigour potential, viability and storage.

Few elite seed companies in India came forward for seed colouring with imported technologies. They import both the colouring materials and sophisticated machinery at a huge cost. In other cases, the colouring materials and processes adopted are very primitive with out any scientific ground. A wide reformation gap is existing in our country in this promising area of recent interest.

Although vast technological development has occurred in colouring of seeds in developed countries, the know-how of the technology remains to be a secret and proprietary critical perusal of available literature revealed the truth.

2. BACKGROUND

Vanderburg (1979) reported that painting grass seeds with fluorescent paint, such as 'Black-Ray' will stay in a good condition for many years. The seeds will not loose their paint and will not give their paint to another seed either.

Tonapi (1988) reported that among the 23 inorganic dyes used for colouring of sorghum seeds methyl red, methyl orange, genetian violet and phenol red drastically affects the viability of seeds. The other dyes, although did not confer adverse effects on vigour and viability of seeds, colouring of seeds using Rhodomine – B, Malachite green, fast green, neutral red and Methylene blue showed promoting effect on seed germination, seedling growth and vigour potential. He suggested that sorghum seeds can be coloured using any of the non toxic dyes of plant origin such a Indigo carmine, Titan yellow, Nigrosine, Rose Bengal to improve their marketability.

Vivekanadhan (1997) reported that sorghum seeds coloured using fresh juice of *Basella rubra* and *Delonix regia* did not give uniform coating due to the lack of adherence. He also reported that yellow colour of turmeric extracted by various methods affected the vigour and viability of sorghum seeds.

Kurdikeri and Vishwaprasad. (2000) reported that amaranthus red, sunset yellow, methylene blue and potassium permanganate either at 0.5 or 1.0 % concentrations could be used for colouring maize seeds without any harmful effect on field performance. He also reported that seeds releated with amaranthus red, sunset yellow and methylene blue at 0.5 or 1.0 % and treated with fungicide showed higher per cent germination, vigour index and low EC as compared to control and $KMNO_4$ treated seeds.

Kurdikeri *et al.* (2001) reported that maize seeds when treated with $KMNO_4$ at 10% concentration showed slightly deleterious effect on seed quality.

3. PRESENT RESEARCH WORK OF THIS DEPARTMENT IN RESPECT TO SEED COLOURING

Tomato seed cv PKM 1 were coloured with different natural colouring botanical and chemical dyes each with three different concentrations and the best concentration was fixed by evaluating the germinated seeds. The seeds were then treated with the best concentrations and were subjected to storage for evaluating the storage potential of coloured seeds.

Botanicals

1. *Curcuma longa* Turmeric
2. *Lawsonia inermis* Henna
3. *Beta vulgaris* Beetroot
4. *Tagetus erectus* Marigold
5. *Zyzgium cumini* Jamun
6. *Basella rubra* Basella
7. *Opuntia sp* Opuntia
8. *Hibiscus rosasinensis* Hibiscus

Chemical dyes

1. Congo red
2. Bromocreol green
3. Jade green
4. Sky blue
5. Turquoise blue
6. Direct chenbaga

Table 1. Effect of different botanicals on germination and vigour index of tomato seeds

Treatments	Germination (%)	Speed of germination	Vigour index	Drymatter production (mg10 seedling^{-1})
Botanicals				
Beetroot				
4 ml	90.00	6.68	1314	14.22
10 ml	96.00	7.25	1591	15.14
Jamun (with seeds)				
6 ml	97.00	6.52	1794	18.01
10 ml	38.00	6.67	1855	18.18
Jamun (without seed)				
8 ml	92.00	6.03	1535	17.95
12 ml	94.00	6.55	1810	19.10
Basella				
6 ml	86.00	6.23	1430	16.70
10 ml	91.00	6.27	1562	18.35
Opuntia				
6 ml	100.00	6.02	1541	15.25
10 ml	86.00	6.31	1444	14.91
Turmeric				
1.0 g	96.00	5.85	1876	18.12
1.5 g	97.00	5.76	1774	17.15
Henna				
6 ml	90.00	5.67	1830	19.11
8 ml	87.00	5.33	1609	17.51
Marigold				
5 ml	70.00	4.94	1322	14.11
15 ml	83.00	4.97	1571	15.11
Hibiscus				
6 ml	88.00	4.00	1420	14.23
8 ml	91.00	4.94	1486	15.11
Chemical dyes				
Jade green				
0.50%	82.00	5.51	1600	19.21
0.75%	84.00	4.22	1319	19.71
Bromocreol green				
0.75%	81.00	3.81	1278	15.15
1.00%	95.00	4.89	1522	17.15

Direct chenbaga				
0.75%	82.00	4.08	1311	17.51
1.0%	93.00	5.52	1673	18.11
Turquoise blue				
0.5%	92.00	5.16	1825	18.52
0.75%	93.00	5.13	1943	19.11
Skyblue				
0.5%	82.00	5.26	1521	18.54
1.0%	87.00	5.24	1712	19.11
Congo red				
0.5%	92.00	5.71	1737	18.51
1.0%	84.00	5.46	1503	17.51
Pink cm				
0.75%	94.00	6.65	1672	18.11
1.0%	94.00	7.45	1858	19.12

REFERENCES

1. Kurdikeri, M.B., M. Vishwaprasad and M. Shekhargouda. (2001). Storage potential of coloured seeds of maize. *Seed Res., 29(1)* : 38-42.
2. Kurdikeri, M.B. and M. Vishwaprasad. (2000). Influence of seed colouring on field performance in maize. *Seed Res., 28 (1)* : 96.
3. Kuridikeri, M.B and G. Basavaraj. (2000). Effect of seed colouring on seed storability in soybean. *Seed Res., 28 (1)*: 39 - 41.
4. Tonapi, A.V.(1998). Studies on the effect of harvesting conditions on vigour potential viability and storability of seeds in sorghum genotypes Ph.D Thesis. Tamil Nadu Agricultural University, Coimbatore - 3.
5. Vanderburg, W.J. (1979). Colouring seeds with fluorescent paint. *Seed Sci. & Technol.*: 565.
6. Vivekanadhan, R. (1997) New approaches in colouring of crop seeds. M.Sc (Ag) Thesis, Tamil Nadu Agricultural University, Coimbatore - 3.

CONCEPTUAL AND UTILITY DIFFERENCES AMONG SEED ENHANCEMENT TECHNOLOGIES VIZ., SEED PELLETING, SEED COATING AND SEED COLOURING

A. BHARATHI[1], P. NATESAN[1], K.VANANGAMUDI[2],
SHERIN SUSAN JOHN[1], H.RAMYA[1] AND P.THANGAVEL[1]

1. INTRODUCTION

Seed is the pivot of Agriculture. Careful farming envisages and enforces first and foremost attention on seed. "Care with the seed – Joy with harvest" is the popular adage which enlightens about the importance incumbent on the seed.

Seed quality is one of the most talked subject among farming communities. To provide higher quality seeds many seed companies have developed new production techniques called seed enhancements.

2. SEED ENHANCEMENTS

Seed enhancements are 'value added' techniques performed on a given seed lot. Seed enhancement is defined as post-harvest treatments that improve germination or seedling growth or facilitate the delivery of seeds and other materials required at the time of sowing. This includes the general areas of enhancements such as coating technologies, which included pelleting and film coating and the coatings may serve as delivery systems. These methods are not mutually exclusive and several techniques can be combined in a particular sequence to obtain an additive effect. Integration of selected individual steps can be used to upgrade seed quality. The two goals of seed enhancement, like seed designing and seed functioning can

1 *Department of Seed Science and Technology, Tamil Nadu Agricultural University, Coimbatore – 641 003*
2 *Deno (Agriculture), Tamil Nadu Agricultural University, Coimbatore - 641 003.*

be improved through seed coatings i.e., seed pelleting, seed coating and film coating.

Seeds vary greatly in size, shape and colour. In many cases, seed size is small or irregular, making singularization and precision placement difficult. In addition, seeds should be protected from a range of pests that attack germinating seeds or seedlings. Thus the seed enhancement technologies can be employed in both situations, to facilitate mechanical sowing to achieve uniformity of plant spacing and also can act as a carrier for plant protectants, so materials can be applied in the target zone with minimal disruption of the soil ecology and environment.

Farmer uses insecticides, herbicides, fertilizers, fungicides and other additive such as bacteria, to grow a better crop and achieve a better economic return on his investment. The way in which we could roll up all things that the farmers wishes to add to the seed to enhance his crop and come out with a product that would be large coated magic seeds. Seed enhancement techniques help in the establishment of crop by the most economical and efficient means possible.

In agriculture, horticulture and forestry to achieve desired plant population and successful establishment of each seed planted, several seed enhancement techniques such as pelleting, coating etc. are advocated.

3. SEED PELLETING

Seed pelleting has been accepted as an important seed enhancement methodology that improves seeding as well as seed performance.

Pelleting is defined as the deposition of a layer of inert materials that may obscure the original shape and size of the seed, resulting in a substantial weight increase and improved plantability.

Pelleting is a process, primarily aimed at creating single seeds to aid precision planting and it is a process of creating more or less spherical units for precision sowing, usually incorporating a single seed with the size and shape of the seed no longer readily evident.

Pelleting is the application of solid materials to seeds in sufficient quantity to make the pelleted seed substantially larger and / or heavier to give a spherical or elliptical shape.

Pelleting is the process of enclosing the seed with foreign material to obtain a standard size and it has its own influence on the micro environment of the seed and influences on seedlings grown at seed soil interface.

Seed pelleting is the process of enclosing a seed into a small quantity of inert material just large enough to produce a globular unit of standard size to facilitate precision planting. The inert material creates natural water holding media and provides small amount of nutrients to young seedlings. The water holding capacity of the pellets can be changed by altering the composition of filler or cementing materials. By adding moisture attracting or repelling materials to the pellet, it is possible to create a perfect seed with specific germination and moisture requirements. Bio-stimulants, microbiological inoculants, micro-nutrients, fungicides and growth regulators can be added to the pellets to improve seed performance, because seed pelleting serves as a mechanism of applying needed materials in such a way that they affect the seed or soil at the seed soil interface. Pellets should always be a stable unit but should then easily disintegrate to release the seed.

In India, seed pelleting is a limited factor. Neither commercial nor government seed organizations realised the importance of seed pelleting whereas in developed countries seed industries have evolved their own patented formulas and most of the seeds are sold with coated only.

The main objective / aim of seed pelleting is precision planting with added advantages of better establishment and increased productivity. Also pelleting of seeds is beneficial to agricultural and horticultural crops for improving the initial establishment and inturn the final yield potential.

3.1. Materials used for pelleting and their criteria of selection

3.1.1. Adhesive materials

The commonly used adhesives are gum Arabic, methyl cellulose, gelatin, caesein, caseinate salts, minerals, plastic resins, polyvinyl acetate, methyl ethyl cellulose, polyurethane, polyvinyl alcohol, poly electrolyte or dextran and poly ethylene oxide. Carboxyl Methyl Cellulose (CMC) is widely used due to the ease of availability, low cost and low rate. Some of the cheap and low cost binding materials are the rice gruel, maida gruel, sago gruel and starch (revive) gruel.

The concentration recommended for CMC is 3% w/v methyl ethyl cellulose is 5% w/v, gum arabic 45%, and nitric coat 4.3 w/v solution in water. The low cost adhesives such as rice gruel, maida gruel etc., are used at the concentration of 5%, 10% depending upon the filler material used for pelleting. Selection of adhesive is also based on selective purposes. Plastic resins, polyvinyl acetate and insoluble poly electrolyte complexes are used to bind pesticides to seeds, polyethylene oxides is to prevent erosion of surface sown seeds, polyurethane to bind lime in a way that resists coat abrasion, blends of polyvinyl alcohol and polyvinyl acetate to bind vermiculite and polyelectrolytes or dextran to aggregate soil around the seeds, there by improving the aeration of sown seeds.

3.1.2. Properties of adhesives

- Must have affinity (combinability) for both seed coat and selected filler material.
- Should have the required degree of water solubility or insolubility for easy emergence.
- Should have the required strength and plasticity to prevent dusting and breakage.
- Should have the most appropriate viscosity for easy application.

3.1.3. Coating materials or the filler materials

The materials used as filler, for pelleting must be beneficial and harmless to both seed and the rhizosphere.

Lime, gypsum, dolomite or rock phosphate and charcoal are the filler materials commonly used for pelleting. Other materials include clay minerals such as montmorillonite and vermiculite. Besides blood, peat, poultry manure, moss and mucilage are also used. The selection criteria are:

- Must be porous to allow movement of air to the seed.
- The coating must weaken or break down easily when it comes in contact with soil moisture to prevent any physical impedance to seed germination.
- Should be non toxic to seeds.
- Able to apply on commercial basis.
- Low cost

3.1.4. Botanical materials

For low cost investment and environmental benefit, botanic leaf powders are used as filler materials. The commonly used botanicals are arappu (*Albizia amara*), pungam (*Pongamia pinnata*), notchi (*Vitex negundo*), prosopis (*Prosopis juliflora*), neem (*Azadirachta indica*), moringa (*Moringa pterygosperma*) and Tamarind (*Tamarindus indica*). They are recommended @ 200-300 g kg^{-1} of seed. The fineness of the powder should be in such a way that it passes through muslin cloth. These leaf powders contain an auxin like substances which regulates the growth of seedling in initial establishment.

3.2. Methodology

The three basic steps involved in pelleting are

1. Stamping
2. Coating and
3. Rolling

The seeds are uniformly coated with adhesive in correct quantity initially. Then the filler materials are sprinkled on the coated seeds and are rolled on the filler material for effective and uniform coating.

The seed is placed in the mixer and moistened with a sticker solution. Enough treating powder is then added to dry the sticker usually in a volume relationship of a four parts powder to one part sticker. The thickness of the seed coating is dependent on the amount of sticker in relation to the amount of seed. The total mixing time should not exceed four minutes since prolonged agitation damages seeds (or) chips of pelleted coat.

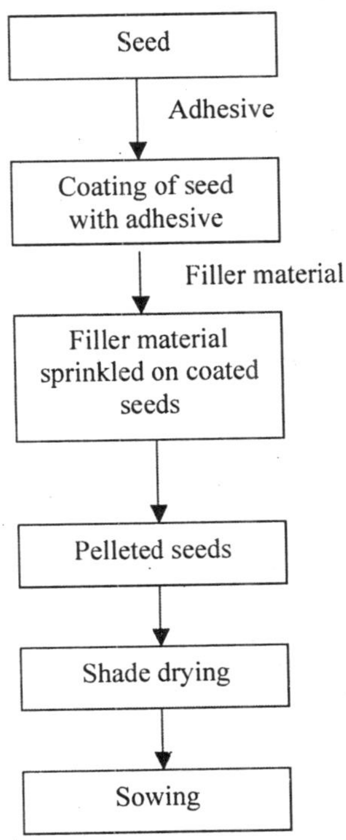

3.2.1. *Precautions*

1. Excess adhesive should not be added. It may lead to clogging of seeds.
2. The adhesive should be with required viscosity, otherwise the coating will not be perfect.
3. The filler material used should be a fine powder. Then only coating will be perfect and coating also will be retained for longer duration.
4. The coated seed should be spread in thin layer of filler material, otherwise more seed will be there in a single pellet. It will also bring the pellets to a ununiform size.

3.3. Advantages of pelleting

1. Singling of seed is achieved, helpful in precision planting in mechanized farming.
2. Seeds with free wing enables aerial seeding by improving ballistic ability.
3. Small seed and irregular shaped seeds are made easy to handle.
4. Accurate dosing of seed with chemical is possible and wastage is prevented.
5. Pelleting acts as inoculant, protective, nutrient and hydrophilic, oxygen suppliers.
6. Lime pelleting of seed, protects the multiplication of rhizobia in the rhizosphere of acid soils.
7. Stress conditions can be overcome by pelleting even in low water holding capacity.
8. Saving of chemicals / fertilizers applied to soil.
9. Uniform field establishment and increased yield.
10. Protection from birds, animals and insects.
11. Increased seed size and reduced seed rate.
12. Stimulation of germination.
13. Influences the micro-environment.

3.4. Disadvantages of pelleting

1. When the seed is pelleted, its cost and weight is more than bare seeds.
2. If proper pelleting techniques / machine is not used, there would be missing of seed during pelleting or sometimes more than one seed may contain in a pellet.
3. More moisture is required to germinate since pellet must be dissolved first.
4. Delayed field emergence at low moisture level.

4. SEED COATING

The seed enhancement technique, seed coating is an effort to increase the effectiveness of the treatments.

Seed coating is the substance applied to the seed that does not obscure its shape and the general size of the raw seed with a minimal weight gain. Seed coating examples are fungicides, microbiological treatments and micronutrients. The major benefit of a seed coating is that the seed enhancement material is placed directly on to the seed. Smaller amounts of chemicals are needed as compared to broadcasting. Seed coating is one of the most economical approaches for improving seed performance.

Factors that are crucial for the success or failure of coating are the particle size, distribution of the coating material, the exact specifications of the adhesive used and the porosity of the coating.

The objectives of seed coating mainly include, the protection of rhizobia, supply of micro and macro nutrients, protection from birds and rodents, supply of growth regulators, attraction of moisture, supply of oxygen, germination stimulation, germination delay, increase in seed weight or size and the supply of selective herbicides or herbicide antidotes.

The coating should result in a more or less continuous coating, which eliminates or minimizes product dust off. Coating methods may contain polymers, pesticides, biologicals, identifying colourants or dyes and other additives. The combination of the elements in the coating give a synergistic effect. The coating material also acts as a wick under less than ideal moisture conditions and draws moisture to the seed for better emergence and survival.

The two goals of seed enhancement, like seed designing and seed functioning can be improved through seed film coating. In view of this, many seed industries have developed specialized seed coatings, seed pellets and film coatings.

Seed coating is defined as a general term for the application of finely ground solids or liquids containing dissolved or suspended solids to form a continuous layer covering the natural seed coat which includes pelleting and many other seed treatments.

It is a mechanism of applying needed materials in such a way that they affect the seed or soil at the seed-soil interface. Thus it provides an opportunity to package effective quantities of materials such that they can influence the micro environment of each seed.

5. FILM COATING

Film coating technology is a sophisticated process of applying precise amount of active ingredients along with a liquid material, directly on to the seed surface without obscuring its shape and total seed weight may increase upto 10 per cenmt. Since the coatings are very thin, multiple coatings of various materials are possible. Most film coating contain a colorant that helps in visual monitoring and placing accuracy during automatic seeding. The coating materials can be designed for specific seed moisture requirements. Fungicides and insecticides can also be added. Film coatings can help to smooth the rough surface of seed, allowing the seed to flow more evenly during automatic seeding. Also provides strong identity for treated seeds.

The film forming formulation consists of a mixture of polymer, plasticizer and colourants. Application of the film-forming mixture results in uniform deposition of materials on each seed with little variation among seeds. The formed film may act as a physical barrier, which reduce the leaching of inhibitors from the seed coverings and may restrict oxygen diffusion to the embryo.

Film coated seeds have better flow characteristics in the planter due to reduced friction between seeds. Film coating may be performed as the final step after pelleting to provide a dust free pellet. Film coating provides an ideal method for the application of chemical and / or biological seed treatments. Relatively high loading rates of plant protectants can be applied with film coating. It also reduces the exposure of workers to chemicals from treated seed.

Coating will continue to be an important delivery system for materials required at the time of sowing. The application of plant protectants, chemicals and / or biologicals, for efficient and economical pest management will be a major emphasis.

The application of polymers to seed serves as an extra exterior shell in order to give the seed characteristics (eg. delay water uptake, enhance germination) that could be beneficial for germination in certain environments.

An ideal polymer would not be permeable to water vapour, but should be water soluble and allow the seed to imbibe sufficient moisture for germination without scarification. Synthetic coats, which restrict water entry, may prevent deterioration. It should be carried in a minimal quantity of solvent to facilitate application and speed drying.

5.1. Materials used for coating

- *Seed*
- *Coating materials*
  Fungicides, insecticides, micro and macro nutrients etc., are used.
- *Adhesives or polymers.*
  Adhesives which are used for pelleting can be used. The polymer used should not degrade when exposed to heat or humidity, be environmentally safe and should be non-toxic to plants or don't inhibit germination.
- *Solvents or diluents*
  Water, fevicol or any other organic solvents can be used.
- *Equipments* – such as pelletizing pan or fluidized bed seed – coating apparatus.

5.2. Methodology

The seeds are usually coated by the following methods.

a. Dry seed dressing

Seeds are treated as such with the active ingredient or coating materials along with the adhesives (liquid form). The seeds are stirred well or rotated in a drum for uniform coating. The coated seeds are shade dried and used for sowing.

b. Slurry seed dressing

Solvents or diluents such as water (@ 5 ml to 20 ml /kg of seed) or fevicol (10% to 20%, @ 30 ml/kg of seed) is used. The concentration of the diluents vary greatly with the seeds. Here the seeds are coated with the active ingredients, simultaneously with the adhesives or polymers diluted in the above diluents or solvents. Then the seeds are stirred uniformly and coated seeds are allowed to shade dry and used for sowing. Care should be taken for the use of optimum concentration of adhesives and diluents for obtaining effective coating.

5.3. Advantages of seed coating

1. Enables accurate and even doses of chemicals and reduces chemical wastage.
2. It makes room for including all the required ingredients, like inoculants, protectants, nutrients, plant growth hormones, hydrophilic substances, herbicides, oxygen suppliers etc.

3. By encasing the seed within a thin film of biodegradable polymer, the adherence of the seed treatment to the seed is improved.
4. It ensures dust free handling, making treated seed, both user and environment friendly.
5. Addition of colorant helps visual monitoring of placement accuracy as well as identification of cultivars.
6. Coating results in more even seeding rate due to the smooth flow of the seeds.
7. Improves the appearance and quality of treated seeds.
8. Major benefits are safety and ease of handling.
9. Polymer acts as a temperature switch and protective coating which regulates the intake of water and subsequent germination of seed.
10. Better plantability and emergence of seeds.
11. Hydrophobic seed coatings are used to delay seed germination, enable synchronous flowering of male and female parental lines in hybrid seed production.

5.4. Disadvantages of seed coating

1. Coated seeds fetch high cost, than the bare seeds.
2. Proper coating equipments and solvents or diluents is needed for effective coating.

6. SEED COLOURING

Seed colouring is colouring of seeds with different naturally colouring dyes and artificial chemical dyes in order to enable brand identification and to give the seed a distinct and attractive look.

Seed colouring is also an enhancement technique mainly done to improve its marketability, brand identity and to enable the farmers for easy identification of the varieties based on colour. It also acts as insect and bird repellant. In storage, seed lots of different years could be easily identified and adulteration of seeds can be checked. The practice of providing an exogenous colour coating to seeds is only of recent interest. But it is very much prevalent in developed countries since a decade. Colouring of seeds along with pelleting, film coating, coating with biologicals is adopted in private sector seed companies in United States, Canada and Europe.

Natural dyes have several advantages on health, safety and ecological point of view. In addition most of them have medicinal and antimicrobial properties.

Nature is full of fascinating colours, without which life would have been dull and monotonous. Colour has been playing a dominant role in the life of man from time immemorial. Seeds are endowed with an array of inherent colours specific to each genera and species. The advancement made in this line during the past 10 years indeed is very great with parallel inventions of sophisticated machineries for coating.

Few elite seed companies in India came forward for seed colouring with reported technologies. They import both the colouring material and sophisticated machinery at a huge cost. In other cases, the colouring materials and processes adopted are very primitive without any scientific ground. A wide information gap is existing in our country in this promising area of recent interest; Processors do this colouring of seeds because it is required by law to avoid accidental use of treated seeds as food or feed. But some companies do this for brand identity, in such case, it is necessary to prove that they are non-toxic with respect to seed germination, vigour potential, viability and storage. Although vast technological development has occurred in colouring of seeds in developed countries, the know-how of the technology remains to be a secret and proprietary critical perusal of available literature revealed the truth.

6.1. Materials required

Seeds, colouring agents either natural dyes or synthetic dyes, water, pulper.

6.2. Methodology

For natural dyes, the root, fruits, leaves, seeds and flowers can be used. For eg. Jamun, Basella, opuntia the fruits were used. The fruits when fully ripened were fed into the pulper and the dye was extracted. In case of Henna the leaf powder was mixed with water and used. In case of beet root, the root portion was cut into pieces and fed into the pulper and the dye was extracted. In case of hibiscus and marigold the flowers were crushed and the dye was extracted. For turmeric the rhizome powder was mixed with water and used. In case of Anatto the seeds were soaked in water @ 1:1 ratio and the dye was extracted and used.

The dyes so extracted were used with different concentrations as below.

1. *Curcuma longa* → 1, 1.5 and 2.0 g per 5 g of seed.
2. *Lawsonia inermis* → 6, 8 and 10 ml per 5 g of seed.
3. *Beta vulgaris* → 4, 6, 8 and 10 ml per 5 g of seed.

4. *Zyzgium cumini* → 6, 8, 10 and 12 ml per 5 g of seed.
5. *Basella rubra* → 6, 8 and 10 ml per 5 g of seed.
6. *Opuntia* sp. → 6, 8 and 10 ml per 5 g of seed.
7. *Tagetes erecta* → 5, 10 and 15 ml per 5 g of seed.
8. *Hibiscus rosasinensis* → 6, 8 and 10 ml per 5 g of seed.
9. *Bixa orellana* → 5, 10 and 15 ml per 5 g of seed.

In case of chemical dyes the dye powder was mixed with water and then they were treated with seed. The different concentrations used were

1. Congo red - 0.25, 0.50, 0.75 and 1.0%.
2. Bromocersol green - 0.25, 0.50, 0.75 and 1.0%
3. Jade green - 0.25, 0.50, 0.75 and 1.0%
4. Sky blue - 0.25, 0.50, 0.75 and 1.0%
5. Turquoise blue - 0.25, 0.50, 0.75 and 1.0%
6. Pink CM - 0.25, 0.50, 0.75 and 1.0%
7. Direct chenbaga - 0.25, 0.50, 0.75 and 1.0%

The seeds were put into a beaker and the dye was slowly added along with constant stirring for uniform mixing of seeds. The seeds were then shade dried then sun dried to original moisture content and then put for germination.

6.3. Advantages

- Prevents inadvertent mixing of high value seeds with other seeds of same crop.
- Gives protection to seed
- Seal the cracks on seed coat
- Improves the physical appearance
- Provides a smooth and even surface enhancing market value and plant ability.
- Gives a distinct and attractive look to the seed..

6.4. Disadvantages

- Involves extra cost
- Time consuming.

7. CONCLUSION

Seed management is an inevitable practice for successful crop production. Establishment is markedly improved when sown species possess good 'establishment ability' and can compete aggressively with weeds. Seed coatings are seen as one way in which seedlings may have an improved 'competitive ability' conferred upon them. In crop establishment, early seedling growth is of great importance in obtaining optimum plant stands and maximizing yields. Seed coatings should be developed as a management tool to be an integral component to enhance seedling establishment under a wide range of environmental conditions. To get an active and efficient uniform growth of seedlings and plants seed pelleting technique may give a wonderful effect. Colouring of seeds is required by law to avoid accidental use of treated seeds as food or feed, the practice which is mainly followed by the seed processors.

PART V

SEED STORAGE, TESTING AND HEALTH

SEED TESTING AND CERTIFICATION PROCEDURES FOR PELLETED SEEDS

S.SAMBANTHAMANI, P.SUNDARARAJAN AND M.V.KRISHNARAJAN

1. INTRODUCTION

Coated or pelleted seeds has to be dealt appropriately in order to make positive identification of all individual seeds and inert matter which is impracticable without destroying the structures presented for testing. A wide range of materials may be used to coat seeds as individuals in discrete units as in pellets or spaced in strips or sheets. Where reference is made to seed pellets the rules also apply to encrusted seed and seed granules and to seed tapes, to seed mats. For seeds thus coated modified techniques are recommended to test the seeds, based on specific instructions.

2. DEFINITIONS

Seed pellets: More or less spherical units developed for precision sowing, usually incorporating a single seed with the size and shape of the seed no longer readily evident. The pellet, in addition to the pelleting material, may contain pesticides, dyes or other additives.

Encrusted seed: Units more or less retaining the shape of the seed with the size and weight changed to a greater or lesser extent. The encrusting material may contain pesticides, fungicides, dyes or other additives.

Seed granules: Units more or less cylindrical, including types with more than one seed joined together. The granule, in addition to the granulating material, may contain pesticides, dyes or other additives.

Directorate of Seed Certification, Coimbatore – 641 040.

Seed tapes: Narrow bands of material, such as paper or other degradable material, with seeds spaced randomly, in groups or in a single row.

Seed mats: Broad sheets of material, such as paper or other degradable material, with seeds placed in rows, groups or at random throughout the sheets.

Treated seed: Seed to which only pesticides, dyes or other additives have been applied which have not resulted in a significant change in size, shape or addition to the weight of the origin seed and which can still be tested according to the methods prescribed in other chapters.

3. SAMPLING

3.1. Size of lot

The lot should be reasonably homogeneous, the maximum weight of lot may be as great as the maximum weight of lot for which sampling procedures are prescribed, subjected to the tolerance of 5% and subjected to the seed number limitation.

Seed tapes or seed mats may contain a maximum number of seeds of 1,000,000,000 (10,000) except that the weight of lot, including the coating material may not exceed 42,000 kg (40,000 kg plus 5%). When lot size is expressed in units the total weight of the lot must be given on the Certificate.

3.2. Sampling intensity

Sampling the lot of seed pellets should be done according to the intensity appropriate to the particular lot. Sampling the lot of seed tapes should be done by taking packets or (from reels) pieces of tape at random analogously, following the prescriptions of 2.6.2. and 2.6.2.A., provided that packets or reels containing up to 2,000,000 (20 units of 100,000) seeds, may be combined as a basic unit and therefore are to be considered as one container.

3.3. Size of submitted sample

Submitted samples shall contain not less than the number of pellets or seeds indicated in column 2 of Table 1, Part 1 and Part 2. If a sample is used the following statement must be inserted on the Certificate: "The sample submitted contained only pellets (seeds) and is not in accordance with the International Rules for Seed Testing".

3.4. Drawing and disposal of submitted sample

As submitted samples of coated seeds normally contain fewer seeds than corresponding samples of uncoated seeds, special care is necessary in

Table 1. Part 1. Sample sizes of pelleted seeds in number of pellets

Determinations 1	Submitted samples not less than 2	Working samples not less than 3
Purity analysis (including verification of species)	7500	2500
Weight determination	7500	Pure pellet fraction
Germination	7500	400
Determination of other seeds	10000	7500
Determination of other seeds (encrusted seeds and seed granules)	25000	25000
Size grading	10000	2000

Table 1. Part 2. Sample sizes of seed tapes

Determinations 1	Submitted samples not less than 2	Working samples not less than 3
Verification of species	2500 seeds	100 seeds
Germination	2500 seeds	400 seeds
Purity analysis (if required)	2500 seeds	2500 seeds
Determination of other seeds	10000 seeds	7500 seeds

drawing the sample to ensure that it is representative of the lot. Precautions are necessary to avoid damage to or change in the pellets or seed tape during drawing, handling and transport, and samples must be submitted in suitable containers.

3.5. Size of working sample

Working samples shall contain not less than the number of pellets or seeds indicated in column 3 of Table 11A, Part 1 and Part 2. If a smaller sample is used the actual number of pellets or seeds in the sample shall be reported on the Analysis Certificate.

3.6. Obtaining the working sample

For pelleting seeds used one of the dividers described *viz.*, conical divider, soil divider or centrifugal divider of fall must never exceed 250 mm. For seed tape take pieces of tape at random, to provide sufficient seeds for the test.

4. PURITY ANALYSIS

4.1. Objective

A purity analysis in the strict sense (i.e. of the seeds inside the pellets and tapes) is not obligatory though, if requested by the sender, a purity analysis on depelleted seeds or seed removed from tape may be carried out to separate the sample to determine the percentage of each part by weight. Separations for pelleted seed are defined but for taped seed no separation is made.

4.2. Definitions for pelleted seed

4.2.1. Pure pellets

Pure pellets shall include:

(a) Entire pellets regardless of whether or not they contain seed,

(b) Broken and damaged pellets in which more than half the surface of the seed is covered by pelleting material, except when it is obvious that either the seed is not of the species stated by the sender or there is no seed present.

4.2.2. Unpelleted seed

Unpelleted seed shall include:

(a) Free seeds of any species,

(b) Broken pellets containing a seed that is recognizably not of the species stated by the sender.

(c) Broken pellets containing seed recognizable as being of the species stated by the sender but not included in the pure pellets fraction.

4.2.3. Inert matter

Inert matter shall include:

(a) Loose pelleting material

(b) Broken pellets in which it is obvious that there is no seed

(c) Any other material defined as inert matter in Rule 3.2.3. of Chapter 3

4.3. Verification of species

In order to check that the seed in the pellets is largely of the species stated by the sender, it is obligatory to remove the pelleting material from 100 pellets taken from the pure pellet fraction of the purity test and determine the species of each seed. The pelleting material may be washed off o removed in the dry state. Similarly 100 seeds must be removed from tapes and the identity of each seed was determined.

4.3.1. *Procedure for verification of species*

The pelleting material may be washed off or removed in the dry state. For taped seed, depending on the material the tape is made of, strip off or dissolve away the tape so that 100 seeds can be examined. When the seeds in the tape are also pelleted remove the pelleting material as indicated above.

4.4. Working sample

For pellets the purity analysis shall be made on a working sample taken from the submitted sample. The size of the working sample shall be that indicated in column 3 of Table 11A, Part 1. The analysis may be made on one working sample of this number of pellets or on two sub-samples of at least half this number each independently drawn. The working sample (or each sub-sample) shall be weighed in grams to the minimum number of decimal places necessary to calculate the percentage of its component parts to one decimal place.

4.5. Procedures for purity tests on depelleted seeds and seeds removed from tapes

When a purity test on depelleted seeds is to be undertaken at the request of the sender the working sample of not less than 2500 pellets is depelled by shaking in fine mesh sieves immersed in water. A sieve of 1.00 mm mesh above a sieve of 0.5 mm is recommended. The pelleting material is dispersed in the water and the remaining seed material is dried overnight on filter paper and then in an air oven at the temperature indicated for the species under test. After drying, the material must be subjected to a purity analysis. The component parts (Pure seed, other seeds and inert matter) shall be reported as percentages of their total weight, ignoring the pelleting material. The percentage of pelleting material shall be reported separately only on request.

When a purity test on seeds removed from tapes is requested, the tape material of the working sample with paper tapes is cautiously separated and stripped off. Water soluble tape material is moistened until the seeds come free. When pelleted seeds are found in the tapes follow the procedure in the paragraph above. The moistened seeds must be dried and the free seed material must be subjected to a purity test as above. The component parts (pure seed, other seeds and inert matter) shall be reported as percentages of their total weight ignoring the tape material.

The results of these tests are to be reported under 'Other Determinations' and endorsed 'weight of the material excluded'.

4.6. Separation

The working sample of pellets (or sub-sample) after weighing shall be separated into its components.

4.7. Calculation and expression of results

The percentage by weight of each of the component parts shall be calculated to one decimal place. Percentages must be based on the sum of the weights of the components, not on the original weight of the working sample, but the sum of the weights of the components must be compared to the original weight as a check against loss of material or other error.

4.8. Reporting results

The result of a purity analysis shall be given to one decimal place and the percentage of all components must total 100. Components of less than 0.05% shall be reported as trace. The name and number of seeds of each species found in the examination of the 100 seeds removed from pellets or tapes shall be reported on the International Seed Testing Association International Seed Analysis Certificate under 'Other Determinations'. The percentages of pure pellets, unpelleted seed and inert matter must be reported in the spaces provided on the International Seed Analysis Certificate for the components of the purity test on unpelleted seed.

4.9. Determination of number of other seeds

4.9.1. Object

This determination, to estimate the number of seeds of other species is carried out only at the request of the sender.

4.9.2. Definitions

In determining the number of other seeds, the definition prescribed in Rule 3.2. shall be observed. Other seeds refer to species other than that of the pure seed as defined in Rule 3.2.1.

4.9.3. General principles

The determination is made by a count of seeds of the species (or groups of species) designated by the sender, and the result is expressed as a number of seeds found in the weight and approximate number of pellets examined for tapes in the length of tape (or area of mat) examined.

4.9.4. Procedure

4.9.4.1. Working sample

The working sample shall be not less than that prescribed in column 3 of Table 11A, Part 1 and Part 2. The working sample of pellets may be divided into two sub-samples.

4.9.4.2. Determination

The pelleting material and/or tape material shall be removed as described earlier but drying is not obligatory. The working sample is searched for seeds of all other species or of certain designated species, as required by the sender.

4.9.4.3. Calculation and expression of results

The result is expressed as the number of seeds belonging to each designated species or category found in the actual weight and approximate number of pelleted seeds examined and for seed tapes the length of tape (or area of mat) is examined. In addition the number per unit weight, per unit length or per unit area (e.g. per kilogram, per metre or per square metre) may be calculated.

4.9.4.4. Reporting results

The actual weight and approximate number of pelleted seeds and/or the length of tape is (or area of mat) examined and the scientific name and number of seeds of each species sought and found in this weight, length or area shall be reported on the International Seed Testing Association International Seed Analysis Certificate under 'Other Determinations', not withstanding that the result may in addition be expressed in some other way (e.g. number of seeds per kilogram, per metre or per square metre).

5. THE GERMINATION TEST

5.1. Object

To determine the percentage by number of normal seedlings of the kind of seed of which the sample purpose to be, using pellets from the pure pellet fraction or tape without removing the seeds from the tape material. An additional germination test on pure seed taken out of the pellets or tape may be carried out at the request of the sender or as a check on a test of pellets or tapes, but care must be taken that the covering material is removed in such a way as not to affect the germination capacity of the seeds.

5.2. Definitions

A pellet is regarded as having germinated if it produces at least one normal seedling of the species stated by the sender. Seedlings that are obviously not of the species stated by the sender, even if normal for their species, are not included in the germination figure but their number shall be reported separately.

5.3. General principles

Germination tests on pelleted seeds shall be made with pellets from the pure pellet fraction of a purity test. The pellets shall be placed on the substrate in the condition in which they are received (e.g. without rinsing or soaking). Germination tests on seed tapes are made on the tape without removing the seeds from the tape material or in any way pre-treating the tape.

5.4. Materials

Paper, sand and in certain situations soil are permissible as substrates. For pelleted seed the use of pleated paper, and for seed tapes between paper method of which the upright rolled towel has proved satisfactory in many cases, is recommended.

The pleated paper recommended for tests on pellets has a weight of 100-120 per square metre and a water absorption of 220-240%. The pleated filter papers are enveloped by cover strips, of weight 70 g per square metre, and water absorption of 220-240%.

5.5. Counting equipment *viz.*, counting beards, vacuum counters

Germination apparatus *viz.*, the bell jar or jacobsen apparatus, the germination cabinet, and room germinators.

5.6. Procedure

5.6.1. Working sample

The pure pellets shall be well mixed and 400 pellets counted at random in replicates of 100. The working sample from seed tapes shall consist of randomly taken pieces of tape to make up four replicates of atleast 100 seeds each.

5.6.2. Test conditions

Methods, substrates, temperatures, light conditions and special treatment as for particular species should be used. Where substrates prescribed are found not to give satisfactory results, pleated paper should be used for pellets and between paper method for tapes.

5.6.3. *Moisture and aeration*

The water supply may be varied according to the pelleting material and the kind of seed so as to achieve optimum conditions for germination. If pelleting material adheres to the cotyledons, water may be sprayed cautiously on to the seedlings at the time of counting.

5.6.4. *Special treatments for breaking dormancy*

When fresh ungerminated seeds remain at the end of the test period a retest may be made using one of the special treatments prescribed.

5.6.5. *Duration of test*

Extension beyond the period prescribed may be necessary. However, slow germination may be an indication that test conditions are not optimum and a germination test of seeds removed from the covering material may be made as a check.

5.6.6. *Evaluation*

Evaluation of seedlings as normal or abnormal shall be in accordance with the definitions and details of the essential stretchers. Abnormality may on occasion be due to the pelleting or tape material and when this is suspected a retest shall be carried out in soil of good quality.

5.6.7. *Multiple seed structures*

Multiple seed structures may occur in pellets or in tapes or more than one seed may be found in a pellet. In either case these shall be tested as single seeds. The result of the test indicates the percentage of structures or pellets which have produced at least one normal seedling. Pellets or seeds in tapes producing two or more such seedlings are counted and their number recorded.

When pellets are tested for monogermity the number of pellets which have produced either one, two or more than two normal seedlings are determined in the germination test and each is expressed as a percentage of the total number of pellets producing at least one normal seedling.

5.6.7. *Calculation and expression of results*

Results are expressed as percentage by number. In addition, for taped seeds the total length of tape (or area of mat) used in the germination test is measured and the total number of normal seedlings is noted. From these data the number of normal seedlings per metre (or square metre) is calculated.

5.6.8. Reporting results

The percentage of pellets or seed in tapes with normal seedlings, with abnormal seedlings and without seedlings shall be reported on the International Seed Testing Association International Seed Analysis Certificate. The method used for the germination test and the duration of the test must be indicated. In addition, for seed tapes the number of normal seedlings per metre of tape (or square metre of mat) shall be reported.

6. WEIGHT DETERMINATION AND SIZE GRADING OF PELLETED SEED

Because of the technical requirements of precision drilling, weight determination or size grading may be necessary.

6.1. Object

The object is to determine the weight per 1000 pellets and/or size grading of the samples as submitted.

6.2. Principles

For weight determination the number of pellets in a weighed quantity of pure pellets is counted and the weight per 1000 pellets is calculated. For size determination a sample of the size specified in Appendix A is screened as specified and the percentage of each screening fraction is determined.

6.3. Apparatus

For weight determination a suitable counting machine or counting equipment for germination tests may be used. For size determination a suitable screening machine is used.

7. CERTIFICATES

7.1. Reporting results

International Seed Testing Association International Seed Analysis Certificates for coated seeds should be clearly marked in the space following the heading ANALYSIS RESULTS with the words 'SEED PELLETS, ENCRUSTED SEEDS, SEED GRANULES, SEED TAPES or SEED MATS'. The name and number of seeds of each species found in the verification of species examination shall be reported under 'Other Determinations'.

STORABILITY OF HARDENED AND PELLETED SEED

V.MANONMANI

1. INTRODUCTION

Seed viability is a major factor in crop stand establishment and subsequent productivity in many parts of the world. Losses in seed quality occur during field weathering, harvesting and storage. Seeds get damaged if seeds are exposed to high temperature or high humidity. Seed coating provides an opportunity to package effective quantities of material such that they can influence the seed or soil at the seed soil interface. Seed coating materials were reported to improve the germination and increase the seedling emergence at changing soil moisture, especially in the sub optimal range (Mucke, 1987). This nutrient environment could be created by adding variety of substances to the seed to enhance the seed quality. Seed pelleting is a technique of seed encapsulation with organic, inorganic nutrients, water absorbants and pesticides. Soil may be inert or limiting in nutrient supply due to continued cropping. Pelleting of seeds with nutrients may overcome the soil deficiency atleast in the early plant growth (Khan, 1980). Seed pelleting is an important key for successful agriculture through which uniform and vigorous field stand can be achieved. Normally, the treated seeds are used for sowing immediately. But there are instances that non-availability of labour and failure of rainfall. Under such situations, the treated seeds have to be stored.

Seed pelleting claimed to play a vital role in modern agriculture for precision planting and to supplement nutrition through seed for uniform and vigorous seedling growth and safeguarding storage life of seed.

Agricultural Research Station, Bhavanisagar – 638 451.

According to Roos (1979), storage behaviour of pelleted seed was very much depended upon the pelleting material, kind of seed and storage container.

2. EFFECT OF HARDENING AND PELLETING ON STORABILITY OF SEEDS

According to Vijaya (1996), fortified and pelleted seeds of blaçkgram and cowpea could be stored safely for two months only. Nargis (1995) indicated that pelleted seeds of tomato could be stored safely upto five months. Saraswathi (1993) demonstrated that pelleting affected storability in cotton. But hardening treatment which involves soaking and drying of seeds maintained viability better as compared to the control and pelleted seeds, probably because of better membrane stability caused by quenching of free radicals (Basu, 1976).

The study conducted by Basaria Begum (2001) revealed that hardening cum coating with DAP + Trichoderma + Rhizobium had not enhanced moisture absorption but slightly reduced it compared to control of Blackgram variety CO-5. In contrast, Roos (1979) and Saraswathi (1993) found absorption of moisture was more in pelleted seed. Initially upto 3 months the pelleted seeds recorded more viability and vigour. Thereafter these parameters were greatly reduced and after six months, they lost their viability. This might be due to the phytotoxic effect of DAP on the coated seed.

She also reported that during storage periods seed borne fungi such as *Aspergillus* and *Rhizopus* were observed. The percentage of infection increased with storage period but it was minimum with seeds hardened with leaf extract and pelleted with DAP + *Trichoderma* + *Rhizobium*. In this study, seeds subjected to different treatment sequences registered significant different in moisture content, rate of germination, germination %, seedling length, dry matter production, vigour index, field emergence and fungal infection during storage. Among the different treatments, seeds hardened with pungam leaf extract (1%) maintained their superiority during storage in terms of viability and vigour.

The hardened seeds could be stored safely upto a period of six months under ambient conditions without deleterious effect on seed quality. However, pelleted seeds could be stored only for three months.

The hardened cum pelleted seeds of Ragi cv. CO 13, could be stored upto five months (Punithavathy, 1997). In contrast, Jegathambal (1996) reported that the hardened cum pelleted seeds could be stored upto 14 days with higher germination. In paddy, seed hardening with KCl (Dhamalingam *et al.*, 1988) and with a mixture of KCl and $CaCl_2$ (Nagaraj, 1996) could be stored for more than 30 days without appreciable loss in vigour and viability. The decline in germination might be due to depletion of food reserves, loss of membrane integrity, denaturation of proteins and enzyme system, ultra structural changes and increased chromosomal abnormalities. Whereas, Vigneshwari (2002) reported that hardened and pelleted seeds maintained their viability for one month and thereafter the viability declined in ragi.

The decrease in germination, seedling length, and vigour index also reduced during storage in hardened cum pelleted seeds. Physiological deterioration of seed vigour might be the outcome of deterioration at various levels in the enzyme activity and seed composition. Seed moisture content was maximum where seeds are hardened with KCl and $CaCl_2$ followed by pelleting with pungam leaf power. Increase in seed moisture during storage was due to the absorption of moisture from atmosphere due to hygroscopic nature of seed.

Study conducted by Saraswathi (1993) proved that the germination of pelleted cotton seed was on par with unpelleted seeds upto 4 months of storage. Beyond 4 months, germination declined significantly in the pelleted seeds clearly indicating that the pelleted cotton seeds could be stored for 4 months without any deleterious effect.

After 4 months of storage, decline in germination was significantly more in the pelleted seeds than in the unpelleted seeds. This might be due to more absorption of atmospheric moisture by the pelleted seeds than unpelleted seeds under ambient storage.

Pelleted seeds of clover could be stored for four months without loss of viability. Vartha and Clifford (1973) stated that pelleted grass seeds were found to store well for more than two months without reduction in germination.

Basu *et al.* (1976) found that pre-sowing soaking and drying treatment with water and dilute solutions of cinnamic acid, tannic acid and thiosulphate significantly checked the loss of viability during storage under high humidity conditions in rice and jute.

According to Roos (1979), pelleted carrot and onion seeds were storable equal to or even better than the unpelleted seeds.

Pelleted onion seeds maintained their viability better compared to non pelleted seeds after five years of storage (Konstantinov and Petkov, 1982).

Subbarao (1984) found that oat and brinjal seed pelleted with nutrients performed better over unpelleted seeds during storage. Sevanan (1988) studied an increase in the field emergence potential after six months of storage in the pelleted seeds than in the unpelleted seeds in gingelly cv. TMV3. Sabir Ahmed (1989) concluded that pelleting of soybean seeds with notchi leaf powder and sambangi seed powder maintained 70 per cent germination over eight months of ambient storage. Selvaraju (1992) reported that the pelleted seeds of sorghum could be stored for more than four months under ambient storage conditions. Jegathambal (1996) found that the hardened and pelleted seeds could be stored upto two weeks with higher germination and seedling vigour. Nagaraj (1996) confirmed that the paddy seeds soaked in potassium chloride (one per cent) + calcium chloride (one per cent) + arappu leaf powder performed on par with unpelleted seeds upto three months of storage under ambient condition. Cowpea and blackgram seeds fortified with micronutrients and pelleted with macronutrients could be stored in cloth bag upto two months without reduction in vigour and viability (Vijaya, 1996). Whereas Kavitha (2002) stated that blackgram seed hardened with prosopis leaf extract and pelleted with nutrients maintained their viability upto six months (Plate 1).

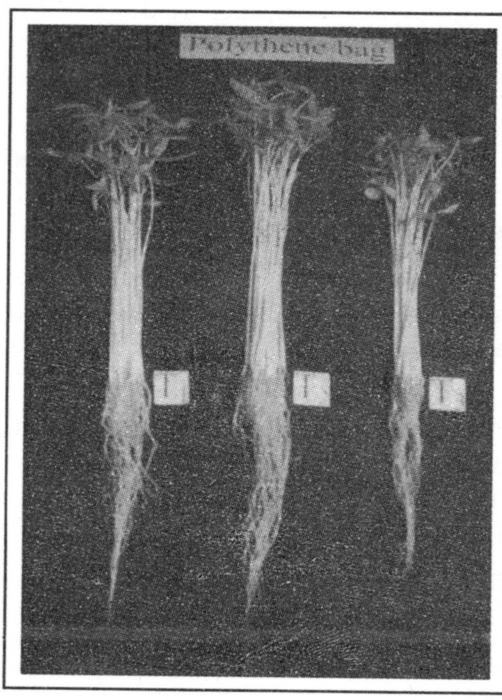

Plate1. Effect of hardening and pelleting and storage container on seedlimg growth and vigour at 7 DAS AT 6 months after storage

Manonmani (1997) conducted study to investigate the storability of teak drupes after coating with biocides *viz.*, captan, thiram, carbaryl and botanicals *viz.*, leaf powders of *Azadirachta indica, Vitex negundo, Albizia amara* and treating with kernel oil of *Azadirachta indica*. From the fourth month of storage upto the end of the study period *A. amara* leaf powder recorded higher germination percentage followed by *Vitex negundo*. The observation on fungal infection percentage and insect damage percentage revealed that among the botanicals, the lowest of 18.6% was recorded by the *Albizia amara* and *Vitex negundo* leaf powders. The result of the experiment underscores the effectiveness of coating the teak drupes with leaf powders of *Albizia amara* in reducing the fungal and insect infection percentage, besides recording high germination percentage during storage. It is hence recommended that teak drupes should be coated with *Albizia amara* leaf powder @ 25 g/100 g of teak drupes, in order to improve the storability of teak drupes. The presence of gibberellin like substance in addition to saponins and nutrients especially the micronutrients (Zn) in *A. amara* leaf powder which might have synergistically interacted to amino acid tryptophan to form Indole Acetic Acid (IAA).

A study conducted by Srimathi *et al.* (2002) to trace the influence of nutrients *viz.*, DAP, $ZnSO_4$ borax added to the pellet on the initial seed quality, field emergence, filled potential and storability upto 3 months in soybean cv. CO 1. Pelleting the seed with $ZnSO_4$ @ 250 mg kg^{-1} of seed improved the initial seed quality and its production potentiality. The performance of pelleted seed including pelleting without nutrients was better than unpelleted seed. The germinability of the seeds was maintained upto three months without much reduction in initial quality.

The germination of the pelleted seeds after ambient storage were better than unpelleted seeds upto the evaluated period of 3 months. Even here, the storability of $ZnSO_4$ pelleted seed was better than other nutrients, simple pelleting and unpelleted control seeds.

Manonmani (2002) conducted study to invgestigate the storability of soybean seeds after coating with botanicals *viz.*, leaf powders of *Albizia amara, Vitex negundo* and *Azadirachta indica* @ 1:100. Among the botanicals, seeds coated with *Albizia amara* recorded 58.0 per cent germination after 12 months of storage. This was followed by *Vitex negundo* leaf powder (55.3%) whereas control recorded germination after 12 months of storage. The superiority of *Albizia amara* was clearly showed by increased root length, shoot length and vigour index over control (Table 1).

Table 1. Effect of botanicals on the germination, root length, shoot length and vigour index of soybean cv. CO 2 in storage

Treat-ment	Germination (%)		Root length (cm)		Shoot length (cm)		Vigour index	
	P_0	P_{12}	P_0	P_{12}	P_0	P_{12}	P_0	P_{12}
Arappu (*Albizia amara*)	87.3	58.0	20.76	15.00	20.53	12.16	3607	1576
Notchi (*Vitex negundo*)	87.3	55.3	18.40	14.23	19.76	12.46	3499	1477
Neem (*Azadirachta indica*)	86.7	51.0	17.96	14.36	21.23	10.26	3358	1273
Control	84.0	0.00	16.73	0.00	20.50	0.00	3103	0.00
Mean	86.4	34.0	18.39	10.90	20.50	8.73	3392	10.81
SEd	0.763		0.763		0.763		0.763	
CD (P=0.05)	1.536	1	10.977		0.296		133.5	

Seed coating with botanicals prior to storage showed significant influence on the seed storability of soybean. The beneficial effect of *Albizia amara* leaf powder in improving seed germination has been reported by Angamuthu (1991) who ascribed this to the presence of gibberellin like substance in addition to saponins and nutrients especially the micronutrients (Zn) which might have synergistically interacted into amino acids tryptophan to form Indole Acetic Acid (IAA).

Shenbaganathan (2001) conducted storage experiment in greengram CV. CO 6 after imposing hardening and pelleting treatments. The seeds hardened with 1 per cent prosopis leaf extract and pelleted with DAP 40 g + $ZnSO_4$ 100 mg + $FeSO_4$ mg + Ammonium molybdate 250 mg + Gypsum 300 g kg^{-1} of seed. Among the seed treatments, the hardened seeds stored better than the hardened cum pelleted seed (Plate 2).

The better efficiency of hardened seed might be due to the adequate drying given to the seed after hardening. In pelleted seed, the enclosed filler material might have absorbed more of the moisture from the surrounding atmosphere and should have acclerated the moisture absorption by the seed which deteriorates the germinability of the seeds at faster level. In addition, the macro and micro nutrients in combination with

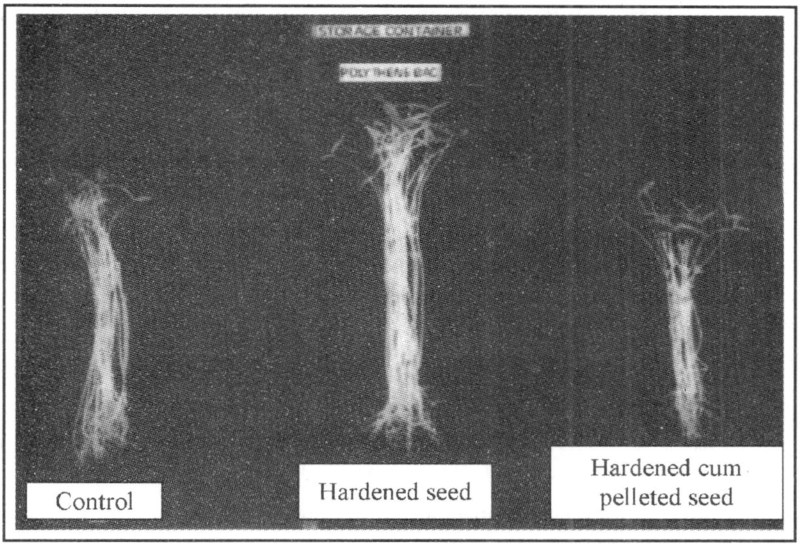

Plate 2. Seedlings characteristics after 12 months of storage as influenced by containers and seed treatments

gypsum could also have created an adverse toxic environment to the enclosed seed which could have damaged the seed coat.

Among the treatments, hardened cum pelleted seeds were poorer in maintaining seed quality attributes like shoot length, root length, dry matter percentage and vigour index while hardened seed performed better in maintenance. The seedling length recorded by hardened seeds were higher compared to untreated and hardened cum pelleted seed even 12 months of storage. Punithavathi (1997) in ragi, Jegathambal (1996) in sorghum and Nagaraj (1996) in rice also stored the hardened cum pelleted seed upto 5, 2 and 3 months respectively under ambient storage.

Based on the above proven facts, we can come to the conclusion that, seed pelleting provide an extra nutrients to uniform early germination and early vigour under field conditions and maintain seed viability in storage.

REFERENCES

1. Angamuthu, K. (1991). Studies on seed pelleting, physiological maturity, sowing quality and storage in small millets. M.Sc. (Ag.) Thesis, Tamil Nadu Agricultural University, Coimbatore.
2. Basaria Begam, J. (2001). Seed hardening cum pelleting studies in Blackgram cv. CO-5. M.Sc.(Ag.) Thesis, TNAU, Coimbatore-3.

3. Basu, R.N. (1976). Physico-chemical control of seed deteroration. *Seed Res., 4(1)* : 15-23.

4. Dharmalingam, C., K.Paramasivam and V.Sivasubramanian. (1988). Seed hardening to overcome adversity. The Hindu, No.16 (Wed.).

5. Jegathambal, R. (1996). Pre-sowing seed treatment to augment productivity of sorghum cv. CO 26 under rainfed agriculture. Ph.D. Thesis, Tamil Nadu Agric. Univ., Coimbatore.

6. Khan, A.A. (1980). In: The physiology and biochemistry of seed dormancy and germination. (Ed. A.A.Khan), North Holland Publishing Company, Amsterdam, New York, Oxford, p.127.

7. Kostantinov, G. and P.Petkov. (1982). Effect of pelleting on direct sown onion seeds. *Grandinarshai Lozarska Nauka, naak, 6*: 26-30.

8. Manonmani, V. (1997). Investigation on seed collection, storage and germination teak and sandal. Ph.D. Thesis, TNAU, Coimbatore-3.

9. Manonmani, V. (2002). Effect of coating with botanicals on control of seed quality deterioration in soybean (*Glysine max* L.). XI National seed seminar on Quality seed to enhance Agricultural profitability, January 18-20, 2002.

10. Mucke, J.D. (1987). The regulation of water transport in pelleted sugarbeet seed. *J. Agron. and Crop Sci., 161*: 79-83.

11. Nagaraj, S. (1996). Seed hardening and coating studies in upland rice (*Oryza sativa* L.). M.Sc.(Ag.) Thesis, Tamil Nadu Agric. Univ., Coimbatore.

12. Nargis, S. (1995). Influence of pelleting, magnetic treatment and radiation on the performance of differentially aged seeds in tomato (*Lycopersicon esculentum* Mill.) cv. PKM 1. M.Sc.(Ag.I) Thesis, TNAU, Coimbatore-3.

13. Punithavathy, N. (1997). Seed hhardening and pelleting studies in Ragi cv. CO 13. M.Sc.(Ag.) Thesis, TNAU, Coimbatore-3.

14. Roos, E. (1979). Storage behaviour of pelleted, tableted and taped lettuce seed. *Hort. Sci., 104(2)*: 283-288.

15. Sabir Ahamed, A. (1989). Studies on the production of quality seed and storage in soybean *Glycine max* (L.) Merrill. M.Sc. Thesis, Tamil Nadu Agricultural University, Coimbatore.

16. Saraswathi, G. (1993). Studies on seed pelleting in relation to sowing of cotton cv. LRA 5166 (*Gossypium hirsutum* L.). M.Sc.(Ag.) Thesis, Tamil Nadu Agric. Univ., Coimbatore.

17. Selvaraju, K. (1992). Studies on certain aspects of seed management practices for seed production under moisture stress condition in sorghum cv. CO 26. M.Sc.(Ag.) Thesis, Tamil Nadu Agric. Univ., Coimbatore.

18. Sevanan, J. (1988). Seed treatments on quality and storability on gingelly (*Sesamum indicum* L.). M.Sc.(Ag.) Thesis, Tamil Nadu Agric. Univ., Coimbatore.

19. Shenbaganathan. (2001). Seed management techniques for quality seed production of greengram (*Vigna radiata*) cv. CO 6 under dryland condition. M.Sc.(Ag.) Thesis, Tamil Nadu Agric. Univ., Coimbatore.

20. Srimathi, P., K.Malarkodi, R.Geetha and V.Krishnasamy. 2002. Nutrient pelleting to augment quality seed production in soybean. *Seed Research,30(2).*

21. Subbarao, L.V. (1984). Seed management for storage and production of transplant in brinjal. M.Sc.(Ag.) Thesis, Tamil Nadu Agric. Univ., Coimbatore.

22. Vartha, E.W. and P.T.Clifford. (1973). Effects of seed coating on establishment and survival of grasses surface sown on tussock grassland. *New Zealand J. Expt. Agric., 1(1)*: 39-43.

23. Vigneshwari. (2002). Seed hardening and pelleting in ragi cv. CO 13 to induce drought resistance and to improve productivity under rainfed condition. M.Sc.(Ag.) Thesis, Tamil Nadu Agric. Univ., Coimbatore- 3.

24. Vijaya, J. (1996). Standardization of pre-sowing seed management techniques for pulses. M.Sc.(Ag.) Thesis, Tamil Nadu Agric. Univ., Coimbatore.

25. Woodraff, D.R. (1969). Studies on pre-sowing drought hardening of wheat. *Aust. J. Agric. Res., 20*: 13-24.

PEST MANAGEMENT FOR STORAGE OF HARDENED AND PELLETED SEED

V. BALASUBRAMANI

1. INTRODUCTION

Insect pests of stored seeds are the descents of earlier insects which came into existence 250 to 500 million years ago. Insects that are associated with stored grain products constitute beetles, weevils and moths. Several hundred species of insects are associated with stored seeds and stored grain products but only 14 species are well adopted for living in stored grain and responsible for most of the damage. Minor pests constitute approximately 175 species of insects and mites that they rise to populations that are injurious to grain and grain products. The group of insects that damage stored seed can broadly be placed in two groups. The first group, internal feeders spend the most of their life, feeding within the kernel of grain and the second group, external feeders spend their life feeding on the surface of seed or on the finished products of grain (Table 1).

2. POST-HARVEST MANAGEMENT OF STORED SEED INSECT

Amongst the present methods of insect control, following are the important methods which can help in storage of seeds particularly at farmer's level (Gahukar, 1994).

1. Preventive measures
2. Curative measures

2.1. Preventive measures

"Prevention is better than cure". Hence the following preventive measures are recommended.

Centre for Plant Molecular Biology, Tamil Nadu Agricultyural University, Coimbatore – 641 003.

Table 1. Insects attacking stored seeds

Common name	Scientific name	Family	Order
A. Internal feeders			
Rice weevil	*Stiphilus oryzae*	Curulionidae	Coleoptera
Maize weevil	*Stiphilus oryzae*	Curulionidae	Coleoptera
Granary weevil	*itophilus zeamais*	Curulionidae	Coleoptera
Lesser grain borer	*Rhizopertha dominica* Fab	Bostrichidae	Coleoptera
Pulse beetle	*Callosobruchus maculatus*, (Fab) *C. chinensis* L.	Bruchidae	Coleoptera
Cigaratee beetle	*Lasioderma serricorne* (F)	Anobiidae	Coleoptera
Drug store beetle	*Stegobium paniceum* L.	Anobiidae	Coleoptera
Tamarind beetle or Groundnut bruchid	*Caryedon serratus* (Oliv.)	Bruchidae	Coleoptera
Angomois grain moth	*Sitotroga cerealella*	Gelechiidae	Lepidoptera
Grain lice	*Liposcelis divintorius L. transvallensis* (Enderlein)	Liposcelidae	Psocoptera
B. External feeders			
Khapra beetle (Everts)	*Trogoderma granarium*	Dermestidae	Coleoptera
Red floru beelte	*Tribolium castaneum* (Herbst). T. confusum	Tenebrionidae	Coleoptera
Saw toothed beetle	*Orzaephilus surinamensis*	Silvanidae (Linn)	Coleoptera
Flat headed flour beetle	*Latheticus oryzae* (Water)	Tenebrionidae	Coleoptera

Warehouse moth	*Ephestia cautella*	Phycitidae	Lepidoptera
Almond moth	*Plodia interpunctella* (Hub)	Phycitidae	Lepidoptera
Rice moth	*Corcyra cephalonica* (Staint)	Galleriidae	Lepidoptera

(i) Hygiene or sanitation
(ii) Disinfection of stores / receptacles
(iii) Legal method

Entry of an insect which is not found in the particular area can be prevented by the imposition of Destructive Insect Pests Act 1914.

2.2. Curative measures

The infestation of stored seed insect pests can be controlled by the following methods.

(i) Non-chemical control measures
(ii) Chemical control measures

2.2.1. *Non-chemical control measures*

The measures where chemicals are not used for control of insect pests of stored Seeds are:

a. Ecological control measures

The infestation of stored seeds by insect pests largely depend on the three factors *viz.*,

(i) Temperature
(ii) Moisture content of grain
(iii) Availability of oxygen

All these factors are required for normal rapid development and multiplication of insects. Hence, they have to be properly manipulated through design and construction of storage structures / godowns and storage practices so as to create ecological conditions unfavourable for attack by insects.

b. Mechanical control measures

Among other methods, mechanical methods are quite practicable. Several mechanical devices have been designed and developed both for monitoring and mass trapping stored product insects.The bahaviour of the stored product insects is exploited here. The various devices are discussed below.

Entoler

A direct method of controlling insect infestation is by removing insects from the infested commodity. Equipments used to remove dockages should also remove insects that are outside the seeds. This would not work for the immature stages of *Sitophilus* spp. or *Rhizopertha dominica*. Flour mills often pass flour through the sieves that remove foreign matter including insects, just before it is packaged. By far the most extensively used direct means of mechanical control are the Entolers, which use centrifugal force to impact insects or seeds containing insects. Entolers are used primarily in flour mills. Kernels infested with primary feeders such as *Sitophilus* sp. *R. dominica* break apart and bare separated from intact kernels. Apart from the above device, there are several traps which can be used for detection of insect infestation. They are:

1. Probe trap (Barak, 1989; Mohan and Gopalan, 1992)
2. Pulse Beetle Trap (Mohan, *et al.*, 1993)
3. Light traps (Kirtkpatrik, *et al.*, 1970; Mohan, 1993)
4. Sticky traps (Cogburn and Vick, 1981)
5. Bait traps (Pinniger, 1975)
6. Pheromone traps (Campion *et al.*, 1975; Pavan Kumar, 2001)
7. TNAU Automatic Insect Removal Bin (Mohan, 1994).

Recently, storage container models of 2 kg, 25 kg, 100g and 500 kg capacity which can remove insect automatically have been designed and developed by Tamil Nadu Agricultural University, Coimbatore, Tamil Nadu, India. The structure has four major parts namely outer container, innrt performed container, collection vessel and the lid. This device exploits wandering behaviour of stored product insects as well as the movement of these towards well aerated regions. The seeds are held in the specially designed inner perforated container, the space between inner and outer container provides good aeration for the insects.

Insects, while wandering, enter the perforation to reach the aerated part and while doing so, get slipped off and fall into the collection vessel through a pitfall mechanism in the collection vessel. In order to quickly collect the insects, as and when they emerge from seeds, perforated (2 mm) rods are fixed in the inner container (Mohan, 1994).

(i) Heat treatment
(ii) Control atmosphere
(iii) Mixing of inert dust
(iv) Use of plant products
(v) Use of activated clay

d. Cultural control measures

(i) Pre-harvest spray (Endosulfan 0.07 or Malathion 0.1%)
(ii) Time of harvesting
(iii) Resistant cultivars

e. Engineering control measures

(i) Hermatic or airtight storage

Seeds kept in airtight sealed structures remain insect free, reason being that during storage seeds respires and liberates carbon dioxide. Excessive CO_2 accumulation and depletion of oxygen do no allow the insects to survive.

(ii) Drying of Seeds

In grain having moisture content below 10% most of the insect sp. do not survive/ multiply. In villages the seeds are dried by spreading a thin layer of Seeds in the sun or by dryers. However, it should be kept in mind that too rapid drying or over drying may damage seeds, develop wrinkles, cause cracking of seed coat and destroy vitamins and seed viability.

2.2.2. *Chemical control measures*

Amongst the present methods of insect control, chemical control is the most and perhaps most effective one. They may be used for both type of treatments.

(a) Prophylactic treatment
(b) Curative treatment

(a) Prophylactic treatment

1. If the people is meant for seed purpose, mix 1 mg of activated kaolin or malathion 5% D for every / 100 kg of seed and store / pack in gunny or polythene lined bags.
2. Apply one of the following pesticides at the specified dosage over the bags. Malathion 50 EC : 10 ml per litre of water and 3 litres of spray

solution per 100 sq.m. DDVP 76% SC: 7 ml per litre of water and 3 litres of spray solution per 100 sq.m

3. Air change alleways or gang ways with one of the following chemicals. Malathion 50 EC : 10 ml/litre of water

   DDVP 76% SC : 7 ml/litre of water

   Apply one litre of spray solution for every 270 cu.m. or 10,000 cu feet.

   Spray the chemicals on the walls and floors and repeat the treatment based on the extent of flying and crawling insects.

4. Gunny bag impregnation : Empty bags are soaked in 0.1% malathion emulsion for 10 minutes and dried before using for seed storage.

(b) Curative treatment

1. Draw samples of seeds of fortnightly intervals and classify the infestation as follows.

   When there is not pest nil infestation

   Upto 2 insects - mid infestation

   More than 2 insects - severe infestation

2. Decide the need for shed fumigation (entire store house or godown) or cover fumigation (only selected blocks of bags)

   Choose the fumigant and work out the requirement on the following guidelines.

   I. Aluminium phosphate

   For cover fumigation 3 table of 3 grams each per tonne of grain

   For shed fumigation 21 tablets of 3 grams each for 28 cu. metres

   Period of fumigation 5 days

REFERENCES

1. Barak, A.V. (1989). Development of a new trap to detect and monitor khapra beetle (coleoptera : Dermestidae). *J.Econ. Entomol., 82* : 1470-1477.
2. Campion, D.G., D.R. Hall and P.F. Prevett. (1975). Use of pheromones in crops and stored products pests management control and monitoring. *Insect Sci. Appl. 8*: 737-741.
3. Cogbum, R.R. and K.W. Vick. (1981). Distribution of Angoumois grain moth, almond moth and Indian meal moth in rice fields and rice storages in Taxes as indicated by pheromone baited adhesive trap. *Environ. Entomol. 10*: 1003-1007.

4. Gahukar, R.T. (1994). Storage of food Seeds and insect control in developing countries. *Insect Sci. Applic. 15 (4/5)* : 383-400.
5. Krikpatrick, R.L., D.L. Yancey and F.O. Markze. (1970). Effectiveness of green and UV light in attracting stored product insects to traps. *J. Econ. Entomol., 63*: 1853-1855.
6. Mohan, S. (1993). Studies on the detection and management of stored product insects of rice. Ph.D. Thesis, Tamil Nadu Agricultural University, Coimbatore.
7. Mohan, S. (1994). A storage container model for automatic insect removal of stored product insect from seeds. *Bulletin of grain technology, 32(2)*: 178-181.
8. Mohan, S. and Gopalan. (1992). Stored grain insect trap. *Rice India, 2(8)*: 15-16.
9. Mohan, S.A Janagarajan and M. Rangarajan. (1993). A device to trap pulse beetle trap. *Indian Food Ind., 12 (3)* : 45.
10. Pavan Kumar, S. (2001). Studies on the detection techniques for management of stored product insect pests. M.Sc., Thesis, Tamil Nadu Agrl. University, Coimbatore.
11. Pinniger, D.B. (1975). A bait trap technique for assessment of stored product beetle production. *Trop Stored Prod. Infn, 9*: 9-11.

DISEASE MANAGEMENT FOR STORAGE OF HARDENED AND PELLETED SEED

K. PRABAKAR

1. INTRODUCTION

Seeds are basic input for crop production. The major world food crops *viz.*, barely, beans, maize, millet, peanut, pulses, rice, sorghum, sugarbeat, wheat and fibre crop such as cotton and vegetable crop such as cabbage, carrot, tomato are attacked by a number of seed borne pathogens and has significance as far as seed health is concerned.

Plant diseases cause significant yield losses and the global loss due to plant disease is estimated to be 12 per cent of potential production, which is equivalent to a monetary loss of $ 50 billion at the producer level; interms of quantity, this is approximately 550 million tones inspite of the use of fungicides costing $ 1 billion. The loss of cereals is 135 million tonnes. Losses to disease in different regions of the world vary from 30 per cent in the agriculturally developing countries of Asia to 25 per cent in Europe and 15 per cent in North America. Losses caused by the seed borne inoculum or diseases have not been quantified.

2. DISCOLOURATION AND SHRIVELLING

Seed borne pathogens cause discolouration of a part or whole of the seed. Due to discolouration the market value of the seed is reduced.

3. REDUCTION AND SHRIVELLING

Seeds after sowing start germinating and with the activity of germination of seeds, seed borne pathogens becomes active which may

Department of Seed Science and Technology, Tamil Nadu Agricultural University, Coimbatore – 641 003.

cause seed decay or pre or post emergence damping – off. Eventually the plant stand is poor in the field. The reduction in germination depends upon the variety / cultivar, type, amount and location of inoculum , environmental condition and other factors.

4. DISEASE DEVELOPMENT

Seeds are most important means of perpetuation of plant pathogens. Pathogens of some diseases exclusively survives in seeds. Pathogens remain viable longer in seed than in vegetable plant parts or in soil. Transmission of a pathogen through seed is considered more important than other survival means. The host parasite relationship within the seed also forms the earliest possible infection in field. Since the pathogens are in direct contact with the seeds, the changes of seedling infection are enhanced. The inoculum of the pathogen from seed may spread fast under favourable conditions and cause an epidemic.

On farm, where diseased plants / contaminated seeds are used as cattle feed and diseased plants for bedding, the spreading of manure can introduce the pathogen to uncontaminated fields. A pathogen can be introduced to new areas through seeds. In early times the movement of pathogens was restricted to certain areas because of limited means of travel, but today the changes of spread through seeds to large areas have increased due to expanded and rapid modes of transport. In exchange of germplasm / plant materials, the risk of introducing of new strain / physiologic races of a pathogen or new pathogen from one country to other country has increased.

5. BIOCHEMICAL CHANGES IN SEEDS

Various seed borne pathogens brings about qualitative changes in the physical and chemical properties of seeds such as colour, odour, oil content, iodine value, refractive index and protein content. These changes ultimately affect the commercial value of seeds.

6. DETECTION METHODS OF SEED BORNE PATHOGENS

Healthy and viable seed is the first pre-requisite for increasing seed production and to reduce possible seed crop failures. Until and unless we do not know the health status of the seed, it is not possible to manage the disease. To know health status of the seed, different testing methods for different pathogens have been developed.

6.1. Fungi

Examination of dry seeds

Seed washing test
NaoH seed soak method
Blotter method
Agar plate method
Seedling symptom test
Embryo count method
Fluorescence method

6.2. Bacteria

Examination of dry seeds
'Growing on' test
Isolation in agar medium
Indicator test
Dome test
Bacteriophage method
Serological techniques

6.3. Virus

Examination of dry seeds
'Growing on' test
Indicator plant test
Serological tests
a. Microprecipitin test
b. Double diffusion
Latex agglutination
ELSIA
Dotblot
ISEM (Immunosorbent Electron Microscopy)
Electron microscopic examination
PCR technique

PART VI

NETWORK EXPERIMENTS / ON STATION TRIALS / ON FARM TRIALS-RESULTS AND EXPERIENCES

SEED HARDENING AND PELLETING IN SORGHUM AND BAJRA

K. VANANGAMUDI[1], R. JERLIN[2], R. UMARANI[3] AND C. MENAKA[3]

1. INTRODUCTION

Ever increasing anthropoid pressure in the country necessitates manifold increase food production. This is possible, if Indian Agricultural Scientists strive towards a second green revolution. The "Second Green Revolution" could be achieved by rainfed farming only. For making this "Green Revolution under Rainfed Agriculture" a reality,the development of pre-sowing seed treatments for better survival and establishment of crop plant under rainfed conditions is highly essential and appropriate. Pre-sowing hardening (or) imbibition and drying of seeds is one of the methods which result in modifying physiological and biochemical nature of seed so as to get the characters that are favourable nutrients, water absorbants and pesticides. These techniques have been tried in sorghum and cumbu with the following objectives.

1. To study the effect of hardening and pelleting on seed germination and seedling vigour
2. To evaluate the effect of seed hardening and pelleting on productivity of sorghum and cumbu under rainfed conditions
3. To predict the storability of hardened and pelled seeds.

Based on the results on farm trail was conducted with the processed seeds of sorghum and cumbu were soaked in different chemicals and

1 *Dean (Agriculture), Tamil Nadu Agricultural University, Coimbatore - 641 003.*
2 *Agricultural Research Station, Bhavanisagar – 638 451.*
3 *Department of Seed Science and Technology, Tamil Nadu Agricultural University, Coimbatore- 641003*

botanicals of various concentration for different durations *viz.*, 6 hours for cumbu and 16 hours for sorghum. For soaking the seeds, a seed to solution ratio of 1:0.6 was adopted. After soaking, the seeds were shade dried to its original moisture content of 8%. The hardened seeds were pelleted using $ZnSO_4$, DAP powder and gypsum. The DAP granules were powdered and sieved to get fine powder for proper pelleting. As an adhesive for pelleting, 10% maida gruel @ 50 ml per kg of seeds was used. DAP, $ZnSO_4$ and gypsum were used @ 100 g, 100mg and 300g for sorghum and 250 g for cumbu respectively. The seeds were pelleted thoroughly without forming any aggregates using seed pelletizer. The pelleted seeds were shade dried for 24 hours and used for sowing. The treatment details of hardening and pelleting for sorghum and cumbu seeds are as follows

T_1	:	No seed soaking
T_2	:	Seed hardening with 2% KH_2PO_4 (20 g in 1 litre of water) for 15 hrs followed by 5 hours shade drying / 2% KCl for 6 h
T_3	:	Hardening with 0.1 ppm brassinolide for 16 hours
T_4	:	Hardening with 200 ppm salicylic acid for 16 hours
T_5	:	Hardening with 2% KH_2PO_4 for 16 hours in equal volume of water + seed coating with DAP 100 g/kg
T_6	:	Hardening with 1% *Prosopis* leaf extract for 16 hours + seed coating with DAP 100 g + $ZnSO_4$ 100 mg/kg of seed
T_7	:	Hardening with brassinolide + KH_2PO_4 and seed coating with DAP 100g + $ZnSO_4$ 100 mg/kg of seed
T_8	:	Hardening with brassinolide + KH_2PO_4 and seed coating with DAP 100g + $ZnSO_4$ 100 mg/kg of seed
T_9	:	Hardening with brassinolide + KH_2PO_4 *Prospis* + seed coating with DAP 100g + $ZnSO_4$ 100 mg/kg of seed

2. SORGHUM

Seed hardened with KH_2PO_4 recorded the maximum value in physiological parameters. Similar results were reported by Nagaraj (1996) in paddy. The potassium present in the chemicals improves the oxidative phosporylation, utilization of sugars along with formation of $NADPH_2$ and ribose 5. Phosphate synthesis of mitochondria (Kursanor *et al.*, 1995) and the activity of ATPase enzyme in sugar beet. This beneficiary effect of nutrients might be the reason for the better seedling growth. It might be also due to enhanced metabolic activity and earliness in germination. Seed hardened with KH_2PO_4 significantly increased the enzyme activity and also chlorophyll content. Similar results were reported in wheat.

3. CUMBU

All the hardening and pelleting treatments showed significant improvement in seed germination and vigour over control. Among them, seeds hardened with 0.1 ppm brassiolide gave higher germination (96%) and enhanced seedling growth. Seed hardening with brassinolide might have stimulated the production of auxin and ethylene which have positive influence on seed germination (Clouse and Zureta, 1991). Brassinolide also enhanced the seedling growth. The probable reason for elite seedling growth might be due to cells elongation and cell division and enhancement in enzyme activities as reported by Hayal *et al.* (2001). The increase in seedling growth reflects in higher drymatter production and vigour. Similar findings were reported by Paul and Choudhury (1993) in wheat. Seed hardened with brassinoilide exhibited the maximum and uniform field emergence are survival. This might be due to the stimulation of auxin by brassiolide which in turn helps the induction of cell differentiation, cell elongation, cell division and other enzyme activities which promotes germination. Improvement in field emergence by brassinolide was also reported by Vigneshwari (2001) in ragi. Seed hardening with brassinolide significantly increased the plant height (155.30 m). The increase in plant height by the use of growth regulators has been reported by Kannaiyan (1987) and Sandhu and Hussain (1961) in pearl millet. The increase in plant height by brassinolide might be attributed to its powerful synergistive interaction with available indigenous auxin and it could be observed in terms of increased cell wall plasticity and cell elongation (Yopp *et al.*, 1981). Beneficial effects of brassinolide on yield was noticed and the increase was 13 per cent higher than the control. This is in agreement with the results of Misra and Reddy (1985) in wheat and Kannaiyan (1987) in pearlmillet. The increase in yield might be due to more number of grains per ear and higher grain weight.

As a network experiment, same trial was conducted at 4 different locations viz., Coimbatore, Aruppukkottai, Kovilpatti and Paiyur. In all the centres, cv. CO 26 was used as a test variety except Kovilpatti where the variety tested was K8. The experiment was conducted during Sep. 2000 adopting RBD with four replications. The treatments tried were hardening with 2% KH_2PO_4 (T_2), hardening with 1% *Prosopis* leaf extract (T_3), seed coating with DAP 100g Kg^{-1} of seed (T_4), combination of T_2+T_4 (T_5) and hardening with 1% *Prosopis* leaf extract followed by seed coating with DAP (100g kg^{-1} of seed and $ZnSO_4$ (100 mg kg^{-1} of seed) (T_6). The untreated seed formed the control (T_1). The crop was raised adopting the recommended package of practices. Immediately after sowing, field emergence percentage was recorded and at harvest, grain yield was recorded (kg $plot^{-1}$) and computed to grain yield (kg ha^{-1}). Studies conducted during Sep. 2001 at Coimbatore expressed that seed hardening with 0.1 ppm brassinolide + 2% KH_2SO_4 + 1% *Prosopis* and seed coating with DAP + $ZnSO_4$ (T_9) improved the seed yield by 24.2 per cent. At Aruppukkottai seed hardening with 2% KH_2SO_4 followed by pelleting with DAP @ 100g/kg of seed (T_5) maximized the seed yield (1950 kg/ha) compared to control (1458 kg/ha). The percentage increase was 33.1.

The network experiment was repeated during Sep. 2001 at Aruppukkottai, Kovilpatti, Coimbatore, Virudhachalam and Paiyur to confirm the findings of 1 year experiment. The recommendation give in the Crop Production Guide was included as a check. Due to unfavourable weather condition, crop raised in Kovilpatti, Virudhachalam and Paiyur have failed to establish. Hence, the experiment could be conducted only at Aruppukkottai and Coimbatore. Sorghum cv. K8 and cv. COS 28 were used as test varieties in Aruppukkottai and Coimbatoe, respectively. the experiment was conducted during Sep-Oct, 2001 adopting RBD design. The experiment details are furnished below.

The grain yield was the highest in 2% KH_2PO_4 hardened seed (1170 kg ha^{-1}) at Coimbatore and Paiyur (1379 kg ha^{-1}), while at Kovilpatti the higher yield was recorded by hardening with 1% *Prosopis* leaf extract followed by DAP pelleting (1850 kg ha^{-1}) which was on par with 2% KH_2PO_4 hardening (1810 kg ha^{-1}). At Aruppukottai, the yield was higher in 1% *Prosopis* leaf extract hardening followed by DAP + $ZnSO_4$ pelleting (1620 kg ha^{-1}). The next best treatment was 1% *Prosopis* leaf extract hardening + DAP pelleting (1530 kg ha^{-1}). From the present study it has been understood that seed hardening with 2% KH_2PO_4 improved the grain

T_1	:	No seed soaking (Control)
T_2	:	Seed hardening with 2% KH_2PO_4 (20 g in 1 litre of water) for 16 hours followed by 5 hours shade drying
T_3	:	Hardening with 0.1 ppm brassinolide for 16 hours
T_4	:	Hardening with 200 ppm salicylic acid for 16 hours
T_5	:	Hardening with 2% KH_2PO_4 for 16 hours in equal volume of water + seed coating with DAP 100 g/kg
T_6	:	Hardening with 1% *Prosopis* leaf extract for 16 hours + seed co ating with DAP 100 g + $ZnSO_4$ 100 mg/kg of seed
T_7	:	Hardening with brassinolide + KH_2PO_4 and seed coating with DAP 100g + $ZnSO_4$ 100 mg/kg of seed
T_8	:	Hardening with brassinolide + KH_2PO_4 and seed coating with DAP 100g + $ZnSO_4$ 100 mg/kg of seed
T_9	:	Hardening with brassinolide + KH_2PO_4 *Prospis* + seed coating with DAP 100g + $ZnSO_4$ 100 mg/kg of seed

yield of sorghum in all the centers except at Aruppukottai, where seeds hardened with 1% *Prosopis* leaf extract followed by pelleting with DAP (100g kg of seed^{-1}) and $ZnSO_4$ (100mg kg of seed^{-1}) excelled well. However, in this centre also the KH_2PO_4 hardened seeds was found to bye superior over control recording an increased yield in all the centres. Hence, from this study, it is concluded that seeds hardened with 2% KH_2PO_4 to maximize the yield of sorghum under dryland condition. The following two treatments were appeared to be the best treatment. (i) Seed hardening with 2% KH_2PO_4 for 16 h followed by 5 h shade drying (ii). Seed hardening wiht2% KH_2PO_4 + seed coating with DAP 100g/kg.

The study conducted at Coimbatore, Kovilpatti and Aruppukottai behaved differently for cumbu. At Coimbatore seed hardening with brassinolide (T3) improved the seed yield by15 per cent (Plate 1) while at

Kovilpatti seed hardening with 0.1 ppm brassinolide followed by seed coating with DAP and $ZnSO_4$, (T9) improved the seed yield by 26.7 per cent. At Aruppukkottai, seed hardening with 1% *prosopis* leaf extract followed by seed pelleting with DAP 100g + $ZnSO_4$ 100mg / kg seed (T6) maximized the seed yield by 31.07 per cent.

Plate 1. Field performance of cumbu seeds hardened with 0.1 ppm brassinolide for 16 h

REFERENCES

1. Clouse, S.D., and D. Zureta. (1991). Molecular analysis of brassinolide action in plant growth and development. In B cassinosteriods chemicals. Bioactivity and application. Cutter, H.G.T. Yohata and G. Adam (eds) Washingon, D.C.AM., *Chem. Soci.,* : 120-140.
2. Hayal, S., A. Ahmad, A. Hussain and M. Mobin. (2001). Gropwth of wheat seedlings raised from grains treated with 28 Brassinolide. *Acta Physiologie Plantarum, 23 (1)*: 23-30.
3. Kanniyan, M. (1987). Effects of dates of sowing and seed hardening on yield of pearlmiller (*Pennisetum typhoides* staff and Hubb) under rainfed condition. M.Sc. (Ag.), Thesis, TNAU, Coimbatore.
4. Kursnor, A.L. E. Byskrebentsera, I. Sveshikova and M. Krasavina. (1995). Disorganisation of energy metabilism in roots during potash metabolism Doklady *Akead Nauk SSR, 162* : 9.
5. Misra N.M. and M.G.R.K. Reddy. (1985). Effect of cycocel, ascoirbic acid and zinc sulphate on the yield attributes and yield of rainfed wheat. *Andhra Agric. J., 32*: 32-33.

6. Nagaraj, S. (1996). Seed hardening and coating studies in upland rice (*Oryza sativa* L.) M.Sc. (Ag.) thesis, Tamil Nadu Agricultural University, Coimbatore.

7. Paul, S.R. and A.K. Choudhury. (1993). Effect of seed hardening with potassium salts of different concentrations and soaking durations on germination and seedling vigour of wheat. *Ann. Agric. Res., 14(3)*: 354-359.

8. Sandhu, A.S. and A. Hussain. (1961). Effect of seed treatment with gibberellic acid on germination and growth of bajra. *Indian J. Agron., 5* : 269-272.

9. Vigneshwari. (2002). Seed hardening and pelleting in ragi cv. CO 13 to induce drought resistance and to improve productivity under rainfed condition. M.Sc.(Ag.) Thesis, Tamil Nadu Agric. Univ., Coimbatore- 3.

10. Yopp, J.H. N.B.Mandara and J.M. Sasse. (1981). Brassinolide a growth processing steroidal lactone. *Physiol. Plant.,* : 53 – 55.

SEED HARDENING AND PELLETING IN RAGI

P. NATESAN[1], R. VIGNESWARI[1], A. BHARATHI[1],
K. VANANGAMUDI[2], C. MENAKA[1]
AND M. BHASKARAN[3]

1. INTRODUCTION

Finger millet (*Eleusine coracana* (L) Gaextn) is the most important among the small or minor millets. In India, small millets occupy an area of around 4.5 m ha, accounting for nearly 15-18 per cent of area under coarse cereals. Among small millets, India has 2 m ha under ragi cultivation with a production of 2.6 million tonnes. In Tamil Nadu, it is being cultivated in 1,26,955 ha, which contributes to a total production of 2,59,490 tonnes. Among the total ragi cultivated area in Tamil Nadu, 92,495 ha is under rainfed condition. The yield potential of any crop in dryland condition is decided by various factors like climate, agronomic practices, nutrient status of the soil and quality of input materials especially seed. To boost the productivity under rainfed conditions, all the above said factors had to be managed carefully. Since agronomic practices and nutrient status can be easily managed, much emphasis has to be given to over come moisture stress which is a major constraint in dryland farming. Moisture stress during initial establishment affects the crop stand and so the productivity. Seed hardening and pelleting is a presowing seed treatment which makes the seed to withstand drought during early phase of germination and seedling growth. Seed hardening using chemicals and botanicals were recommended normally. There are several physiological studies which enumerate the role of growth hormones in increasing the productivity of crops. So, an attempt was made with growth regulators

1 *Department of Seed Science and Technology, Tamil Nadu Agricultural University, Coimbatore - 641 003.*
2 *Dean (Agriculture), Tamil Nadu Agricultural University, Coimbatore - 641 003.*
3 *National Seed Research and Training Centre, Varanasi -221 106.*

like brassinolide and salicylic acid and their efficacy as hardening materials and compared with chemicals like calcium chloride, potassium dihydrogen phosphate and botanicals like prosopis leaf extract. Dry farming relies mainly on soil moisture and seasonal rain often encounters deficits in soil moisture and struggles for survival and germination. Several studies have proved the importance of seed hardening and pelleting in drought resistance. But the studies on storage potential of hardened and pelleted seeds were less. Hence, studies were conducted on hardening and pelleting in ragi seeds to trace the influence of these treatments both on productivity and storability. The genetically pure seeds were thoroughly cleaned and processed using BSS 12 x 12 sieve. The retained seeds were used for experimentation. The processed seeds were soaked in different chemicals and botanicals of various concentrations for 6h in a seed to solution ratio of 1: 06. After soaking, the seeds were shade dried to its original moisture content of 10%. The hardened seeds were pelleted using $ZnSO_4$ and DAP powder. The DAP granules were powdered and sieved to get fine powder for proper pelleting. As an adhesive for pelleting, 10% maida gruel @ 50 ml per kg of seeds was used. DAP and $ZnSO_4$ were used @ 100g and 100mg respectively. The seeds were pelleted thoroughly without forming any aggregates using seed pelletizer. The peleted seeds were shade dried for 24 h and used for sowing.

T_1	:	Control (Unsoaked seeds)
T_2	:	Seeds hardened with calcium chloride (0.5%)
T_3	:	Seeds hardened with Brassinolide (0.1 ppm)
T_4	:	Seeds hardened with salicylic acid (200 ppm)
T_5	:	Seeds hardened with KH_2SO_4 (2%) + DAP (100g) powder pelleting
T_6	:	Hardening with 1% *Prosopis* leaf extract (1%) +DAP (100g) and $ZnSO_4$ (100 mg) pelleting.
T_7	:	Seeds hardened with Brassin olide (0.1ppm) and *Prosopis* leaf extract (%) + DAP (100g) and $ZnSO_4$ (100mg) pelleting
T_8	:	Seeds hardened with Brassinolide (0.1ppm) and *Prosopis* leaf extract (%) + DAP (100g) and $ZnSO_4$ (100mg) pelleting
T_9	:	Seeds hardened with Brassinolide (0.1 ppm), KH_2PO_4 (2%) and *Prosopis* leaf extract (%) + DAP (100g) and $ZnSO_4$ (100mg) pelleting.

The treated seeds were sown under rainfed conditions of Coimbatore to evaluate the productivity. In addition, the seeds were also stored under ambient conditions of Coimbatore in both cloth and 700 gauge polyethylene bags to evaluate their storability.

2. HARDENING AND PELLETING ON PHYSIOLOGICAL AND BIOCHEMICAL CHARACTERISTICS

All the hardening and pelleting treatments showed significant improvement in seed germination and vigour over control. Among which, seed hardened with 0.1 ppm brassinolide gave higher germination (95%) and vigour index value (1549). The improvement in germination by this treatment was 11 per cent over control. Seed treatment with brassinolide might have stimulated the production of auxin the ethylene which have positive influence in seed germination as reported. The rate of germination of hardened seeds were higher than pelleted seeds. The quick initiation of germination of hardened seeds might be due to advancement in the physiology of germination during soaking. The reduction in germination rate of pelleted seeds might be due to the restriction imposed by the pelleting material which causes slow adsorption of moisture and so delayed germination. The soluble protein and seedling chlorophyll content were higher than the control. Similar results were also recorded in wheat. Improvement in soluble protein content favours drought resistance. The increased chlorophyll content increases the photosynthetic activity which aids in higher yield. Improvement in physiological and biochemical parameters of hardened and pelleted seeds over control exhibits its superiority in drought resistance capacity.

3. SEED HARDENING AND PELLETING ON PRODUCTIVITY

Seeds hardened with brassinolide exhibited the maximum and uniform field emergence and survival (Plate 1). This might be due to the stimulation of auxin by brassinolide which in turn helps the induction of cell differentiation, cell elongation, cell division and other enzyme activities which promotes germination. The maximum drymatter accumulation and plant height were recorded by the seeds hardened with brassinolide. Increase in drymatter production by the use of brassinolide was also reported in wheat seeds. Use of growth regulators increase the uptake of nutrients like N and P which aid in more effective root growth and leaf area which has positive influence on dry matter production. The increase in plant height by brassinolide might be attributed to its powerful synergistic interaction with available endogenous auxin and it could be observed in terms of increased cell wall plasticity and cell elongation (Table 1).

Plate 1. Field view of ragi seed hardened with Brassinolide (0.1 ppm)

Table 1. Effect of hardening and pelleting on field emergence, survival percentage drymatter production and plant height in ragi cv CO 13.

Treatments	Field emergence (%) (4 DAS)	Survival (%) 30 DAS	Seed yield (kg/plot)	Seed yield (kg/ha)
T1	49	68	3.483	2322
T2	52	78	3.683	2455
T3	82	84	4.184	2789
T4	74	72	3.531	2354
T5	58	76	3.826	2551
T6	56	74	3.742	2495
T7	58	76	3.862	2575
T8	62	78	3.983	2655
T9	62	82	4.128	2751
Mean	61	76	3.825	2450
SEd	2.193	0.972	0.090	59.695
CD (P = 0.05)	6.176	1.636	0.191	127.123

Beneficial effects of brassinolide on yield was noticed in all the treatments with brassinolide as one of the components. This is in agreement with the results, in wheat and pearlmillet. The increase in yield might be due to more number of grains per ear and higher grain weight. The increased carbohydrate content at flowering due to brassinolide application was related to higher rate of photosynthesis per unit leaf area and regulation of source sink relation for assimilate transport.

4. STORABILITY OF HARDENED AND PELLETED SEEDS

The importance of storage potential of hardened seeds arises in the case of expected monsoon failure. The performance of hardened and pelleted seeds under storage revealed that the reduction in germination (Table 2) was considerably higher in hardened and pelleted seeds when compared to control and similar results were reported in paddy. The hardened seeds reached the minimum certification standard of 75 per cent germination with

Table 2. Influence of hardening and pelleting and containers on germination (per cent) of seeds during storage in ragi cv. CO 13

Treatments	Cloth Bag (C1)			Polythene bag (C2)			T Mean
	P0	P6	Mean	P0	P6	Mean	
T1	84 (66.45)	77 (62.03)	82 (64.53)	84 (66.45)	80 (63.44)	82 (65.03)	82 (64.78)
T2	93 (74.76)	62 (51.95)	76 (61.50)	93 (74.76)	67 (64.95)	80 (63.83)	78 (62.66)
T3	95 (77.12)	60 (50.77)	75 (51.95)	95 (38.09)	65 (52.55)	78 (57.52)	72 (59.24)
T4	86 (68.04)	52 (46.15)	69 (56.24)	86 (68.04)	56 (48.45)	72 (58.22)	70 (57.23)
T5	92 (73.59)	33 (33.21)	60 (51.37)	92 (73.59)	39 (38.64)	63 (53.97)	62 (52.67)
T6	91 (72.56)	28 (31.94)	57 (49.85)	91 (72.56)	40 (39.23)	63 (53.97)	60 (51.49)
T7	92 (73.59)	30 (33.21)	59 (50.84)	92 (73.59)	41 (39.81)	64 (54.19)	62 (52.67)
T8	94 (75.86)	29 (32.58)	60 (51.82)	94 (75.86)	42 (40.40)	57 (55.82)	63 (53.82)
T9	94 (75.86)	28 (31.94)	59 (51.13)	94 (75.86)	40 (39.23)	66 (55.15)	62 (52.67)
Mean	91 (73.10)	44 (41.55)	66 (54.34)	81 (72.56)	52 (46.30)	71 (57.43)	

in two months of storage in cloth bag and three months in 700 gauge polythene bag while pelleted seeds reached the 75 per cent germination one month before hardened seeds in both the containers. The imbibition process that occurred during hardening might have lead to faster biochemical degradation of hardened seeds. Seed deterioration alters the semi permeability properties of the membrane and membrane integrity. Increase in electrical conductivity is a good indicator of loss of membrane integrity which attributes to the seed viability. Higher electrical conductivity of seed steep water of hardened and pelleted seeds might be due to its faster deterioration. Thus the study highlighted that

1. Among different treatment combinations, seeds hardened with brassinolide 0.1 ppm has positive influence on drought resistance which was evident from its increased physiological and biochemical aspects of seed. Seeds hardened with 0.1 ppm brassinolide produced higher seed yield with high quality. The hardened and pelleted seeds can be stored upto 1 month in cloth bag and 2 months in polythene bag with the minimum certification standard of 75% germination under ambient condition.

SEED HARDENING AND PELLETING IN GREENGRAM

R. UMARANI[1], G. SASTHRI[2] AND P. SRIMATHI[3]

1. INTRODUCTION

Greengram, the major pulse is extensively grown in all types of soil under varying climatic conditions. It is presently grown in all the states of India. Greengram seeds serve as main resource in many food products. The sprouted seeds are consumed as salad which are rich in vitamins particularly ascorbic acid, niacin, riboflavin and thiamine and also rich in minerals particularly calcium, iron and potassium. The average protein content of greengram is 24 per cent, more importantly, greengram has higher digestibility (79 per cent) and biological value (72 per cent). The estimated global production of greengram is 2.5 to 3 million tonnes contributed from an area of 5 million hectares. In India, the area under greengram is 2.73 million hectares with an annual production of 1.1 million tonnes. The average yield is 0.41 tonnes per hectare. The breeder seed production in India during 2000-2001 is 391.42 and for Tamil Nadu it is 3.382. The meagre yield in pulses is due to the fact that pulses are cultivated almost in marginal and rainfed areas. The major constraints for higher productivity of pulses in drylands are the inadequacy of soil moisture and poor fertility status of the soil. To increase the productivity, various presowing physiological seed treatments through chemical fortification were suggested.

Seed hardening is one of the presowing seed management practice recommended for dryland agriculture to resist the seed against the adverse

1 *Department of Seed Science and Technology, Tamil Nadu Agricultural University, Coimbatore – 641 003.*
2 *Department of Plant Breeding and Genetics, National Pulses Research Centre, Vamban – 622 303*
3 *Department of Seed Science and Technology, Forest College and Research Institute, Mettupalayam – 641 301.*

agro ecological conditions, the seed faces at the time of sowing until the seed put forth seedling on availability of moisture from rain, the natural source of irrigation under dryland conditions. During the hardening treatment, the drought resistance in seed is created by water or chemicals or botanicals, based on its suitability for invigouration and productivity which varies with crop and lot used for sowing. Hardening of seeds also said to result in the absorption of more water due to increase in the elasticity of cell wall and development of a stronger and efficient root system. Seed pelleting, the process of enclosing the seed into a small quantity of inert material just large enough to produce a globular unit of standard size to facilitate precision planting is another presowing seed management practice towards productivity. But pellets should always be a stable unit that easily disintegrate to release the seed. Birader (1988) narrated the importance of pelleting, in regulating the size of seeds for precision planting in reducing the amount of seeds required to plant and in elimination of thinning the crop. He also stated that along with pelleting materials, bacterial inoculants, nutrients, fungicides, pesticides etc., could be embedded with inert material so that cost of production could be reduced while the yield and the quality of the crop could be improved. The bioactive materials used as seed coating materials were reported to improve the germination and increase the seedling emergence at changing soil moisture, especially in the suboptimal range in greengram.

Therefore, network experiments on seed hardening and pelleting in greengram cultivars *viz.*, CO6, CO4, K1, Paiyur 1, were carried out in different centres *viz.*, Coimbatore, Madurai, Aruppukottai and Paiyur respectively during 2001. The treatmental details are

T_1	-	Untreated seeds
T_2	-	Seed hardening with $MnSO_4$ 100 ppm
T_3	-	Seed hardening with 1 per cent prosopis (*Prosopis juliflora*) leaf extract and seed pelleting with DAP 40 g + $ZnSO_4$ 100 mg + $FeSO_4$ 100 mg + Ammonium molybdate 250 mg

2. SEED HARDENING

Seeds were preconditioned in between moist gunny bags for 2 h in water and soaked in 1/3 volume of 100 ppm of $MnSO_4$ and one per cent prosopis leaf extract for 4h and shade dried to the original moisture content.

3. SEED PELLETING

The filler materials selected for pelleting were finely ground. Using maida flour, 10 per cent gruel was prepared as adhesive, on boiling the flour with water. Using this adhesive the seeds hardened with one per cent prosopis leaf extract were pelleted with nutrients using seed pelletizer (Plate 1).

Plate 1. Unpelleted and pelleted seeds of greengram

The hardened cum pelleted seeds along with untreated seeds were observed for seed and seedling quality parameters in the laboratory. The hardened cum pelleted seeds took lesser days to complete their germination than rest of the seeds. Evaluated germination and vigour parameters expressed that the above seeds recorded 2.1 per cent higher germination than untreated seed. The hardened cum pelleted seeds also produced the seedling with maximum length and drymatter production. The computed values for vigour index were higher in hardened cum pelleted seeds by 17 per cent than control. Similary, field emergence was also higher in the hardened cum pelleted seeds. This could be due to the modification of physiological and biochemical nature of the seeds so as to get the characters that were favourable for drought resistance. The probable reason for the higher germination of prosopis leaf extract hardened cum pelleted seed could be of greater hydration of colloids, higher viscosity and elasticity of protoplasm, increase in bound water content, lower water deficit and more efficient root system (May *et al.*, 1962). The increased seedling growth and dry weight observed in this treatment might also be due to greater early vigour and higher percentage of germination because of which the seedlings had reached autotropic stage well in advance the others. The increase in dry weight was claimed to be due to enhanced lipid utilization through glyoxalate cycle, a primitive pathway leading to faster growth and development of seedling to reach autotropic stage well in advance of others and enabling them to produce relatively more quantity of drymatter (Jayaraj, 1977). The amylase (hydrolase) and peroxidase (desmolase), enzyme activity were higher in hardened cum pelleted seeds which could be attributed to bioactive chemicals present in prosopis leaf extract (Rathinavel and Dharmalingam, 1999). Then the hardened and pelleted seed along with control were sown in field and recommended package of practices were followed. At Coimbatore centre in the CO 6 cultivar the following characters were evaluated (Table 1).

Table 1. Influence of hardening cum pelleting treatment on seed yield and quality characters of greengram

Characters	Untreated	Hardening	Hardening cum pelleting
100 seed weight (G)	3.470	3.483	3.774
Germination (%)	87	88	89
Field emergence (%)	89	90	92
No. of seeds / plant	16	22	25
Pod yield plant^{-1} (gm)	6.320	7.735	8.340
Seed yield plant^{-1} (g)	5.203	5.482	6.537
Seed yield (kg ha^{-1})	710	724	862

The pod characters, seed yield and seed quality parameters measured through 100 seed weight, germination and field emergence were higher in hardened cum pelleted seeds. As far as the increase in the seed yield is concerned, the physico-chemical treatments could have triggered the biosynthesis, enhanced the cell division and the metabolic activity of the plants resulting on the increased uptake of nutrients which could have been possibly accounted for improvement in crop performance. The prosopis hardening cum pelleting treatments might have improved the growth of the plant during early stages of the crop with increased vigour and associated stronger root system which inturn might have favoured the derivation of available soil moisture and nutrients enabling better growth that resulted in higher yield (Jegathambal, 1996). In all the other centres (Madurai, Aruppukottai and Paiyur) seed hardening with 1% prosopis leaf extract for 1 h followed by seed pelleted with DAP 40 g + $ZnSO_4$ 100 mg + $FeSO_4$ 100 mg + Ammonium molybdate 250 mg using 10% maida as an adhesive improved the seed yield and other quality characteristics (Plate 2 and 3; Table 2).

Plate 2. Field view of hardened cum pelleted seeds of greengram

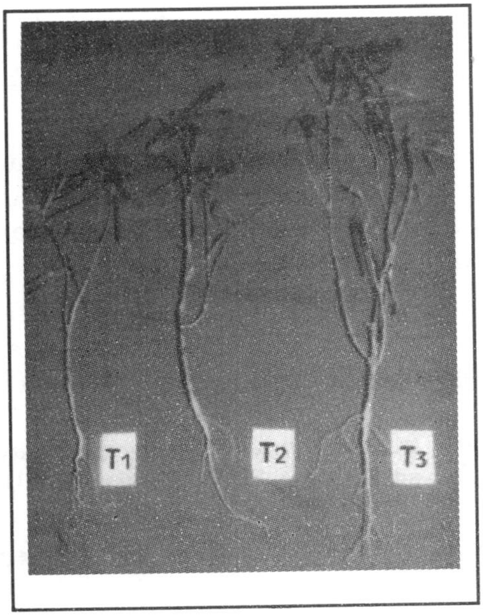

Plate 3. Influence of hardening and hardening cum pelleting on number of pods plants^{-1}

Table 2. Influence of seed hardening cum pelleting on seed yield of greengram

Characters	Untreated	Hardening	Hardening cum pelleting	CD (P = 0.05)
Madurai				
Days to first flowering	26	25	24	-
Seed yield kg plot^{-1} (20 m^2)	2.55	2.85	3.27	0.31
Resultant seed germination	77	79	82	2.51
Aruppukottai				
Plant height (cm)	82.6	93.8	102.8	5.83
No. of pods plant^{-1}	19.0	22.2	26.4	1.96
Seed yield (kg ha^{-1})	681	694	813	58
Paiyur				
Seed yield kg ha^{-1}	911	927	953	
Stover yield kg ha^{-1}	1118	1134	1168	

The results of these experiments reveal that hardening of greengram seeds with 1 per cent prosopis leaf extract for 1h and pelleting with DAP 40g + $ZnSO_4$ 100 mg + $FeSO_4$ 100 mg + Ammonium molybdate 250 mg can be recommended to defy the drought condition.

REFERENCES

1. Biradar, B.B. (1988). Precision planting by seed pelleting. *Seeds & Farms, 14(5)* : 32-34.
2. Jagathambal, R. (1996). Presowing seed treatment to augment productivity of sorghum Cv. Co 26 under rainfed agriculture. Ph.D. Thesis, TNAU, Coimbatore.
3. Jayaraj, T. (1977). Study on the effect of plant protection chemicals on seed quality in sesame cv. KRR 2 and TMV 3. M.Sc. (Ag.) Thesis, Tamil Nadu Agricultural University, Coimbatore.

4. May, L.M., E.J. Milthorpe and F.L. Milthorpe. (1962). Presowing hardening of plants to drought. *Field crop Abstr., 15*: 93-98.

5. Rathinavel, K. and C. Dharmalingam. (1999). Seed hardening to augment the productivity of cotton cv. LRA 5166 (*Gossypium hirsutum* L.). *Madras Agric. J., 86(1-3)* : 68-72.

SEED HARDENING AND PELLETING IN GROUNDNUT

M. BHASKARAN[1], V. VAKESWARAN[2] AND K. VANANGAMUDI[2]

1. INTRODUCTION

Groundnut, the king of oilseed crop of India accounts for 28 and 36% of the total area and production under oilseeds, respectively. It is grown both in kharif and rabi season. The area under groundnut cultivation especially in summer season is increasing steadily in command areas due to low water requirement. On the other hand, its productivity is decreasing, this could be due to non availability of quality seeds of improved varieties at proper time and place and poor nutrient management. Further, it is grown mostly as rainfed crop during kharif. Although considerable work has been carried out on management of rainfed crop through seed hardening and pelleting technology, the same thing has not reached the farmers as expected. It is proved that use of quality seed coupled with seed management strategies *viz.,* seed hardening and pelleting can enhance groundnut productivity. Sustainability of enhanced groundnut production is as important as enhancing the production itself. Given the current level of technology exists a wide commercially untapped yield reservoir in groundnut. The productivity potentials of the currently available improved groundnut production technologies could be exploited under real farm situations.

The overall productivity of groundnut is quite low i.e. 750 and 1500 kg/ha in kharif and summer respectively in India. Among many factors,

1 *National Seed Research and Training Centre, Varanasi – 221 106*
2 *J.K.Seeds Pvt. Ltd, Hyderabad*
2 *Dean (Agriculture), Tamil Nadu Agricultural University, Coimbatore - 641 003.*

seedling emergence, establishment and age of the seed play a major role in getting higher yields. Pre-sowing seed treatment with growth stimulants, seed protectants or even mere soaking in water helps to tide over the adverse soil conditions during initial establishment (Austin *et al.*, 1973).

2. PRE-SOWING SEED MANAGEMENT TECHNIQUE

Among several factors which affect groundnut yield, maintenance of adequate population (32 plants/m^2) in the field is the main problem which reduced the yield upto 30-40 per cent. There are several management practices to improve the yield of groundnut crop, but none of the practices help in maintaining the desired plant population per unit area. Since groundnut seed contain more lipids, the maintenance of higher germination for more than six to eight months is a problem. At farmer's level, the germination of sowing seeds will be 60-70 per cent. When such seeds are sown, there is every chance of reduction in field population to the tune of 25-30 per cent from the beginning itself. Increasing the seed rate by 30-40 per cent more is also not economical due to higher seed cost. Moreoever, in groundnut, mostly single seed is sown per hill and also gap filling is not practicable. Under such conditions, when poor quality seeds with low germination of 50-55 per cent are used, the field stand will be affected by 40-50 per cent. Hence, all the other applied inputs are wasted, then optimum plant population is not achieved. Hence, a seed hardening technique has to be adopted and followed at the field level, so as to maintain the plant population of about 90-95 per cent and there by we can increase the yield by 15-25 per cent without much additional cost.

The pre-sowing seed hardening technique involves three steps *viz.*, i) soaking the seeds for water absorption, ii) Separation of dead seeds and (iii) Sowing of radicle emerged seeds. For example, when the seed lot has lower germination of 60-70 per cent, then this technology may be adopted. 150 kg of kernels per hectare may be taken in gunny bags and tied very loosely at the top and fully immersed in fresh water for four hours. After that, the gunny bags with seeds are taken out from the water and kept in dark, covering with another moist gunny bag for 14-16 hours. When the radicle emergence is found in 30-40 per cent of seed, the kernels are spread over dry gunny bags under shade and the sprouted kernels are hand picked and air dried under moist condition by covering with the same moist gunny bag for another two to three hours and again the process of separation and shade drying of sprouted viable seeds may be repeated. Such seeds can be sown immediately (or) even one or two days after treatment. Before sowing the seeds, seed treatment with chemical should be carried out carefully without pealing the test and breaking the merged radicle. This seed management technique in groundnut is a boon to the farmers in getting higher yield by maintaining required population. If we look into the economics of this technology a net additional profit of Rs.4975/- can be realized (Narayanasamy, 2003) also revealed that pre-sowing treatment of groundnut seeds with 0.5% calcium chloride for 6 h followed by shade drying for 16 h resulted in significantly higher germination, field emergence, number of graded pods/ plant, test weight and also registered an increase in graded pod yield of 1614 and 1607 kg/ha over control of 1260 and 1292 kg /ha during summer and kharif seasons, respectively. However, soaking seeds directly in born solution my cause adverse effect on groundnut seed germination and an over all performance, thus, this treatment is not recommendable.

3. SEED PELLETING IN GROUNDNUT

Seed pelleting or coating is the most applicable technique in direct sown crops, which need initial vigour for sustained crop growth and development. Coating provides an opportunity to package effective quantities of materials such that they can influence micro environment of each seed which supplies not only micro and macronutrients but also protects the crop from pest and diseases from the earlier stages to the inclusion of pesticides. Because of increased awareness of organic farming, use of plant products in agricultural research is gaining importance. Seed coating is an important area in which plant materials could be widely used. Organic farming is gaining momentum now-a-days. Use of plant products as manure and for pest and disease control has become common. Hence attempt was made to eclucidate information on the use of leaf powder, organic and inorganic substances for seed coating.

Seed coating is a technique of seed encapsulation with organic and inorganic nutrients, water absorbents and pesticides importance of seed coating was well recognized and farmers may be able to save on the inputs require and the associate cost applying them (Lonergan *et al.*, 1995 and Scott, 1989). Seed coating materials were reported to improve the germination ability and to increase seedling emergence at changing soil moisture especially in the sub-optimal range (Mucke, 1987). This technique will be relevant to direct sown paddy wheat, pulses and oil seed crops, which require initial vigour for sustained crop growth and development (Kuppusamy *et al.*, 1992). Dried leaf powders of certain plant species could also be used for seed coating (Reddy, 1995). The leaf materials might have got certain physiologically active substances causing the beneficial effect

(Nargis, 1995). Further studies to decipher the chemical nature of substance would be rewarding. However, studies are plenty on allelopathic effect of plant materials which normally are deleterious to seed (Hasur,1993). Different adhesives at different concentrations exerted profound influence on dry and we stability of pelleting material (Scott, 1975). In line with the above findings studies made with groundnut kernels of the variety CO2 on the usage of adhesives namely, rice gruel, maida paste and gum at 5 per cent concentration of 2 and 5 ml, respectively had good binding capacity, however the binding materials significantly reduced the germination and vigour. Cracking of the test of the groundnut due to hard binding of the adhesives and subsequent entry and activation of the adhesives on the seed contents and embryo might be the reason for the low germination of the seeds subjected to adhesive treatments.

Normally chalk powder, dolomite and gypsum are used as filler materials along with nutrients, pesticides and biofertilizers. Present dry trend is to use leaf power'ds or coating and pelleting (Sabir Ahamed, 1989; Saraswathi, 1994; Nargis, 1995). Prosopis leaf powder, arappu leaf powder, vermicompost, cowdung ash, gypsum and groundnut micronutrient mixture. Used as filler materials in the present study adversely affected the germination potential of the groundnut kernels. Among the filler materials, gymsum treated seeds performed better than the other materials. The vigour potential of the pelleted seeds should the same bind. The reduction in germination and vigour of the pelleted kernels, observed in the present study is due to mechanical change and rapid absorption of water during pelleting and subsequent rapidity in the activation of hengus as reported by Roos (1979) and Saraswathi (1994).

REFERENCES

1. Austin, R.B., P. C. Longdon, T.W. Hegarty, T.H. Thomas and R.A. Mande. (1973). Pre-sowing seed treatments. **In:** Seed Biology (Ed. W. Heydecker), Butter Worths, London. pp. 521-531.
2. Hasur, S. (1993). Effect of Acioa basteri, Cassia siamea, Flemingia macrophylla and Gmelina arborea leaves on germination and early development maize and cassva. *Agriculture, Ecosystem and Environment., 45(3-4)* : 263-273.
3. Kuppusamy, G., A.R. Lakshmanan and A. JeyabaL. (1992). Effect of seed pelleting with bidigested slurry and macronutrients on rice fallow green gram cv. CO 2. *Madras Agric. J., 76 (4)* : 183-187.
4. Lonergan, J.F., D. Meyar., R.G. Fawcett and A. J. Anderson. (1995). Lime pelleted clover seeds for nodulation in acid soils. *J. Aust. Inst. Agric. Sci., 21*: 240-250.

5. Mucke, J.D. (1987). The regulation of water transport in pelleted sugar beet seed. *J. Agron. Crop. Sci., 161(1)*: 79-83.
6. Nargis, S. (1995). Influence of pelleting, magnetic treatment and radiation on the performance of differentially aged seeds in tomato (*Lycopersicon esculentum* mill.) cv PKM 1. M.Sc. (Ag.) Thesis, Tamil Nadu Agric. Univ., Coimbatore.
7. Reddy, K.V. (1995). Studies on pre-sowing seed treatment in differentially aged seeds of beinjal (*Solanum melogena* L.) cv. PKM 1. Msc. (Ag.) Thesis Tamil Nadu Agric. Univ., Coimbatore.
8. Roos, E. (1979). Storage behaviour of pelleted tableted and taped lettuce seed. *J. Amer. Sco. Hort. Sci., 104(2)*: 283-28.
9. Sabirahmed, A. (1989). Studies on the production of quality seed and storage in soybean (Glycine max (L.) Merrill). M.Sc. (Ag.) Thesis, Tamil Nadu Agric. Univ., Coimbatore.
10. Saraswathi, G. (1994). Studies on seed pelleted in relation to sowing of cotton cv LRA 5166 (Gossypium hirsutum L.) M.Sc. (Ag.) Thesis, Tamil Nadu Agric, Univ., Coimbatore.
11. Scott, J.M. (1989). Seed coatings and treatments and their effects on plant establishment - *Advan. Agron.,* 12: 43-83.

NUTRIENT COATING AND FOLIAR APPLICATION ON SEED YIELD AND QUALITY IN SESAME - A NET WORK EXPERIMENT

P. BALAMURUGAN[1], V. BALASUBRAMANI[2] AND K. SUNDARALINGAM[3]

1. INTRODUCTION

Establishment of seedling under dryland is an important factor for realizing satisfactory yield. Presowing seed coating / pelleting is one such method modifying physiological and biochemical nature of seed and is recommended for precision planting as it results in singling of seeds. Sesame is an important oil seed crop often grown as rainfed crop with low productivity in dry land cultivation. In dryland farming, basal application of fertilizer is not feasible due to restricted moisture supply. Hence, seed pelleting with nutrients and water absorbants offers vast scope for improved crop stand and in turn the productivity. Studies were conducted at the Department of Seed Science and Technology, Tamil Nadu Agricultural University, Coimbatore to improve seed yield in sesame cv CO1 through pelleting and foliar application under rainfed conditions as a net work experiment. Sesame cv. CO1 seeds were treated with thiram 2 g kg^{-1} and pelleted with gypsum 300 g + Ammonium molybdate + $ZnSO_4$ + $MnSO_4$ borax (300 mg each kg^{-1} of seed). After treatment seeds were evaluated for quality parameters. Seed pelleting and thiram treatment showed significant improvement in seed germination and seedling vigour by 2 and 6 per cent respectively.

2. PELLETING WITH NUTRIENTS AND FOLIAR SPRAY ON QUALITY PARAMETERS

Sesame cv. CO1 seeds pelleted (Plate 1 and 2) were sown in a field trial at the Department of Seed Science and Technology, Coimbatore during

1 *Department of Seed Science and Technology, Tamil Nadu Agricultural University, Coimbatore – 641 003.*
2 *Centre for Plant Molecular Biology, Tamil Nadu Agricultural University, Coimbatore – 641 003.*
3 *Krishi Vigyan Kendra, Sirugamani – 636 203.*

September 2001. During crop growth, DAP (0.5%), ammonium sulphate (0.2%) boric acid 0.1%, NAA (40 ppm), salicylic acid (100 ppm) were sprayed as foliar application on 25th and 40th DAS. All other recommended package of practices were followed. The treatment combinations and the results are furnished in table 1.

Table 1. Effect of seed pelleting and foliar sparay on quality parameters of sesam oil cv. CO1

Treatments	Dry matter production (g/plant)	Number of capsule plant^{-1} (g)	Capsule weight plant (g)	Seed yield Per Plant (g)	Seed yield (kg/ha)
T_1 : Control	7.06	71.8	18.67	7.23	
T_2 : Seed treatment with thiram 2g kg^{-1} (300g)	7.10	74.7	19.73	7.76	550
T_3 : Seed pelleting with Gypsum (300g) + Amm. Molybdate + $ZnSO_4$ + $MnSO_4$ + Borax (300 mg each kg -1 seed)	7.80	80.4	21.00	8.26	571
T_4 : T1 + foliar spray of DAP (0.5%) + Amm. Sulphate (0.2%) + boric acid (0.1%) + NAA 40 ppm on 25th and 40th DAS	7.13	78.1	20.70	8.03	567
T_5 : T1 + foliar spray of DAP 0.5% + Amm. Sulphate 0.2% + boric acid 0.1% + salicylic acid 100 ppm on 25th and 40th DAS	7.15	83.6	21.67	8.49	602

T_6 :	T3 + foliar spray of DAP 0.5% + Amm. Sulphate 0.2% + boric acid 0.1% + AA 40 ppm on 25th and 40th DAS	8.0	86.2	22.47	8.93	642
T_7 :	T3 + foliar spray of DAP 0.5% + Amm. Sulphate 0.2% + boric acid 0.1 % + Salicylic acid (100 ppm) on 25th and 40th DAS	8.0	90.3	22.83	9.24	671
	CD (P = 0.05)	0.23	0.26	1.027	0.18	19.33

Plate 1. Field View of Experimental Plot

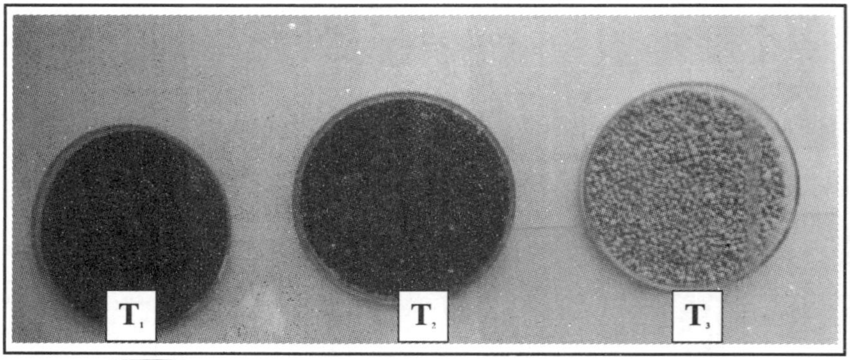

Plate 2. Thiram treated and pelleted seeds

T_1- Control
T_2 – Seed treatment with thiram (2g kg^{-1} of seed)
T_3 – Seed pelleting with gypsum (300g) + Ammonium molybdate + $ZnSo_4$ + $MnSo_4$ + Borax (300 mg each kg^{-1} of seed)

The increase in plant height and drymatter production of sesame due to pelleting could be due to enhanced seedling establishment and possibly due to activation of metabolic activity of seed. Significant differences were recorded due to the effect of pelleting of seed and subsequent foliar nutrition for days to first flowering, 50% flowering and days to maturity. The earliness in flower formation and seed maturity was noticed in seed pelleting and foliar application of nutrients. Seed pelleting combined with foliar spray of salicylic acid increased the weight of capsules $plant^{-1}$ by recording 25.7% increase over control. This may be due to mobilization of reverse food materials to the sink for grain filling process through increase of hydrolyzing and oxidizing enzyme activities. The experiment was conducted also at Agricultural Research Station, Aliyarnagar and Oil Seeds Research Station, Tindivanam in sesame cv. TMV 3 and TMV 6 respectively. Seed yield increased by the application of pelleting treatment and foliar spray of nutrients. Seed yield of sesame was increased by 20.6% over control by pelleting and foliar treatment of DAP 0.5%, Ammonium molybdate 0.2%, boric acid 0.1% and salicylic and 100 ppm at 25th and 40th day after sowing (Plate 3). The results obtained from different centre as follows.

3. CONCLUSION

Seed pelleting with Gypsum (300 g/kg) + Ammonium molybdate + $ZnSO_4$ + $MnSO_4$ + borax (300 mg each per kg of seed) increased the seed germination and seedling growth while seed pelleting combined with foliar spray of DAP 0.5% + Ammonium sulphate 0.2% + boric acid 0.1% + Salicylic acid 100 ppm on 25th and 40th days after sowing improved the seed yield.

Plate 3. Efficacy of treatments on growth of sesame

Treatment	Seed yield (kg / ha)			Mean
	Coimbatore	Tindivanam	Aliyarnagar	
Recommended dose of fertilizer	525	308	497	452
Seed coating	608 (10)	341 (11)	489 (-2)	479 (6)
Foliar spray of DAP 0.5%, Amm, Molybdate 0.2%, boric acid 0.1% and NAA 40 ppm	575 (4)	336 (9)	500 (0.6)	453 (0.1)
Foliar spray of DAP 0.5% Amm. Molybdate 0.2%, boric acid 0.1% and salicylic acid 100 ppm	642 (16)	380 (23)	507 (2)	509 (13)

Seed coating + Foliar spray of DAP 0.5%, Amm. Molybdate 0.2%, boric acid 0.1% and salicylic acid 100 ppm	689 (26)	402 (31)	516 (4)	545 (21)
CD (P=0.05)	12.0	16.9	14.2	

(Figures in the paranthesis are % of increased seed yield)

NUTRIENT MANAGEMENT THROUGH SEED COATING AND FOLIAR APPLICATION IN CASTOR UNDER RAINFED CONDITIONS

P. SRIMATHI[1], D. M. SHETE[2], V. PARAMASIVAM[3],
R. JEGATHAMBAL[4] AND M. BHASKARAN[5]

1. INTRODUCTION

Rainfed agriculture is one of the adverse agro-ecological situations where the success of the crop depends on proper management practices to resist against existing drought. Karivaratharaju (1999) expressed drought resistance as the capacity of the plant to develop normally in dry habitat yielding maximum crop with endureness of drought without injuring the crop. Agronomic methods *viz.*, contour bunding, summer ploughing, crop rotation and water spray are recommended normally to tide over the adverse environmental condition the crop seeks during their growth period. Henkel (1964) and Hegarty (1970) based on their research, recommended seed hardening as a basic technique for better establishment of crop under rainfed conditions, while Karivaratharaju *et al.* (1999) reported that enclosing the seed with easily soluble nutrients will favour the growth of crop at field level. Rashma (2001) reported that foliar spray of nutrients could improve the photoshynthetic efficiency of the crop and could favour better crop growth and yield. In rainfed situation also, foliar sprays with nutrients were recommended to mitigate the ill effects of stress (Balamurugan, 2002).

Castor *Ricinus communis* is an important oil seed crop of multipurpose utility which is highly used in medicinal and cosmetic industries. Castor

1 *Department of Tree Breeding, Forest College and Research Institute, Mettupalayam - 641 301.*
2 *Department of Seed Science and Technology, Tamil Nadu Agricultural University, Coimbatore- 641 003.*
3 *Oilseed Research Station, Tindivanam – 604 001.*
4 *Agricultural Research Station, Vaigai Dam – 625 512.*
5 *National Seed Research and Training Centre, Varanasi – 221 106.*

belongs to the family Euphorbiaceae and is highly cross pollinated as the inflorescence is monocieous in nature. This long duration crop is mostly grown as rainfed crop than as an irrigated crop. Evaluation of seed and crop management practices in this crop may lead to identification of a suitable dryland technique to boost up the yield of castor under rainfed conditions.

2. NUTRIENT MANAGEMENT IN CASTOR THROUGH SEED COATING AND FOLIAR APPLICATION

During 2000-2001 to evolve a suitable seed management practice studies were initiated at Department of Seed Science and Technology, TNAU in three different centres *viz.*, Yethapur, Tindivanum and Coimbatore with following treatment combinations.

T_1	:	Recommended fertilizer schedule of NPK as per Crop Production Guide (RDF)
T_2	:	RDF + Seed coating with gypsum 300 g + Ammonium molybdate + $ZnSO_4$ + $MnSO_4$ + Borax each @ 300 mg/kg of seed using maida 10% as adhesive (SC)
T_3	:	RDF + Foliar spray with DAP 0.5%, Ammonium molybdate 0.2% Boric acid 0.1% and NAA 40 ppm on 25th and 40th days after sowing (FA-NAA)
T_4	:	RDF + Foliar spray with DAP 0.5%, ammonium molybdate 0.2%, Boric acid 0.1%, Salicylic acid @ 100 ppm at 25 th and 40 th days after sowing (FA - SA)
T_5	:	RDF + Seed coating with gypsum 300 g+ Ammonium molybdate + $ZnSO_4$ + $MnSO_4$ + Borax each @ 300 mg/kg of seed using maida 10% as adhesive + RDF + Foliar spray with DAP 0.5%, Ammonium molybdate 0.2%, Boric acid 0.1% and NAA 40 ppm and 25 th and 40 th days after s owing (SC-FA NAA)
T6	:	RDF + Foliar spray with DAP 0.5%, Ammonium molybdate 0.2%, Boric acid 0.1%, Salicylic acid @ 100 ppm at 25th and 40 th days after sowing (SC-FA-SA)

At all the 3 centres the crop was raised during Sep- Oct 2001 under the rainfed conditions of Tamil Nadu adopting the normal package of practices under the statistical design RBD with 3 replications. During crop growth, observations were made on field emergence (%), yield attributing characters and seed yield. Based on the results, the best treatments were identified and again raised as a confirmity trial during 2002-2003 at Coimbatore (October 2002) and observations were made as per the previous year. The results were analysed statistically to evaluate the significance of the technology for rainfed agriculture (Panse and Sukhatme, 1965).

Highly significant variations were observed due to the treatmental influence in all the 3 locations during first year. The observation on field emergence revealed that the field emergence had a non-significant influence due to seed coating treatment as it improved the seed germination only by 2% at Coimbatore and Tindivanum. But at Yethapur, the seed coating reduced the germination by 5% indicating the adverse effect caused by the seed coating on initial field emergence. Srimathi and Malarkodi (2000) also reported a non-significant influence due to seed coating in soybean as per the results of the pooled analysis of the present study. But Sevanan (1998) reported in gingelly that seed coating influence the germination of seeds in additive manner. The yield attributing characters were recorded only at Yethapur and Tindivanam (Table 1).The pooled analysis on yield data from 3 locations expressed that the combinational treatment where the seed coating was followed with foliar spray with salicylic acid @ 100 ppm improved the seed yield by 17% expressing the adaptability of the treatments for boosting the yield of castor under rainfed conditions (Table 2). The results of the present study might be due to the influence of micronutrients added to the rhizosphere due to nutrient pelleting (Karivaratharaju *et al.*, 1999) and the nutrients added to the foliar spray which might have accelerated the photosynthetic (Sasthri, 2001) efficiency of the growing crop even under the stress condition which restrict the translocation of nutrients from soil to sink. The observations on seed quality characters conducted at Tindivanum expressed that the above combinationl treatment also improved the germination and vigour of resultant seeds. Sasthri (2001) and Reshma (2001) also expressed that foliar spray with salicylic acid improved the yield and quality of seed due to their influence in activating the efficiency of photosynthesis. Hence, the results of the present network experiment might be due to additive effect of nutrients added to the soil through seed coating and the foliar application that had improved the seed set and seed yield.

The conformity trial was carried out at Coimbatore during 2002-2003 (October, 2002) with 3 different treatments *viz.*, untreated, seed coating and

Table 1. Influence of seed coating on field emergence of castor at 3 different locations

Treatment	Location			Pooled
	Coimbatore	Yethapur	Tindivanum	
RDF	91	80	78	83.0
RDF + SC	76	82	80	79.0
CD (P=0.05)	2.2	NS	NS	NS

Table 2. Influence of seed coating and foliar spray treatment on seed yield (kg ha^{-1}) of castor.

Treatment	Location			Pooled analysis Mean
	Yethapur	Tindivanum	Coimbarore	
RDF	618	770	889	759
RDF + SC*	615	1005	930	850
RDF + FA** (NAA)	685	1040	948	891
RDF + FA*** (SA)	688	960	1004	884
RDF + SC + FA (NAA)	353	972	985	865
RDF + SC + FA (SA)	640	980	1096	905
CD (P = 0.05)	5.61	25.2	2.3	

seed coating + foliar spray with salicylic acid (as in the year 2001-2003) under rainfed conditions. The results also revealed that the combinational treatment *viz.* nutrient seed coating and foliar application improved the seed yield and also the resultant seed quality attributes. The performance might also be due to the influence of micronutrients on crop growth, seed set and yield attributing characters as discussed earlier. The quality of the resultant seeds was also significant by higher due to the treatmental influence as that of the previous years. Thus, the network experiment on seed pelleting and foliar spray revealed that sowing of nutrient coated seeds followed with foliar application of micronutrients twice during crop growth could improve the yield of rainfed castor by 13%.

REFERENCES

1. Balamurugan, P. (2002). Nutrient Management for Seed Yield Maximization in Rainfed Sesame (*Sesamum indicum*) cv. CO 1.
2. Hegarty, T.W.(1970). The possibilities of increasing field establishment by seed hardening. Hort. Res., 10 : 54-64.
3. Henkel, P.A. (1964). Interrelated effects of imbibition temperature and oxygen on seed germination. In: Seed Ecology (ed.) Heydecker, W. Butterworths, London, pp. 157-169.
4. Karivarathaaraju et al. (1999). A Treatise on Tree Seed Technology, Published by TNAU, pp. 75-80.
5. Panse and Sukhatme. (1965). Seed Production Technology Manual. Release No. 19/88-89. pp.153-164.
6. Rashma, C. (2001). Studies in seed maturation, production, processing and storage in Hedge luceme (*Desmanthus virgatus*) cv. TNDV1 M.Sc. (Ag.) Thesis, Tamil Nadu Agricultural University, Coimbatore -3.
7. Sasthri, G. (2001). Investigation on Seed Production and Storage in Hybrid Cotton TCHB 213, Ph.D. Thesis, Tamil Nadu Agricultural University, Coimbatore - 3.
8. Sevanan, J. (1998). Seed treatment as quality and storability in gingelly (*Sesanum indicum* L.) M.Sc. (Ag.) Thesis, Tamil Nadu Agricultural University, Coimbatore - 3.
9. Srimathi, P. and K. Marlakodi. (2000). Influence of nutrient pelleting on seed yield of soybean cv. CO 1. *Indian Agric., 44(1)* : 79-82.

HARDENING AND PELLETING IN COTTON

K.RATHINAVEL AND K.RAJA

1. INTRODUCTION

Cotton is the chief fibre crops grown in India on commercial scale, as this is the prime supplier of raw material to the Indian textile industry. The area under cotton cultivation is 73.90 lakh hectares with production of 140 lakh bales during 2002-03. The productivity of lint is 322 kg/ha. This is far below the global average of 627 kg/ha. To rise up to the global standards, newer approaches on crop husbandry are warranted. In the states of Punjab, Haryana and Northern parts of Rajasthan, cotton is cultivated under intensive irrigated condition, in the southern parts of Rajasthan, Gujarat, Maharashtra, and Madhya Pradesh and Karnataka mostly under rainfed condition. In all these areas, suitable seed management techniques are required to encounter abiotic stresses such as excess soil moisture, salt stress, depletion of soil moisture during seed germination *etc.*, that are anticipated to occur during cotton seed sowing. Seed quality enhancement to the challenges during seed germination and early seedling growth phases are considered as the paramount importance for assured crop growth and successful production. The major methods suggested for enhancing seed quality are seed hydration, biological seed treatments and seed coating.

2. SEED HARDENING

The principle is that hydrated seed initiate early germination events, but not sufficient enough to permit protrusion. The germinating seedlings could emerge and produce deep root systems before the upper layers of

---

Central Institute for Cotton Research, Regional Station, Coimbatore-641 003.

the soil dry out, harden or become dangerously hot. Since, for a large proportion of the time seeds spend in soil they are simply imbibing water, often very slowly, to avert this situation seed hydration is proposed. In this process, seed is allowed to take up water for a period that is less than the 'safe limit'. Once removed from the water and surface-dried, the seed will not continue to germinate unless extra water is supplied, e.g., by sowing into moist soil. If sown into dry soil (to await the rains, for example) the seed will not germinate. It will dry out slowly but will germinate once the soil is wetted by rain. Benefits of seed priming will be greatest if seed is sown into moist soil because it will have a 'head start' (Harris, 1996). Seed hardening also confers high degree of drought resistance and excessive water loss could be prevented by seed hardening (Crafts *et al.*, 1949). Hardening is the physiological induction as a cause of seed conditioning towards increased drought resistance in crops (Henkel, 1961).

During seed hardening, a number of physico-chemical changes *viz.*, greater hydration of colloids, higher viscosity and elasticity of protoplasm, higher bound water content, increase in temperature, lower water deficit and more efficient root system occurring in plants from seeds given pre-sowing seed hardening treatment (May *et al.*, 1962). In cotton, increased seed yield could be obtained by seed soaking treatment with cycocel (Chlormequat) at 200 or 400 ppm solution. Hardening of cotton seed by soaking in 1000 ppm CCC (Chlormequat) for 6 h followed by drying to the original moisture content before sowing significantly increased the root length; decreased shoot growth and transpiration. The seedlings maintained increased turgidity and showed high water use efficiency (Pothiraj and Sankaran, 1984). Cotton seed soaked in solutions of $KMnSO_4$ @ 2 g lit^{-1} for 18 h increased seed cotton yield (Ragah *et al.*, 1991). In another study, soaking cotton seed in solutions of $ZnSO_4$ @ 2 or 4 g lit^{-1} for 18 h did not affect seed cotton or lint yield (Salwau *et al.*, 1991). Kariev (1981) observed higher 1000 seed weight, seed oil and seed cotton yield in plants from seed soaked in Na salt and d phenylbutric acid at 0.001 per cent concentrations. Soaking cotton seeds in chelated Zn, Cu, Mn, Byfolan (mixed trace elements) 24 h before sowing reduced germination and increased seedling vigour, whereas highest germination and seedling vigour was recorded by soaking for 48 h (Shaban and Eid, 1982). Seed soaking treatment given to cotton with 550 ppm Ascorbic acid or GA_3 or 50 ppm TIBA increased the speed of germination, percentage germination and seedling growth, whereas 2,4,5-T inhibited the growth (Sharma *et al.*, 1984). A study towards the optimization of concentration of seed hardening solutions showed the superiority of potassium chloride (1.0%) with respect to germination and speed of germination. On the other hand, the aqueous leaf extracts of pungam (*Pongamia pinnata*) and prosopis (*Prosopis juliflora*) 1.0 and 0.5% solutions were superior to potassium chloride in all seed quality characteristics such as germination, speed of germination, seedling growth, dry matter production and computed vigour index (Rathinavel and Dharmalingam, 1999).

Field experiments conducted in summer and winter 1996 seasons to assess the productivity of hardened seeds highlighted that seeds given seed hardening treatment with 1.0% pungam leaf extract striking an increase of 7.9, 35.2, 29.0 and 18.0% respectively over control in two seasons trials. The seed cotton yield and seed yield were higher by 31.7 and 35.7%, respectively in the summer crop and 43.6 and 48.0% respectively in the winter crop in the same treatment over control. In the resultant seeds, better quality vested with the seeds from botanical hardening (Rathinavel and Dharmalingam, 1999). Cotton seed germination and seedling growth were inhibited by the extracts from ripe fruits of *Rubia tintorum* (Agakishiev *et al.*, 1978). But soaking of cotton seed in *Nostoc muscorum* suspension increased the seed yield by 400 kg ha^{-1} (Umarova and Urmanov, 1978). Abdallah and Mahmoud (1978) observed that CCC (chlormequat) treatment @ 50, 100 or 200 ppm decreased plant height, internode length, number of fruiting branches, top: root ratio; increased number of nodes $plant^{-1}$, highest dry weight of roots, leaves, stem; increased flower production, shedding percentage, seed index; reduced number of open bolls $plant^{-1}$, number of seeds $boll^{-1}$, lint percentage and subsequently yield of seed cotton in Egyptian cotton. Thandapani and Subharayalu (1986) reported that cotton seed hardening with CCC or KCl increased seed cotton yield from 1.48 to 1.84 and 1.72 tonnes $hectare^{-1}$, respectively.

3. SEED PELLETING

The earliest method adopted for seed "Singulation" was seed pelleting, done to increase the size of the individual seed by coating it with a layer of bentonite clay and other foreign materials, which can regulate the size of the seed for precision planting. Pelleting is done for increased flowability, improved handling, accurate distribution, better seed to soil contact and improved seedling nutrition. During pelleting, seeds are first mixed with adhesives and then coated with filler materials on rolling for uniformity. Materials used for pelleting are adhesives, coating material and bioactive chemicals such as nutrition, growth regulators, pesticides etc. It was reported that simple coating cotton seeds with starch gave good flow characteristics. This makes it easier to handle than seeds have got acid delinting treatment

(Ghosh and Elawady, 1973). When cotton seeds were coated with pesticide disyston using clay as coating material, the release was greater. If methyl cellulose is used instead of clay, the release was reduced substantially. According to Patil and Dighe (1983), coating cotton seed with indigenous material such as mud, ash, cow-dung or their combination did not result in better seed germination. Pelleting with mud + cow-dung reduced seed vigour to the maximum. Cotton seeds pelleted with Gypsum + DAP + Micronutrient recorded higher germination, speed of germination, root length, shoot length and vigour index (Saraswathi, 1993). Nehra *et al.* (1990) opined that cotton seed inoculated with Azotobacter replaced 2/3 recommended dose of nitrogen (53 kg N ha^{-1}) and resulted in yield attributes and seed cotton yield at par with recommended dose of nitrogen (80 kg N ha^{-1}). Full dose of nitrogen with seed inoculation of Azotobacter recorded the maximum of 63 per cent of seed cotton yield over control.

3.1. Effect of pelleting on cotton seed quality

Cotton seeds coated with aldicarb did not affect germination opined by Regupathy (1976). However, seeds coated with Penthiuram and Polymers initially inhibited the activity of catalase and peroxidase and rate of seedling growth but stimulated them later on (Ruban *et al.*, 1983). According to Patil and Dighe (1983), coating of cotton seeds with indigenous materials such as mud, ash, cowdung and their combination did not result in better seed germination. Bordovsky *et al.* (1991) reported that upland cotton seeds stored in gel slurry more than 24 h resulted in poor germination whereas, gell slurry containing starch-based material improved the germination. Dhindwal *et al.* (1991) noticed no advantage with respect to germination, dry matter of seedling, germination rate and vigour index in undelinted cotton seeds coated with fresh cowdung or ash.

3.2. Effect of pelleting on cotton productivity

Increased germination and cotton seed yield due to pelleting of cotton seed with urea and formaldehyde has been reported by Usmanov *et al.* (1981). Umarov *et al.* (1978) reported that cotton seeds pelleted or soaked in butyric acid derivatives enabled better growth formation of fruiting organs and seed cotton yields by 360-570 kg ha^{-1}. Lakatkin and Ibragimov (1983) reported that pelleting of delinted cotton seed with 1.0 per cent solution of Na-Carboxymethyl cellulose after treatment with cytisine increased the germination percentage, dry matter accumulation and seed cotton yield. Wankhade *et al.* (1994) recorded the highest seed cotton yield in seeds treated with Biozyme (containing cytokinin, auxin enzymes and chelated trace elements) than the control. Zhao *et al.* (1994) reported in cotton that seed coating delayed initial seedling development by 1-2 days and decreased seedling emergence, increased seedling growth reduced the incidence of insects and diseases and increased the squaring and boll setting rate and lint yield.

The performance of resultant seeds in terms of germination, speed of germination, seedling growth, dry matter production, computed vigour index, electrical conductivity of seed leachate dehydrogenase enzyme activity and oil content were in favour of the seeds given pelleting treatments with arappu leaf powder + DAP + micronutrient mixture (Rathinavel *et al.*, 2000). A new approach for elite seedling production in protrays for transplanting yielded encouraging results. The fuzzy and delinted seeds of MCU 7 pelleted with arappu leaf powder + DAP + micronutrient mixture and sown in protrays filled with soil moisture composed of sand + red soil + FYm at 1:0.5:1.5 ratio and kept under open condition provided higher germination associated with better seedling growth, higher dry matter production and vigour index. The twenty-day-old seedlings attained optimum growth for transplanting with better field establishment (Rathinavel and Dharmalingam, 1999). Cotton productivity depends to a larger extent on the sowing quality of seeds that are planted. The role of high quality seeds needs no further emphasis and the seed management technique becomes all the more important to upgrade the planting value of seeds. The seeds used for sowing should have maximum viability and vigour so that it will emerge rapidly and uniformly in the field even if the conditions are unfavourable. Seeds are to be pre-conditioned and prepared to meet the unfavourable situation in the field during sowing. Hardening seeds either by hydration and dehydration, chemical fortification, osmoconditioning or soaking with botanicals make them to equilibrate with unfavourable situation with subsequent good germination and better establishment of seedling. It is also evident that coating of cotton seeds with different nutrients, biofertilizers, chemicals, biocides, hydrophilic substances and botanicals were found useful as reported by several workers to solve problems like soil pH, nutrient deficiency, protection against soil microbes and development of better seed soil relationship.

REFERENCES

1. Agakishiev, D., N.A.Palavova and D.K.Akhmedyarova. (1978). The presence of *Rubia tinctorum* fruits of compounds inhibiting cotton seed germination. *Referativnyi Zhumal, 8(55)*: 552.

2. Agakishiev, D. and CH.Gurbanov. (1992). Effect of heteroauxins on seed germination and growth of small roots of seedlings of fine fibred cotton. *Seed Abstr., 18(5)*: 1892.

3. Bordovsky, J.P., W.M.Lyle, and A.B.Onken. (1991). Cotton seed preparation for hydraulic planter. *Agron. J., 83(2)* : 275-278.

4. Crafts, A.S., H.B.Currier and C.R.Stocking. (1949). Water. In: The Physiology of plants. Chronica Botanica Co., Waltham, Mass, USA, p.205.

5. Dhindwal, A.S., B.P.S.Lather and Jagdev Singh. (1991). Efficacy of seed treatments on germination, seedling emergence and vigour of cotton (*Gossypium hirsutum*) genotypes. *Seed Res., 19(1)*: 59-61.

6. Ghosh, B.N. and M.N.Elawady. (1973). The effect of starch coating cotton seed on manual planter operation. *J. Agric. Eng. Res., 18(4)*: 393-395.

7. Harris, D. (1996). The effects of manure, enotype, seed priming depth and date of sowing on the emergence and early growth of *Sorghum bicolor* (L.) Moench in semi-arid Botswana. *Soil and Tillage Research, 40*: 73-88.

8. Henkel, P.A. (1961). Drought resistance of plants. Methods of recognition and investigation. In: Plant water relationship in arid and semi arid conditions. Proc. Madrid. Symp., Paris, UNESCO/NS/A2/421 Paris.

9. Karive, A. (1981). Growth substance increases the seed oil content. Khlopkovodstvo. No.3: 37-38.

10. Lakatkin, I.P. and SH.I.Ibragimov. (1983). Effect of stimulators and Na-Carboxy methylcellulose on germination of delinted cotton seeds. *Seed Abstr., 6(10)*: 2727.

11. May, L.H., E.J.Milthrope and F.L.Milthrope. (1962). Pre-sowing hardening of plants to drought. An appraisal of the contribution by P.A.Henkel. *Field Crop Abstr., 15*: 9398.

12. Nehra, D.S., L.K.Bishnoi and M.S,Kairon. (1990). Effect of nitrogen and Azotobacter on yield of seed cotton (*Gossypium hirsutum* L.). *J. Cotton Res. & Dev., 4(1)*: 61-64.

13. Patil, V.N. and R.D.Dighe. (1983). Effect of coating on cotton seed quality. *Seeds and Farms, 9(11)*: 15-16.

14. Pothiraj, P. and S.Sankaran. (1984). Effect of seed hardening on the physiological vigour of the cotton seedlings. *Ind. J. Plant Physiol., 27(3)*: 300-302.

15. Ragah, M.T., M.I.Salwau and A.A.Hosny. (1991). Effect of seed soaking in $KMNO_4$ and P fertilization on yield and yield components and fibre quality of cotton cv. Giza 75. *Annals of Agric. Sci., 29(4)*: 1345-1357.

16. Rathinavel, K. and C.Dharmalingam. (1999). Effect of seed pelleting on elite seedling production in cotton cv. MCU.7 (*Gossypium hirsutum* L.). *Crop Research, 18(1)*: 137-141.

17. Rathinavel, K., C.Dharmalingam and S.Paneerselvam. (2000). Effect of seed pelleting on the productivity of rice fallow cotton (*Gossypium hirsutum* L.) cv. MCU 7. *Advances in Plant Sciences, 13(1)*: 213-217.

18. Regupathy, A. (1976). Factors affecting the performance of Aldicarb applied to cotton. Ph.D. Thesis, Tamil Nadu Agric. Univ., Coimbatore-3.

19. Ruban, I.N., G.I.Borodin and SH.S.Rashidova. (1983). Effect of polymer composition on some physiological indices of growth and development of cotton seedlings. *Doklady Akademi Nauk Uzbekskoissr, 11*: 35-37.

20. Salwau, M.N., A.A.Hosny and M.Ragah. (1991). Effect of seed soaking in Zn solutions and irrigation intervals on yield and yield components and fibre properties of Giza 75 cotton variety. *Annals of Agric. Sci., 29(4)*: 1299-1312.

21. Saraswathi, G. (1993). Studies on seed pelleting in relation to sowing qualities of cotton cv. LRA 5166 ((*Gossypium hirsutum* L.). M.Sc.(Ag.) Thesis, Tamil Nadu Agric. Univ., Coimbatore-3.

22. Shaban, S.A. and A.Eid. (1982). Effect of some micronutrients on seed germination and seedling growth in maize and cotton. *Research Bulletin, Faculty of Agriculture. Ainshams Univ., No.2047* :14.

23. Sharma, R.K., Vidhu and Y.S.Murthy. (1984). Influence of different growth substances on seed germination and seedling growth of *Gossypium hirsutum* var. H777. *Ind. Bot. Reptr., 3(1)*: 60-73.

24. Thandapani, V. and M.Subharayalu. (1986). Effect of pre-sowing seed treatment with chemical and growth regulators for drought tolerance and yield of cotton (*G. hirsutum* L.) under rainfed condition. *Madras Agric. J., 73(12)*: 668-675.

25. Umarova, SH.U. and Z.Urmanov. (1978). Application of blue green algae of cotton under farm conditions. *Referativnyi Zhumal, 7(55)*: 532.

26. Usmanov, S.V., V.A.Budhkov, B.G.Ostrobroo, A.K.Kulighev and V.A.Akamedov. (1981). Pelleting of cotton seeds with urea and formaldehyde compounds. *Seed Abstr., 4(7)*: 2192.

27. Wankhade, S.T., R.M.Deshpande and H.K.Kene. (1994). Effect of Biozyme on yield of cotton. *PKV Res. J., 18(1)*: 108-109.

28. Zhao, H.J., X.G.Liu and H.D.Wang. (1994). The yield increasing effect of delinting and coating of cotton seed. *China Cottons, 21(3)*: 10-11.

HARDENING AND PELLETING IN HORTICULTURAL CROPS

A.VIJAYAKUMAR[1], V.VAKESWARAN[2]
AND S.SUNDARESWARAN[1]

1. INTRODUCTION

Horticultural crops are more nutritious and valuable in terms of health and industrial development. Most of the crops are grown as cash crops and some of the crop are having more export potential. In addition to these, botanically horticultural crops having more significance. The floral biology and seed production in horticultural crops differ significantly from field crops. Starting from planting i.e., almost all the horticultural crops are transplanted from nursery, harvesting by different pickings, seed extraction and drying methods, etc. Botanically, there is a large difference in seed size, shape and even in structure. The horticultural crops mostly belongs to dicotyledons. Even though they are dicots, the seeds possess predominantly other structures also *viz.*, endosperm, perisperm, etc. Considering the seed size, it varies from 0.5 mm to 25 cm dia. If the seed storage is concerned, there exist both orthodox and recalcitrant behaviours. Most of the fruits and plantation crop seeds belongs to recalcitrant behaviour. Because of these reasons maintenance of moisture at higher level is must for germination. In case of small size seeds, to improve the size and precision planting, pelleting is highly useful. Hence, in this paper some of the work done in horticultural crops are discussed.

2. SEED PELLETING

Coating of tomato seeds with Gypsum + DAP mixture found to enhance crop performance in polyhouse tomato. Pelleting of different seeds require

1 *Department of Seed Science and Technology, Tamil Nadu Agricultural University, Coimbatore 641 003*
2 *J.K.Seeds Pvt Ltd, Hyderabad.*

a prior idea on standardization of adhesive and filler material used and also the concentration and rate to be used respectively in the above components while pelleting. Ahamed Raza (1997) while pelleting onion found both 5% and 10% maida gruel as best adhesive and the quantum of adhesive used did not influence the seed quality. Nargis (1995) working with tomato seeds reported that germination and vigour were enhanced by the leaf powder pelleting not only in fresh seeds but also in seeds aged artificially at 40°C 100% RH for 13 days. All leaf powder pelleting was found to be better in protection from deterioration than the commonly recommended (thiram) fungicidal treatment. The performance of pungam pelleting is even better than vitamin E treatment which is an anti oxidant that control deterioration rate of seeds. Coating of carrot seeds with Gypsum + DAP + micronutrient mixture found to enhance the seedling vigour and crop performance in the field than the uncoated seeds (Ramesh, 1996). Similarly in bellary onion, seed pelleted with halogen, vermicompost, biofertilizer, DAP micronutrient mixture recorded increased plant height, number of leaves and bulb yield (Ahamed Raza, 1997). Anon (1997) reported that in both tomato and soybean nutrient pelleting with $ZnSO_4$ @ 300 mg kg^{-1} using 5% maida gruel as adhesive and arappu leaf powder as filler material enhanced not only the field emergence but also the yield potential. The seed pelleting with *Azospirillum* in combination with the seedling inoculation of VAM @ 5 g $polypot^{-1}$ aided the production of elite seedling in nursery (Parameswari, 1999). Vegetable seeds sown in two different soil types at various soil moisture content exhibited delayed germination (Although pelleted seeds have been widely tested, their performance under soil conditions has not been reliable). Millier and Sooter (1967) found that the rate of emergence in pelleted seeds in soil was much lower than that of non-pelleted seeds; but the adverse effect of pelleting disappeared when the pellets were cracked, indicating the problem to be the micro-environment of the germinating seed.

Seed pelleting in onion increased the yield by stimulating the plant growth process. Longden (1973) stated that pelleting was adopted to make seed round so as to facilitate precision sowing in vegetables besides accurate dosing of seeds with chemicals. According to Moore (1973), the field emergence and efficiency of viability of plant weight at 20 days were equal in both pelleted and unpelleted seeds in lettuce, while Hovadik (1976) explicated delayed germination and reduced emergence of pelleting seeds in parsely, carrot, lettuce, cauliflower, radish, onion and cabbage. Olesen and Elkin (1977) stated that soil acidity inhibited rhizobial multiplication in the rhizosphere of developing seedlings might be overcome by lime pelleting of seed. Adverse effect of high soil moisture content was reduced when tomato seeds were pelleted. Use of pelleted sugarbeet seeds increased germination, and thereby decreased sowing rate; and registered nearly equidistant emergence which decreased labour cost on thinning besides, it increased the average root yields by 1.6 t/ha. Pelleted seeds of carrot and onion can safely be stored for upto three years under low temperature (10°C) and low RH (50%) condition. Earlier ripening of fruits by two days by using seeds pelleted with an organic solvent binder and late ripening of fruits by two days as compared with unpelleted seeds in tomato. Seeds pelleted with N, P, K, Fe, Cu, Mn, B and Zn gave 3-4 per cent higher field germination over unpelleted seeds in tomato (Konstantinov, 1983). Similarly, pelleted onion seeds registered 10-23% higher field emergence over non pelleted ones. Clay pelleting seeds in sugarbeet restored and reduced seedling emergence from wet soil . Seed germination in vegetables especially in gourds are affected by various soil types and conditions such as soil moisture content, soil temperature, salinity and alkalinity of the soil. Soil is quite heterogeneous, germination and seedling establishment may be altered by micro heterogeneous condition in the soil.

Among the vegetables, cucurbitaceous seeds are more sensitive to soil moisture levels hence field emergence in these crops being always problematic even with seeds of high germinability. Either a little high (or) low soil moisture results in poor emergence necessitating resowing. Seed pelleting with arappu powder @ 500 g per kg of seed registered a spectacular improvement in seed quality attributes in terms of germination, speed of germination, root and shoot length, dry matter production and vigour index in the gourds namely ash gourd, bitter gourd and ribbed gourd and the efficacy of the treatment became more pronounced at 60 per cent water holding capacity in the two types of soil i.e., sandy loam and clay soil. The pelleting product namely arappu acting as a butter/wick to establish better seed soil relationship, regulated the moisture supply for better emergence, growth and development (Renuga Devi, 1992).

3. SEED HARDENING

Sujatha Bashyam (1999) reported that hardening of coriander seeds with 2 per cent arappu leaf extract with 24 hrs soaking in 1.0 : 0.6 ratio followed by drying for better germination and vigour. Seed hardening using 2 per cent leaf extract followed by pelleting with $CaCO_3$ registered the maximum germination and vigour index in coriander cv. CO 1 (Sujatha Bashyam, 1999). Sujatha Bashyam (1999) reported that hardened seeds with 2% arappu leaf extract pouched in cloth bag can be stored for 4 months without much reduction in germination this might be due to hygrocopity.

- Hardening bhendi seeds with KH_2PO_4 0.5% for 16 hours found to increase the seed yield and quality.
- Karonda seed hardening with water for overnight found to improve seed germination.
- Four 'o'clock plant seed hardening with water for 24 hours improves viability.
- Isabgol seed hardening for 16-20 hours with water will improve germination.
- Oscimum seed hardening with KCl 0.1% for 20 hours found to enhance the germination potential.
- To maintain viability in Jack, Mango and Avocado seed moisture can be improved by hardening with water for 3-5 hours.
- Seed hardening/soaking for 72-84 hours with water improves the viability status in tea.
- Nutmeg seed hardening/soaking for 96-100 hours with water enhances the germination potential.
- Tomato seed hardening with sodium and potassium phosphate (0.5%) for 16 hours found to increase the germination and vigour.
- Carrot seed hardening with potassium salts will improve the viability and vigour potential.
- Manila tamarind (*Inga delci*) seed viability can be maintained by seed hardening treatment with water.

4. CONCLUSION

Seed hardening and pelleting is very much useful for precision sowing in crops like solanaceous vegetables and medicinal plants, maintenance of viability of recalcitrant seeds, remove dormancy and plant population maintenance. In protected cultivation of vegetables and flowers, it will be helpful to overcome nutrient deficiency. It is more beneficial in smaller seeds as singling resulted through pelleting helps in reducing the cost and wastage of seeds. It also reduces the problem of thinning and gap filling in seeds. In bigger seeds it is preferred for addition of needed bioactive substances to the seed based on demand and requirement. This prevents the materials applied to the field for the defined purpose in any other form (soil or foliar). The requirement of chemical is also very low. It serves in most cases a solution for enhancing establishment in problem soil. Use of botanic as filler material makes this treatment as ecofriendly, low cost and easily adaptable technology to small as well as big entrepreneurs.

REFERENCES

1. Ahamad Raza, M. (1997). Seed technological studies on bellary onion (*Allium cepa* L.) cv. Agri found dark red. M.Sc.(Ag.) Thesis, Tamil Nadu Agril. Univ., Coimbatore.
2. Anon, (1997). Annual Report, Promotion of Research and Development efforts on hybrid in selected crops. Project Co ordinator unit. NSP (crops) held at Mysore.
3. Geetharani, P. (1995). Studies on mid-storage seed treatment and productivity in tomato (*Lycopersicon esculentum* Mill.) and (*Solanum melongena* L.). M.Sc.(Ag.) Thesis, Tamil Nadu Agril. Univ., Coimbatore.
4. Konstantinov, G. (1983). Growing direct sown tomato cultivar Druzh ba from pelleted seed. *Gradinariki Lzarska Nauka., 20 (4)* : 53-57.
5. Longden, P.C. (1975). Sugar beet seed pelleting. *Adas. Quart. Rev., 18*: 73-80.
6. Miller, W.F. and C. Sooter. (1967). Improving emergence of pelleted vegetable seed. *Transactions of the ASAE, 16(5):* 540-549.
7. Moore, B. (1973). Advances in the study of vesicular arbuscular mycorrhizal. *Ann.Rev.Phytopath., 11* : 171-196.
8. Nargis, S. (1995). Seed pelleting, magnetic treatment and radiation on the performance of differentially aged seeds in tomato cv. PKM 1. M.Sc.(Ag.) Thesis, Tamil Nadu Agril. Univ., Coimbatore.
9. Olsen, F.J. and D.M. Elkins. (1977). *Agronomy J., 69* : 871-874.
10. Parameswari, K. (1999). Seed technological studies in tamarind (*Tamarindus Indica* linn) M.Sc. (Ag) Thesis, Tamil Nadu Agricultural University, Coimbatore.
11. Ramesh, D. (1996). Studies on enhancement of seed germination and vigour in carrot (*Daucus carota* L.) cv. Zino. M.Sc.(Ag.) Thesis, Tamil Nadu Agri. Univ., Coimbatore.
12. Renuga Devi, J.(1992). Studies on seed treatment and storage in Ash gourd (*Benincasa hispida* Thumb.), Bitter gourd (*Momordica charantia* Lin.) and Ribbed gourd (*Luffa acutangula* Roxb.) seeds. M.Sc. (Ag.) Thesis, Tamil Nadu Agricultural University, Coimbatore – 3.
13. Sujatha Bashyam, (1999). Seed quality testing and management procedures in three important condiments of coriander (*Coriandrum sativum* Linn.) fennel (*Foeniculum vulgare* Mill) and fennugreek (*Trigonella feonum graecum* Linn.). M. Sc. (Ag.) Thesis, Tamil Nadu Agricultural University, Coimbatore.

PERFORMANCE OF HARDENED AND PELLETED SEEDS OF CROP SPECIES UNDER MINE SPOIL SITUATIONS

S. KAVITHA[1], M. BHASKARAN[2] AND K. VANANGAMUDI[3]

1. INTRODUCTION

Mining industry is next to agriculture and is spread over vast areas throughout the world. But mining and its allied works have wasted and ravaged the land surface, particularly the loss of top soil, which is an integral storage and exchange site for nutrients. Due to mining activity, not only the existing vegetation is destroyed, but also the flora, fauna, hydrological relations and the soil biological systems are affected as a result of soil loss. Revegetation on these spoils to restore the pre-mining ecological equilibrium or to return such areas to an ecologically acceptable post-mining land use, is a challenging problem. Neyveli Lignite Corporation Limited is a leading Government of India Enterprise, pioneered and planned the exploitation of large lignite resources at Neyveli in the Southern part of India, since 1971.

The mined soil of Neyveli Lignite Corporation (NLC) is sandy clayey in texture devoid of organic matter and other essential plant nutrients. Surface crusting, high bulk density, poor buffering capacity, lack of soil structure and inadequate soil microflora are the soil problems that restrict plant growth and productivity of the mine spoil over burdens. Therefore, it is an imperative need to reclaim mine spoil so as to effectively rehabilitate the environment. At present, crop production is ventured in mine spoil soils which is subjected to high degree of uncertainity in productivity. It is

1 *Central Institute for Cotton Research, Coimbatore – 641 003.*
2 *National Seed Research and Training Centre, Varanasi – 221 106.*
3 *Dean (Agriculture), Tamil Nadu Agricultural University, Coimbatore - 641 003.*

imperative to evolve a suitable strategy for augmenting the productivity of such mine spoils. Reclamation and rehabilitation of the mine spoil can be made with selection of suitable and seed management technologies *viz.*, seed hardening and film coating with organic and inorganic substances and soil amendment techniques, which include the addition of red earth, sand and FYM.

Because seed is one of the major input in agriculture and decides the future of crops. The rationale for preplant seed treatment is to mobilize the seeds own resources and to augment them with external resources to maximum improvement in stand establishment and yield. Physiological and non-physiological treatments are needed to remove several soil and hydrological constraints. Physiological seed treatments that improve seed performance, are based primarily on seed hydration and dehydration. Among several non-physiological seed treatments, coating or pelleting can indirectly improve seed germination and stand establishment.

2. EXPERIMENTAL APPROACH

2.1. Description of the study area

Neyveli, a remote hamlet in Cuddalore district occupies a pre-eminent place in industrial map of India. and located 200 kms to south west of Chennai. Neyveli Lignite field is the second important fossil fuel in South India next to the Godavari Valley Coal field. Lignite occurs over an area of 480 sq. km with deposits of 3300 million tonnes.

2.2. Physical, chemical and microbial properties of mine spoil

The lignite mine spoil at Neyveli is characterized to have very pale brown to pink colour, massive structure, sandy clay loam to sandy loam texture, high bulk density, lower Cation Exchange Capacity, poor in the status of exchangeable Ca, Mg and Na, organic carbon, available N, P, K, Fe, Mn, Cu, Zn and S and requires improvement in physical properties like structure, texture and bulk density. The major nutrients, in the study area namely organic carbon (which is a measure of nitrogen availability in soils), available N, P and K were found to be in a very low range (N: 8.4 - 16.8 kg ha^{-1}, P: 1.0 - 3.0 kg ha^{-1} and K: 5.0 - 14.0 kg ha^{-1}) and they played a major role in the growth and productivity of crops. Micronutrients have been identified as essential for crop growth, but they are required in very small quantities by the plants and are seldom deficient in mine spoils (Fe: Trace - 0.266, Mn: Trace - 0.314, Cu: 0.002 - 0.146, Zn: Trace - 1.28 and S: 31.0 - 81.5).

Organic carbon content and Cation Exchange Capacity (CEC) are integrative properties of soil and indicative of soil fertility status. High organic matter status of soil is associated with high CEC. The CEC and organic carbon content of mine spoils ranged from 1.5 to 3.9 Cmol (p+)/kg and 0.029 - 0.176, respectively. The Electrical Conductivity (EC) is a measure of soluble salt concentration or salinity level of soil which can affect the growth of crops, if its value exceeds certain limit.

pH is a measure of acidity or alkalinity level of soil. Any drastic change in pH causes marked changes in the soil environment, especially in the nutrient availability, in the composition and density of soil microflora etc. In the present investigation, the EC and pH values of mine spoils were found to range from 0.025 to 0.135 dSm^{-1} and 6.7 - 7.9, respectively. Thus, the pH level observed in the present study was within the safer limit and was not a growth constraint for crops.

Generally microbial activities are lacking in mine spoils. The results of analysis of mine spoil indicated that the lignite mine spoil at Neyveli is characterized to have very low bacterial (2.9) and fungal (4.5 per gram of soil) population. Actinomycetes were nil. By the mining activity all the microflora and micro fauna were completely lost, but presently the situation is improving due to the application of biofertilizer in the spoil areas. N_2 fixing microbes are constantly multiplying by the reclamation process.

2.3. Treatment details

To evaluate the performance of pre-sowing invigouration treatment on better survival, establishment, growth and productivity of maize, sorghum, ragi, blackgram, cotton, tomato and marigold under NLC mine spoil condition for revegetation of mine spoils, the seeds were given hardening treatments with different chemicals and growth regulators and coated with carbendazim + imidachloprid +DAP + micronutrient mixture + seed coating polymer + biofertilizer. Mine spoils obtained from NLC, Neyveli were utilized for the experiment. The hardened, film coated and pelleted seeds and mine spoil enriching amendments were evaluated under pot culture experiments along with untreated seeds. The treatments tried for each crop are given below.

Maize

Dry seed (Control) (T_1), seed hardening with 2% KH_2PO_4 + seed film coating with pink polykote @ 3g kg^{-1} + carbendazim @ 2g kg^{-1} + imidachloprid @ 1ml in 5ml of water + *Azospirillum* @ 60g kg^{-1} of seed (T_2) and seed hardening with 2% KH_2PO_4 + Seed film coating with pink polykote @ 3g kg^{-1} + carbendazim @ 2g kg^{-1} + imidachloprid @ 1ml in 5ml of water + 30g DAP + 20g micronutrient mixture + *Azospirillum* @ 60g kg^{-1} of seed (T_3).

Sorghum

Dry seed (Control) (T_1), seed hardening with 2% KH_2PO_4 + seed film coating with pink polykote @ 3g kg^{-1} + carbendazim @ 2g kg^{-1} + imidachloprid @ 1ml in 5ml of water + *Azospirillum* @ 40g kg^{-1} of seed (T_2) and seed hardening with 2% KH_2PO_4 + seed film coating with pink polykote @ 3g kg^{-1} + carbendazim @ 2g kg^{-1} + imidachloprid @ 1ml in 5ml of water + seed coating with 30g DAP + 20g micronutrient mixture + *Azospirillum* @ 40g kg^{-1} of seed (T_3).

Ragi

Dry seed (Control) (T_1), seed hardening with 0.1 ppm brassinolide + seed film coating with pink polykote @ 3g kg^{-1} seed + carbendazim @ 2g kg^{-1} + imidachloprid @ 1ml in 5ml of water + *Azospirillum* @ 60g kg^{-1} of seed (T_2) and seed hardening with 0.1 ppm brassinolide + seed film coating with pink polykote @ 3g kg^{-1} seed + carbendazim @ 2g kg^{-1} + imidachloprid @ 1ml in 5ml of water + 30g DAP + 20g micronutrient mixture + *Azospirillum* @ 60g kg^{-1} of seed (T_3).

Blackgram

Dry seed (Control) (T_1), seed hardening with 100 ppm $ZnSO_4$ + seed film coating with black polykote @ 3g kg^{-1} seed + carbendazim @ 2g kg^{-1} + imidachloprid @ 1ml in 5ml of water + *Rhizobium* @ 30g kg^{-1} of seed (T_2) and seed hardening with 100 ppm $ZnSO_4$ + Seed film coating with black polykote @ 3g kg^{-1} seed + carbendazim @ 2g kg^{-1} + imidachloprid @ 1ml in 5ml of water + 30g DAP + 20g micronutrient mixture + *Rhizobium* @ 30g kg^{-1} of seed (T_3).

Cotton

Dry seed (Control) (T_1), seed hardening with 2% KCl + seed film coating with pink polykote @ 3g kg^{-1} seed + carbendazim @ 2g kg^{-1} + imidachloprid @ 1ml in 5ml of water + *Azospirillum* @ 80g kg^{-1} of seed (T_2) and seed hardening with 2% KCl + seed film coating with pink polykote @ 3g kg^{-1} seed + carbendazim @ 2g kg^{-1} + imidachloprid @ 1ml in 5ml of water + 30g DAP + 20g micronutrient mixture + *Azospirillum* @ 80g kg^{-1} of seed (T_3).

Tomato

Dry seed (Control) (T_1), seed hardening with 2% KH_2PO_4 + seed film coating with red polykote @ 3g kg^{-1} seed + carbendazim @ 2g kg^{-1} + imidachloprid @ 1ml in 5ml of water + *Azospirillum* @ 50g kg^{-1} of seed (T_2) and seed hardening with 2% KH_2PO_4 + seed film coating with red polykote

@ 3g kg^{-1} seed + carbendazim @ 2g kg^{-1} + imidachloprid @ 1ml in 5ml of water + 30g DAP + 20g micronutrient mixture + *Azospirillum* @ 50g kg^{-1} of seed (T_3).

Marigold

Dry seed (Control) (T_1), seed hardening with 0.75% KNO_3 + seed film coating with pink polykote @ 3g kg^{-1} seed + carbendazim @ 2g kg^{-1} + imidachloprid @ 1ml in 5ml of water + *Azospirillum* @ 400g kg^{-1} of seed (T_2) and seed hardening with 0.75% KNO_3 + seed film coating with pink polykote @ 3g kg^{-1} seed + carbendazim @ 2g kg^{-1} + imidachloprid @ 1ml in 5ml of water + 30g DAP + 20g micronutrient mixture + *Azospirillum* @ 400g kg^{-1} of seed (T_3).

The mine spoil enriching amendments used for the experiment are given below (i.e.) M_1-Mine spoil alone (MS); M_2-Red soil alone (RS); M_3-MS + RS @ 1:1; M_4 -MS + RS + Sand + FYM @ 1: 2: 1:1 (Enriched mine spoil). Mine spoils were gently powdered, weighed and transferred to each pot. The calculated quantity of soil amendments were mixed thoroughly and added.

3. EXPERIMENTAL FINDINGS

3.1. Emergence

The 2% KH_2PO_4 hardened and carbendazim + imidachloprid + DAP + micronutrient mixture + polykote + *Azospirillum* film coated seeds of maize, sorghum and tomato sown in mine spoil recorded higher emergence, which were 9.0, 8.0 and 8.0 per cent higher, respectively than untreated seeds (Fig. 1). Hardening and film coating of seeds improved field emergence over control seed. During seed hardening, the first phase of germination ends with completion of imbibition process and hence the time taken from sowing to emergence is much reduced (Hegarty, 1970). Seed soaking with organic salt solutions also advanced the germination process in field conditions (Durrant *et al.*, 1983). Hardening of seed increased the field germination of corn by promoting embryo growth (Zubenko, 1959). The improvement in emergence could also be ascribed to activation of cells, which results in the enhancement of mitochondrial activity leading to the formation of more high energy compounds and vital bimolecules, which were made available during the early phase of germination (Dharmalingam *et al.*, 1988). These initial changes culminate in enlargement of the latent embryo. The hardened seeds were dried back to its original moisture content, the triggered germination events were halted. When the seeds are sown, germination event begins from the point where it stopped previously. As a consequence, early emergence and establishment of seedling are achieved before the available soil moisture is depleted.

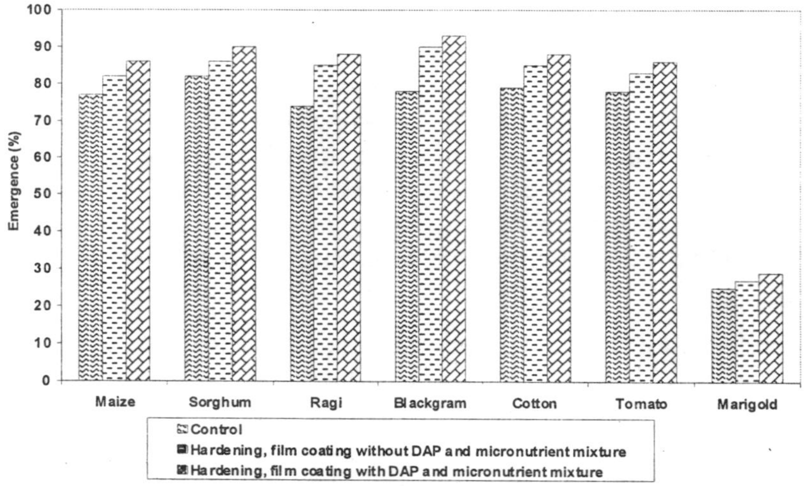

Figure 1. Seed hardening and film coating on emergence (%) under lignite mine spoil condition

In ragi, 0.1 ppm brassinolide hardened and 2g carbendazim + 1ml imidachloprid + 30g DAP + 20g micronutrient mixture + 3g pink polykote + 60g *Azospirillum* film coated seeds recorded 14.0 per cent higher emergence, over untreated seeds (Fig. 1). This might be due to the stimulation of auxin by brassinolide which in turn helps the induction of cell differentiation, cell elongation, cell division and other enzyme activities which promote germination. Improvement in emergence by seed treatment with growth regulators was also reported by Asborno *et al.* (1999) in paddy and Niranjana *et al.* (1999) in maize, paddy and sunflower.

In pulses, the micronutrients play an important role. In blackgram, 100 ppm $ZnSO_4$ hardened and 2g carbendazim + 1ml imidachloprid + 30g DAP + 20g micronutrient mixture + 3g black polykote + 30g *Rhizobium* film coated seeds recorded higher emergence of 15.0 per cent over untreated seeds (Fig. 1; Plate 1). This might be due to the above said treatment contains Zn which is constituent of an enzyme essential for the synthesis of plant hormone indole acetic acid, which is presumed to be capable for stimulating emergence

KCl (2%) hardened and 2g carbendazim + 1ml imidachloprid + 30g DAP + 20g micronutrient mixture + 3g pink polykote + 80g *Azospirillum*

Plate 1. Influence of seed hardening and film coating on emergence (%) of blackgram seeds in NLC mine spoils

film coated seeds of cotton sown recorded higher emergence, which was 9.0 per cent higher than untreated seeds (Fig. 1). In the case of marigold, 0.75% KNO_3 hardened and 2g carbendazim + 1ml imidachloprid + 30g DAP + 20g micronutrient mixture + 3g pink polykote + 400g *Azospirillum* film coated seeds registered higher emergence of 4.0 per cent in mine spoil compared to untreated control seeds (Fig. 1). Polymer film coating serves as an effective dust free delivery system for fungicides and insecticides used to control diseases and pests during seed germination and seedling emergence (Scott, 1989; Robani, 1994). According to Iqbal and Srinivasan (1987), polymers are considered as beneficial in prolonging the moisture supply under stress conditions, since they are hydrophilic and they have a capacity to absorb water about 100 to 1000 times of their weight from the surrounding rhizosphere. These special hydrogel transform themselves in to water laden gel 'chunks' which act as local water reservoirs. Over a period of time, the water from these reservoirs is released gradually to the soil and thereby to plants as the need be. Surface coatings of seed with hydrophilic colloids were effective accelerating water uptake from sand and in accelerating emergence from soil under ordinary planting conditions in the field (Dexter and Takao, 1960).

Interaction effect of the best performing treatment over other treatments are shown in table 1.

3.2. Root length

In case of maize, sorghum and tomato, the 2% KH_2PO_4 hardened and carbendazim + imidachloprid + DAP + micronutrient mixture + polykote

Table 1. The best performing seed treatment for maize, sorghum, ragi, blackgram, cotton and tomato with respect to emergence under pot culture experiment

Sl. No.	Crop	Best performing treatment details	Percentage increase over		
			Control in enriched mine spoil	Treated in mine spoil	Control in mine spoil
1.	Maize	2% KH_2PO_4 hardened + 2g carbendazim + 1ml imidachloprid + 30g DAP + 20g micronutrient mixture + 3g pink polykote + 60g ***Azospirillum*** film coated seed sown in enriched mine spoil (MS+RS+Sand+FYM @1:2:1:1)	7	13	22
2.	Sorghum	2% KH_2PO_4 hardened + 2g carbendazim + 1ml imidachloprid + 30g DAP + 20g micronutrient mixture + 3g pink polykote + 40g ***Azospirillum*** film coated seed sown in enriched mine spoil (MS+RS+Sand+FYM @1:2:1:1)	6	6	14
3.	Ragi	0.1 ppm brassinolide hardened + 2g carbendazim + 1ml imidachloprid + 30g DAP + 20g micronutrient mixture + 3g pink polykote + 60g ***Azospirillum*** film coated seed sown in enriched mine spoil (MS+RS+Sand+FYM @1:2:1:1)	11	10	24
4.	Blackgram	100 ppm $ZnSO_4$ hardened + 2g carbendazim + 1ml imidachloprid + 30g DAP + 20g micronutrient mixture + 3g black polykote + 30g ***Rhizobium*** film coated seed sown in enriched mine spoil (MS+RS+Sand+FYM @1:2:1:1)	12	5	20
5.	Cotton	2% KCl hardened + 2g carbendazim + 1ml imidachloprid + 30g DAP + 20g micronutrient mixture + 3g pink polykote + 80g ***Azospirillum*** film coated seed sown in enriched mine spoil (MS+RS+Sand+FYM @1:2:1:1)	9	8	17
6.	Tomato	2% KH_2PO_4 hardened + 2g carbendazim + 1ml imidachloprid + 30g DAP + 20g micronutrient mixture + 3g red polykote + 50g ***Azospirillum*** film coated seed sown in enriched mine spoil (MS+RS+Sand+FYM @1:2:1:1)	8	10	18
7.	Marigold	2% KH_2PO_4 hardened + 2g carbendazim + 1ml imidachloprid + 30g DAP + 20g micronutrient mixture + 3g pink polykote + 400g ***Azospirillum*** film coated seed sown in enriched mine spoil (MS+RS+Sand+FYM @1:2:1:1)	6	19	23

+ *Azospirillum* film coated seeds sown in mine spoil recorded higher root length of 22.3, 23.2 and 41.0 per cent, respectively over untreated seeds (Fig. 2). In ragi, 0.1 ppm brassinolide hardened and 2g carbendazim + 1ml imidachloprid + 30g DAP + 20g micronutrient mixture + 3g pink polykote + 60g *Azospirillum* film coated seeds recorded higher root length which accounted for 24.4 per cent increase over untreated seed (Fig. 2 ; Plate 2).

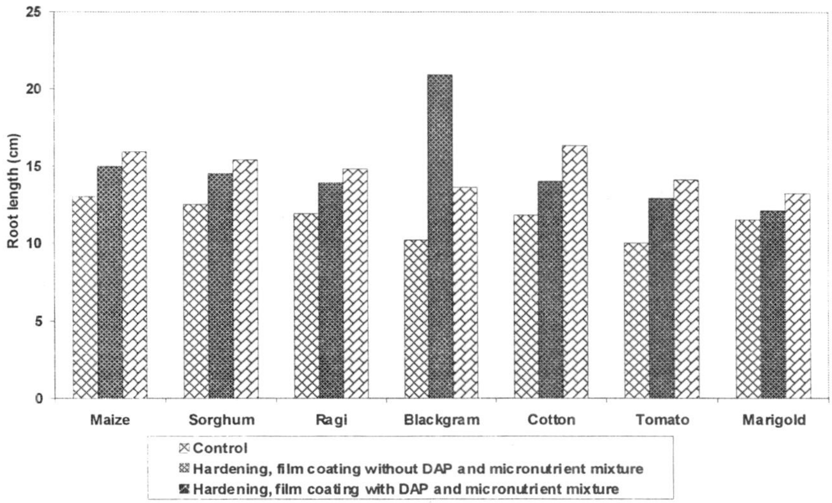

Figure 2. Seed hardening and film coating on root length (cm) under lignite mine spoil condition

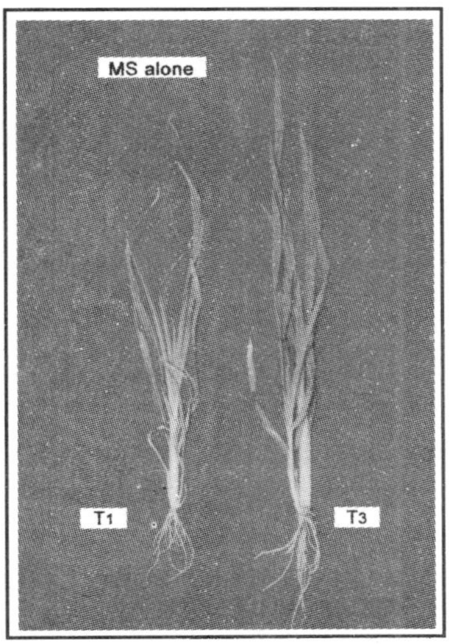

Plate 2. Influence of seed hardening and film coating on root length (cm) of ragi in NLC mine spoils

In case of blackgram, 100 ppm $ZnSO_4$ hardened and 2g carbendazim + 1ml imidachloprid + 30g DAP + 20g micronutrient mixture + 3g black polykote + 30g *Rhizobium* film coated seeds recorded higher root length compared to untreated seeds. The per cent increase over untreated seed was 33.3 per cent (Fig. 2 ; Plate 3).

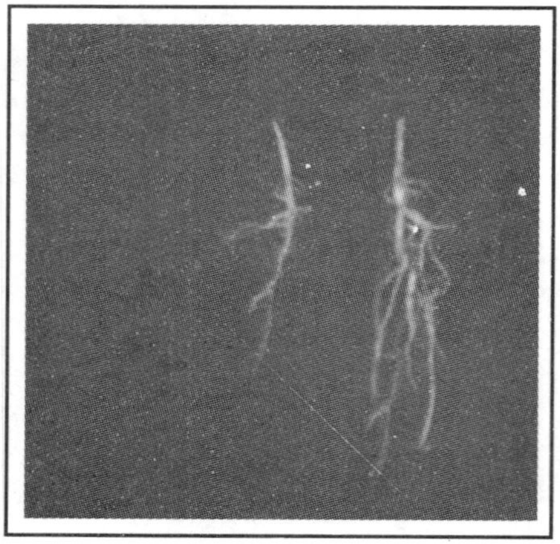

Plate 3. Influence of seed hardening and film coating on root length (cm) of blackgram in NLC mine spoils

In cotton, the 2% KCl hardened and 2g carbendazim + 1ml imidachloprid + 30g DAP + 20g micronutrient mixture + 3g pink polykote + 80g *Azospirillum* film coated seeds recorded higher root length of 38.1 per cent compared to the root length produced by untreated seeds sown in mine spoil (Fig. 2; Plate 4). In marigold, 0.75% KNO_3 hardened and 2g carbendazim + 1ml imidachloprid + 30g DAP + 20g micronutrient mixture + 3g pink polykote + 400g *Azospirillum* film coated seeds recorded higher root length (Fig. 2).

The beneficial effect of nutrients and polymer present in the pelleting material might be the reason for better root growth. It might be also due to enhanced metabolic activity and earliness in emergence. The root growth was increased due to the biofertilizer (*Azospirillum* and *Rhizobium*) treatment. The increase in plant growth attributes might be due to increased uptake of nutrients by microorganisms associated plants and their synergistic effect. The microorganisms that are used as biofertilizers stimulate the plant growth by providing necessary nutrients by their colonization at the rhizosphere or by their symbiotic association. The association may also regulate the physiological processes in the ecosystems by involving in the decomposition of organic matter, fixation of atmospheric nitrogen, secretion of growth promoting substances, increasing the availability of mineral nutrients and protecting the plants from pathogen. Thus, the rhizosphere effect through microbial activity modifies the plant itself by providing the plant growth substances, and increasing the availability of elements to the root zone (Newman *et al.*, 1992; Jakobsen *et al.*, 1994).

Plate 4. Influence of seed hardening and film coating on root length (cm) of cotton in NLC mine spoils

Interaction effect of the best performing treatment over other treatments are shown in table 2.

3.3. Dry matter production and plant height

In case of maize and sorghum, the higher drymatter production was observed in plants from 2% KH_2PO_4 hardening and carbendazim + imidachloprid + DAP + micronutrient mixture + polykote + *Azospirillum* film coating treatment, which was 24.2 and 18.2 per cent higher than control (Fig. 4). The increase in drymatter production due to the above said treatment in the present study is in accordance with the report of Selvaraju (1992) in sorghum. This might be due to the combined effect of hardening and film coating which enhanced the root-shoot ratio and nutrient uptake and so the drymatter production (Rangasamy ***et al.***, 1993). In **maize,** 2% KH_2PO_4 hardened and 2g carbendazim + 1ml imidachloprid + 30g DAP + 20g micronutrient mixture + 3g pink polykote + 60g ***Azospirillum*** film coated seeds recorded the increased plant height (Fig. 3; Plate 5) in mine spoil over untreated control.

In the case of sorghum, the same treatment performed well by recording increased plant height in mine spoil (Fig. 3; Plate 6) over untreated seed.

In tomato, 14.8 per cent increased plant height was recorded in mine spoil (Plate 7) by 2% KH_2PO_4 hardened and 2g carbendazim + 1ml

Table 2. The best performing seed treatment for maize, sorghum, ragi, blackgram, cotton and tomato with respect to root length and root volume at vegetative stage under pot culture experiment

Sl. No.	Crops	Best performing treatment details	Root length at vegetative stage (cm)		
			Percentage increase over		
			Control in enriched mine spoil	Treated in mine spoil	Control in mine spoil
1.	Maize	2% KH_2PO_4 hardened + 2g carbendazim + 1ml imidachloprid + 30g DAP + 20g micronutrient mixture + 3g pink polykote + 60g ***Azospirillum*** film coated seed sown in enriched mine spoil (MS+RS+Sand+FYM @1:2:1:1)	16.3	47.8	80.8
2.	Sorghum	2% KH_2PO_4 hardened + 2g carbendazim + 1ml imidachloprid + 30g DAP + 20g micronutrient mixture + 3g pink polykote + 40g ***Azospirillum*** film coated seed sown in enriched mine spoil (MS+RS+Sand+FYM @1:2:1:1)	17.3	50.0	84.8
3.	Ragi	0.1 ppm brassinolide hardened + 2g carbendazim + 1ml imidachloprid + 30g DAP + 20g micronutrient mixture + 3g pink polykote + 60g ***Azospirillum*** film coated seed sown in enriched mine spoil (MS+RS+Sand+FYM @1:2:1:1)	17.8	52.0	89.1
4.	Blackgram	100 ppm $ZnSO_4$ hardened + 2g carbendazim + 1ml imidachloprid + 30g DAP + 20g micronutrient mixture + 3g black polykote + 30g ***Rhizobium*** film coated seed sown in enriched mine spoil (MS+RS+Sand+FYM @1:2:1:1)	24.8	29.4	72.5
5.	Cotton	2% KCl hardened + 2g carbendazim + 1ml imidachloprid + 30g DAP + 20g micronutrient mixture + 3g pink polykote + 80g ***Azospirillum*** film coated seed sown in enriched mine spoil (MS+RS+Sand+FYM @1:2:1:1)	21.0	58.9	119.5
6.	Tomato	2% KH_2PO_4 hardened + 2g carbendazim + 1ml imidachloprid + 30g DAP + 20g micronutrient mixture + 3g red polykote + 50g ***Azospirillum*** film coated seed sown in enriched mine spoil (MS+RS+Sand+FYM @1:2:1:1)	31.7	29.8	83.0
7.	Marigold	2% KH_2PO_4 hardened + 2g car bendazim + 1ml imidachloprid + 30g DAP + 20g micronutrient mixture + 3g pink polykote + 400g ***Azospirillum*** film coated seed sown in enriched mine spoil (MS+RS+Sand+FYM @1:2:1:1)	19.4	25.8	44.3

Plate 5. Influence of seed hardening and film coating on plant height (cm) of maize in NLC mine spoils

Plate 6. Influence of seed hardening and film coating on plant height (cm) of sorghum in NLC mine spoils

imidachloprid + 30g DAP + 20g micronutrient mixture + 3g red polykote + 50g *Azospirillum* film coated seed at maturity stage over untreated seed (Fig. 3; Plate 7).

Plate 7. Influence of seed hardening and film coating on plant height (cm) of tomato in NLC mine spoils

The increase in plant height due to seed hardening was reported by Mehrotra *et al.* (1968) in wheat. Hardening increased the volume of root system which could have greatly contributed to better exploitation of soil water and nutrition resulting in increased plant height (Kannaiyan, 1987)

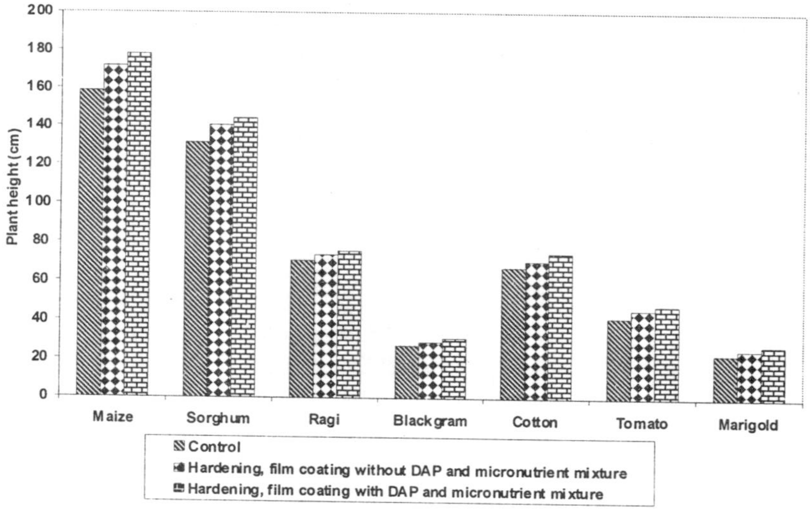

Figure 3. Seed hardening and film coating on plant height (cm) under lignite mine spoil condition

In ragi, the increase in drymatter accumulation at different stages of seedling was observed from the 0.1 ppm brassinolide hardened and 2g carbendazim + 1ml imidachloprid + 30g DAP + 20g micronutrient mixture

+ 3g pink polykote + 60g *Azospirillum* film coated seeds, which was 24.3 per cent higher over control (Fig. 4). Increase in drymatter production by the use of brassinolide was also repored by Hayat *et al.* (2001) in wheat seeds. Use of growth regulators increase the uptake of nutrients like N and P, which aid in more effective root growth and leaf area (Rangasamy, 1986), which has positive influence on drymatter production.

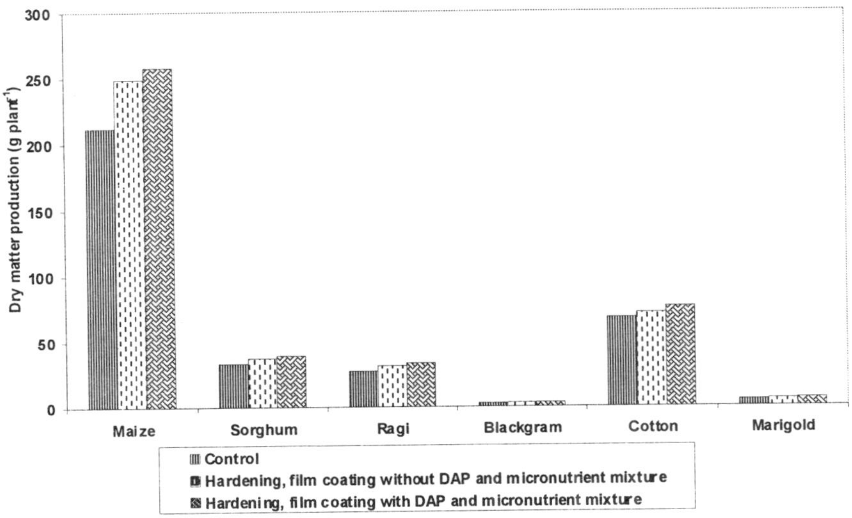

Figure 4. Seed hardening and film coating on dry matter production under lignite mine spoil condition

Brassinolide (0.1 ppm) hardened and 2g carbendazim + 1ml imidachloprid + 30g DAP + 20g micronutrient mixture + 3g pink polykote + 60g *Azospirillum* film coated seeds of ragi recorded higher plant height in mine spoil (Plate 8) at maturity stage compared to control seed. Increase in plant height by the use of growth regulators has been reported by Sandhu and Hussain (1961) in pearl millet. The increase in plant height by brassinolide might be attributed to its powerful synergistic interaction with available endogenous auxin and it could be observed in terms of increased cell wall plasticity and cell elongation (Yopp *et al.*, 1981; Chio *et al.*, 1990).

In blackgram, 100 ppm $ZnSO_4$ hardened and 2g carbendazim + 1ml imidachloprid + 30g DAP + 20g micronutrient mixture + 3g black polykote + 30g *Rhizobium* film coated seeds produced higher drymatter (21.7 per cent), plant height (20.5 per cent) (Fig. 4) and number of nodules (Table 3) (Fig. 4; 3). Similar results have been reported by Kavitha (2002). The increased plant height could be attributed to the presence of ammoniacal

Plate 8. Influence of seed hardening and film coating on plant height (cm) of ragi in NLC mine spoils

nitrogen, phosphorus and other essential micronutrients in the film coating combination. Certain seed coating materials improve seedling vigour at changing soil moisture especially in the sub-optimal range and this was reported by Mucke (1987). However, the increase in the number of nodules plant^{-1} was rapid during the later part of the crop growth between flowering to maturity and it was uniform throughout the growth stages. Higher number of nodules registered after imposing pelleting treatment may be attributed to the favourable effect of micronutrients on growth and multiplication of *Rhizobium*. This increased nodulation during later part of the growth could have promoted higher drymatter accumulation also. Lu *et al.* (1987) opined that the pelleting treatments increased the number of nodules plant^{-1} in legumes. Normally high population of rhizobia in pulses before sowing is required to bring about effective nodulation, because in pulses the productivity mainly depends upon optimum nodulation (Kurundkar *et al.*, 1991).

In cotton, 2% KCl hardened and 2g carbendazim + 1ml imidachloprid + 30g DAP + 20g micronutrient mixture + 3g pink polykote + 80g *Azospirillum* film coated seeds increased the dry matter production and plant height (which was 11.0 per cent higher in mine spoil and 14.7 per cent higher in enriched at maturity stage compared to plant height recorded by untreated seeds) (Fig 3 and 4). In marigold, 0.75% KNO_3 hardened and 2g carbendazim

Table 3. Influence of seed hardening, film coating treatments and mine spoil enriching amendments on number of nodules plant $^{-1}$ at different growth stages in blackgram cv. VBN 4 under pot culture experiment

Treatments	Vegetative stage					Flowering stage					Maturity stage					Mean
	M_1	M_2	M_3	M_4	Mean	M_1	M_2	M_3	M_4	Mean	M_1	M_2	M_3	M_4	Mean	
T_1	1.1	1.6	1.4	1.7	1.5	6.7	7.6	7.2	7.9	7.4	14.4	17.5	16.6	19.0	16.9	8.6
T_2	1.4	1.9	1.7	2.0	1.8	7.4	8.4	8.1	8.7	8.2	16.4	19.4	18.6	20.8	18.8	9.6
T_3	1.6	2.1	1.9	2.3	2.0	8.9	10.2	9.6	10.8	9.9	18.2	21.6	20.4	23.1	20.8	10.9
Mean	1.4	1.9	1.7	2.0	1.7	7.7	8.7	8.3	9.1	8.5	16.3	19.5	18.5	21.0	18.8	9.7

Mine spoil enriching amendments mean	M_1	M_2	M_3	M_4
	8.5	10.0	9.5	10.7

	T	M	S	T x M	M x S	T x S	T x M x S
SEd	0.05	0.06	0.05	0.10	0.10	0.09	0.17
CD(P=0.05)	0.10**	0.11**	0.10**	0.20**	0.20**	0.17**	NS

+ 1ml imidachloprid + 30g DAP + 20g micronutrient mixture + 3g pink polykote + 400g *Azospirillum* film coated seeds recorded higher plant height in mine spoil than untreated seed (Fig. 3). The highest drymatter accumulation due to this treatment at the earlier phase of crop growth indicated the possible increase in seedling vigour. Increase in plant height due to polykote, carbendazim, imidachloprid, DAP, micronutrient mixture and biofertilizer (*Azospirillum* and *Rhizobium*) might be attributed to the effect of nutrients present in the coating materials, which could improve the growth resulting in higher plant height (Farley and Draycott, 1978), or due to enhanced seedling establishment and possibly due to activation of metabolic activity of seed (Ponnuswamy and Vijaya, 1997; Thiyageshwari and Ramanathan, 1998). The activation of metabolic activity of seed could also be due to hydrophilic polymer present in the coating material, which might improve the rate of water uptake by the seed (Baxter and Waters, 1986) leading to early germination and better seedling establishment, which might also help in better plant height. Increased plant height using polymer was reported by Angamuthu (1991) in minor millets. Min (1988) in alfalfa, in maize hybrids; Rana *et al.* (2001) in mustard. Better plant growth could be the possible reason for higher drymatter production (Nayak *et al.*, 1989). Hence, improved plant height have ultimately improved the drymatter production.

The nitrogen containing compounds present in the coating material coupled with the additional nitrogen fixed by *Azospirillum* added in the coating material might have stimulated the seedling growth by increasing the seed cytokinin (Khan, 1980). Beneficial effect of biofertilizers on significant increase in the plant growth, drymatter content, number of bolls, sympodial branches and yield have been reported by Purushothaman and Gunasekaran (1981). *Azospirillum* is reported to be capable of fixing nitrogen at the rate of 0.7 kg N ha^{-1} day^{-1} and is responsible for increase in plate height. N is known to produce excess shoot growth at the expense of root growth (Cornett Zane, 1982). Inoculation of seeds with *Rhizobium* recorded more number of nodules at all the growth stages. The increased nodule number $plant^{-1}$ due to the *Rhizobium* inoculation was reported by many scientists in pulses such as redgram (Prabakaran and Srinivasan, 1995), lablab (Prabakaran and Rangarajan, 1992), chickpea (Namdeo *et al.*, 1989). According to Vaishya and Gajendragadkar (1984), the treatments *Rhizobium*, *Rhizobium* + *Azospirillum* and *Rhizobium* + *phosphobacterium* produced significantly more nodules than the control in redgram. Normally, high population of rhizobia on pulse before sowing is required to ensure survival of optimum number of rhizobia on seed and in soil in order to bring about effective nodulation, because pulse productivity mainly depends upon optimum nodulation (Kurundkar *et al.*, 1991). The increase in plant height and drymatter production might be due to increased supply of nitrogen due to more nodulation by the *Rhizobium* inoculation (Saraswathi *et al.*, 1988).

Interaction effect of the best performing treatment over other treatments are shown in table 4.

Table 4. The best performing seed treatment for maize, sorghum, ragi, blackgram, cotton and tomato with respect to drymatter production and plant height at maturity stage under pot culture experiment

Sl. No.	Crops	Best performing treatment details	Drymatter production at maturity stage (g plant^{-1}) Percentage increase over: Control in enriched mine spoil	Treated in mine spoil	Control in mine spoil	Plant height at maturity stage (cm) Percentage increase over: Control in enriched mine spoil	Treated in mine spoil	Control in mine spoil
1.	Maize	2% KH_2PO_4 hardened + 2g carbendazim + 1ml imidachloprid + 30g DAP + 20g micronutrient mixture + 3g pink polykote + 60g ***Azospirillum*** film coated seed sown in enriched mine spoil (MS+RS+Sand+FYM @1:2:1:1)	14.7	26.2	53.6	9.3	11.2	24.8
2.	Sorghum	2% KH_2PO_4 hardened + 2g carbendazim + 1ml imidachloprid + 30g DAP + 20g micronutrient mixture + 3g pink polykote + 40g ***Azospirillum*** film coated seed sown in enriched mine spoil (MS+RS+Sand+FYM @1:2:1:1)	17.5	15.8	35.9	9.5	6.7	17.1
3.	Ragi	0.1 ppm brassinolide hardened + 2g carbendazim + 1ml imidachloprid + 30g DAP + 20g micronutrient mixture + 3g pink polykote + 60g ***Azospirillum*** film coated seed sown in enriched mine spoil (MS+RS+Sand+FYM @1:2:1:1)	18.6	18.6	45.0	8.4	31.0	39.8
4.	Blackgram	100 ppm $ZnSO_4$ hardened + 2g carbendazim + 1ml imidachloprid + 30g DAP + 20g micronutrient mixture + 3g black polykote + 30g ***Rhizobium*** film coated seed sown in enriched mine spoil (MS+RS+Sand+FYM @1:2:1:1)	17.0	30.3	49.5	14.5	9.0	25.1
5.	Cotton	2% KCl hardened + 2g carbendazim + 1ml imidachloprid + 30g DAP + 20g micronutrient mixture + 3g pink polykote + 80g ***Azospirillum*** film coated seed sown in enriched mine spoil (MS+RS+Sand+FYM @1:2:1:1)	9.4	19.9	34.5	14.7	18.0	31.0
6.	Tomato	2% KH_2PO_4 hardened + 2g carbendazim + 1ml imidachloprid + 30g DAP + 20g micronutrient mixture + 3g red polykote + 50g ***Azospirillum*** film coated seed sown in enriched mine spoil (MS+RS+Sand+FYM @1:2:1:1)	-	-	-	11.1	36.0	56.2
7.	Marigold	2% KH_2PO_4 hardened + 2g carbendazim + 1ml imidachloprid + 30g DAP + 20g micronutrient mixture + 3g pink polykote + 400g ***Azospirillum*** film coated seed sown in enriched mine spoil (MS+RS+Sand+FYM @1:2:1:1)	14.4	73.9	109.1	8.1	254.7	327.5

3.4. Total chlorophyll content

In case of maize, sorghum and tomato, 2% KH_2PO_4 hardened and carbendazim + imidachloprid + DAP + micronutrient mixture + polykote + *Azospirillum* film coated seeds recorded higher total chlorophyll content of 54.7, 38.6 and 36.3 per cent in mine spoil over untreated seed at maturity stage (Fig. 5). It has been shown the K^+ exert control over membrane protein synthesis and binding, which might be the reason for the production of photoassimilatory surface and maintained high chlorophyll content (Usha *et al.*, 1999). In the case of ragi, 0.1 ppm brassinolide hardended and 2g carbendazim + 1ml imidachloprid + 30g DAP + 20g micronutrient mixture + 3g pink polykote + 60g *Azospirillum* film coated seeds recorded the highest total chlorophyll which was 48.1 per cent higher compared untreated seed at maturity stage (Fig. 5). The chlorophyll content showed increased trend upto flowering stage and then it decreased in the present study. Similar trend was noticed for chlorophyll components namely chlorophyll 'a' and 'b' by Afria and Narnolia (1999) who reported an increase in chlorophyll content by seeds hardening with growth regulators in wheat. Kalarani *et al.* (2001) reported that seed treatment with 1% KCl improved the chlorophyll stability index in ragi. The increase in chlorophyll content with plant growth regulators has been attributed to the inhibition of senescence and enhanced uptake of iron (Menon and Srivastava, 1984).

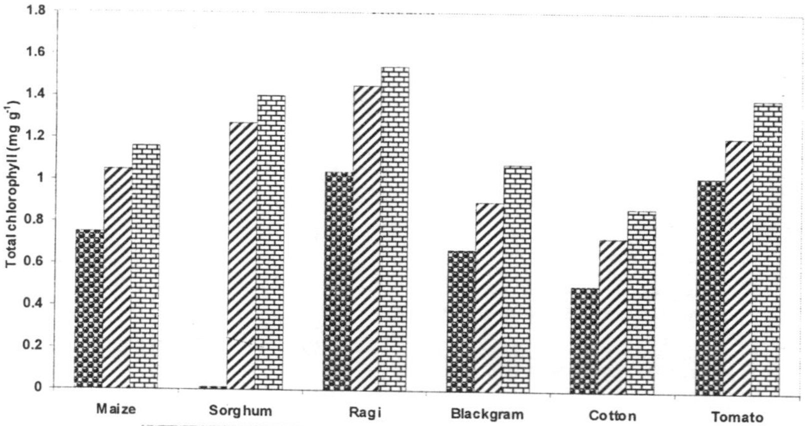

Figure 5. Seed hardening and film coating on total chlorophyll (mg g^{-1}) under lignite mine spoil condition

In blackgram, 100 ppm $ZnSO_4$ hardened and 2g carbendazim + 1ml imidachloprid + 30g DAP + 20g micronutrient mixture + 3g black polykote + 30g *Rhizobium* film coated seeds registered higher total chlorophyll in mine spoil compared to control (Fig. 5). In cotton, 2% KCl hardened and 2g carbendazim + 1ml imidachloprid + 30g DAP + 20g micronutrient mixture + 3g pink polykote + 80g *Azospirillum* film coated seeds produced higher amount of total chlorophyll in mine spoil over untreated seeds. The favourable effect of micronutrients on chlorophyll content and inturn, the photosynthetic efficiency of leaf was reported by Shekhawat *et al.* (1993) (Fig. 5). Total chlorophyll content was high in seed inoculated with *Azospirillum* and *Rhizobium*. This could be attributed to more chlorophyll synthesis due to inoculation of biofertilizers. Mahatim Singh *et al.* (1983) also observed the beneficial effects of biofertilizer inoculation and reported an increase in chlorophyll content, ascribable supply of higher amounts of nitrogen to the growing tissues.

Interaction effect of the best performing treatment over other treatments are shown in table 5.

Table 5. The best performing seed treatment for maize, sorghum, ragi, blackgram, cotton and tomato with respect to total chlorophyll content under pot culture experiment

Sl. No.	Crop	Best performing treatment details	Percentage increase over		
			Control in enriched mine spoil	Treated in mine spoil	Control in mine spoil
1.	Maize	2% KH_2PO_4 hardened + 2g carbendazim + 1ml imidachloprid + 30g DAP + 20g micronutrient mixture + 3g pink polykote + 60g ***Azospirillum*** film coated seed sown in enriched mine spoil (MS+RS+Sand+FYM @1:2:1:1)	30.2	42.5	98.9
2.	Sorghum	2% KH_2PO_4 hardened + 2g carbendazim + 1ml imidachloprid + 30g DAP + 20g micronutrient mixture + 3g pink polykote + 40g ***Azospirillum*** film coated seed sown in enriched mine spoil (MS+RS+Sand+FYM @1:2:1:1)	28.7	31.4	82.2
3.	Ragi	0.1 ppm brassinolide hardened + 2g carbendazim + 1ml imidachloprid + 3 0g DAP + 20g micronutrient mixture + 3g pink polykote + 60g ***Azospirillum*** film coated seed sown in enriched mine spoil (MS+RS+Sand+FYM @1:2:1:1)	39.7	23.4	82.7
4.	Blackgram	100 ppm $ZnSO_4$ hardened + 2g carbendazim + 1ml imidachloprid + 30g DAP + 20g micronutrient mixture + 3g black polykote + 30g ***Rhizobium*** film coated seed sown in enriched mine spoil (MS+RS+Sand+FYM @1:2:1:1)	34.9	37.4	119.4
5.	Cotton	2% KCl hardened + 2g carbendazim + 1ml imidachloprid + 30g DAP + 20g micronutrient mixture + 3g pink poly kote + 80g ***Azospirillum*** film coated seed sown in enriched mine spoil (MS+RS+Sand+FYM @1:2:1:1)	54.7	33.3	132.0
6.	Tomato	2% KH_2PO_4 hardened + 2g carbendazim + 1ml imidachloprid + 30g DAP + 20g micronutrient mixture + 3g red polykote + 50g ***Azospirillum*** film coated seed sown in enriched mine spoil (MS+RS+Sand+FYM @1:2:1:1)	23.2	24.1	62.0

3.5. Grain yield and yield attributing characters

In maize, the important yield attributes *viz.*, cob length, cob girth, cob weight, number of seed rows cob^{-1}, number of seeds row^{-1}, number of seeds cob^{-1} and grain yield (Table 6) were significantly influenced by seed hardening and film coating treatments. The increased grain yield of 61.2 % was recorded by 2% KH_2PO_4 hardened and 2g carbendazim + 1ml imidachloprid + 30g DAP + 20g micronutrient mixture + 3g pink polykote + 60g *Azospirillum* film coated seeds in mine spoil over control (Fig. 6). There was an increase in number of seeds per cob and this could be ascribed to the increase in the number of seed rows per ear and the number of seeds per row. The increase in yield might be due to more number of seeds per cob and higher seed weight. Seed soaking treatment causes higher mitochondrial activity (Henckel, 1964), rapid embryo enlargement (Hegarty, 1970), improved the growth and yield components as well as productivity. The beneficial effects of KH_2PO_4 in improving the productivity of rainfed fingermillet was reported by Maitra *et al.* (1997).

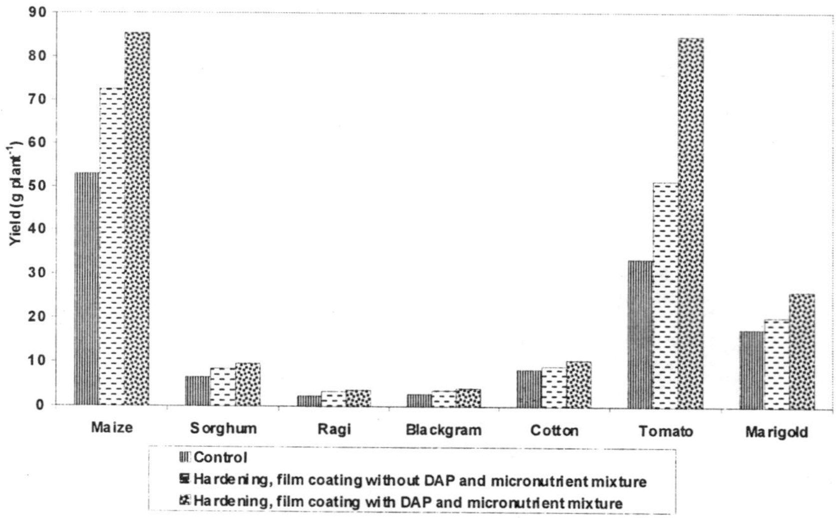

Figure 6. Seed hardening and film coating on yield (g $plant^{-1}$) under lignite mine spoil condition

The yield increase could be attributed to the presence of nutrients in both the treatments, which are the constituent of several dehydrogenase enzymes and also an activator of other enzymes. They are also necessary for the biosynthesis of IAA, the growth regulator, which is essential for normal enlargement of cells. It is also a constituent of aminoacids, from which protein and enzymes are synthesized. The hardening and film coating treatment improved the growth of the plant during early stages of the crop with increased vigour and stronger root system, which inturn derived the available soil moisture and nutrients enabling better growth and higher yield.

Polymer present in the coating material have also helped in higher rate of water uptake in turn, resulted in the early germination with more seedling vigour and better stand establishment, which ultimately led to better growth and productivity of maize. All coating treatments improved the growth of the plant during early stages of the crop with increased vigour and stronger root system, which in turn derived the available soil moisture and nutrients enabling better growth and higher yield. In general, the crop yield depends on the accumulation of photoassimilates during the growing period and the way they are partitioned between storage organs in the plants. The reason for the increased yield might be due to the increased photosynthetic efficiency through stabilisation of chlorophyll, higher production of photosynthates resulting in increased translocation of organic material from the source to sink in the treated plants. Yield increase due to polymer coating was reported by Chikkanna *et al.* (1993) in finger millet, cowpea and groundnut and Padmani *et al.* (1994) in mustard. The beneficial effect of micronutrients on yield attributes and yield of soybean was reported by Chandel *et al.* (1990).

In Sorghum, 2% KH_2PO_4 hardened and 2g carbendazim + 1ml imidachloprid + 30g DAP + 20g micronutrient mixture + 3g pink polykote + 40g *Azospirillum* film coated seeds recorded higher earhead length, earhead weight and grain yield (Fig. 6; Table 6) than untreated control seeds in ragi. In mine spoil the highest grain yield of 46.1 per cent was recorded by the above said treatment over control seeds. Kachapur *et al.* (1987) reported that micronutrients play an important role in increasing the earhead length in sorghum.

In ragi, the number of tillers, number of productive tillers (Table 6), earhead length, earhead weight and grain yield (Table 6) were higher in 0.1 ppm brassinolide hardened and 2g carbendazim + 1ml imidachloprid + 30g DAP + 20g micronutrient mixture + 3g pink polykote + 60g *Azospirillum* film coated seeds. The above said treatment produced more number of productive tillers and grain yield which account for 125.0 and 55.3 per cent increase in mine spoil, respectively over untreated seeds (Fig. 6). Seed soaking treatments caused higher mitochondrial activity (Henckel, 1964), rapid embryo development (Hegarty, 1970), improved the growth and yield components as well as productivity of various crops (Shinde and Bahalerao,

1991). Beneficial effects of brassinolide on yield was noticed in the treatments with brassinolide as one of the components. The increase in yield might be due to more number of grains per ear and higher grain weight. The increased carbohydrate content at flowering due to brassinolide application was related to higher rate of photosynthesis per unit leaf area and regulation of source - sink relation for assimilate transport (Wang and Steffens, 1985). This might be the reason for increased yield due to brassinolide treatment.

In blackgram, number of pods plant^{-1} (Plate 9), number of seeds pod^{-1}, pod yield plant^{-1} and seed yield plant^{-1} (Fig. 6; Table 6) were higher in $ZnSO_4$ (100 ppm) hardened and 2g carbendazim + 1ml imidachloprid + 30g DAP + 20g micronutrient mixture + 3g black polykote + 30g *Rhizobium* film coated seeds compared to untreated seeds. The pod and seed yield was also higher in 100 ppm $ZnSO_4$ hardened and 2g carbendazim + 1ml imidachloprid + 30g DAP + 20g micronutrient mixture + 3g black polykote + 30g *Rhizobium* film coated seeds which were 36.8 and 45.2 per cent increase over control. The sulphur in zinc sulphate was found to increase the level of vitamins, biotins and thiamins and its coenzymes (Srimathi and Malarkodi, 2000). The increased pod yield also due to unaborted reproductive structures could have resulted due to higher photosynthetic activity as the prosopis hardened and pelleted seeds recorded higher chlorophyll content during different growth stages. Moreover, increased level of phosphorus (Durai Singh, 1988) could have promoted adequate vegetative structure to produce more number of reproductive sinks. According to Durai Singh (1988), DAP and molybdenum are able to increase the fertilization efficiency in all the parts of the plant. All these reasons could have attributed to increased pod yield in 100 ppm $ZnSO_4$ hardened + pelleted seeds. Among the seed treatments, yield plant^{-1} was also higher by 35.3 per cent in the above said treatment over control in blackgram. Similar increased yield due to pelleting was reported by Jeyabal *et al.* (1992) and Srimathi and Malarkodi (2000) in soybean, and Jayaseelan (1997) in greengram and Sabir-Ahamed (1999) in blackgram and Ponnuswamy and Vijaya (1997) in cowpea. Rapid, uniformity and percentage of seedling emergence of direct seeded crop have a major impact on final yield and quality. The seeds hardened with 100 ppm $ZnSO_4$ and film coated with various nutrients in the present study enabled quick emergence and established well in the field with higher population that accounted for higher yield per unit area. As far as the increase in the seed yield per plant is concerned, the physio-chemical treatments could have triggered the biosynthesis of nucleic acids, proteins and the consequential enhancement of cell division besides, the enhanced metabolic activity of the plants resulting on the increased uptake of nutrients. This could have possibly accounted for improvement in crop performance.

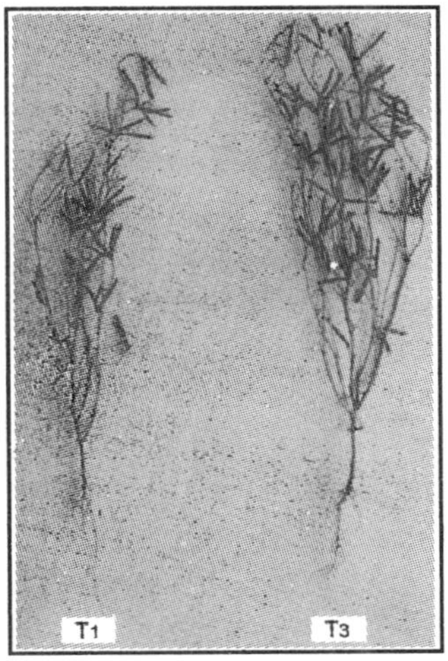

Plate 9. Influence of seed hardening and film coating on yield of blackgram in NLC mine spoils

In cotton, 2% KCl hardened + 2g carbendazim + 1 ml imidachloprid + 30g DAP + 20g micronutrient mixture + 3g pink polykote + 80g *Azospirillum* film coated seeds registered increased number of monopodia and sympodia plant^{-1} (Table 6), number of bolls plant^{-1}, kapas yield and seed yield plant^{-1} compared to control. This treatments recorded higher kapas and seed yield of 25.5 per cent (Fig. 6) higher in mine spoil compared to untreated control. The yield increase was probably due to maintenance of higher leaf area and and less proline content (Kalarani *et al.*, 2001) and the beneficial effects of the N and P present in the pelleting material. The nitrogen containing compounds present in the coating material coupled with the additional nitrogen fixed by *Azospirillum* added in the coating material might have stimulated the seedling growth by increasing the seed cytokinin (Khan, 1980). Punithavathi (1997) reported that KCl and $CaCl_2$ act as an osmoticum and also induce protoplasm hardening which inturn confer drought tolerance in seedlings.

Any technique that can maintain soil moisture at an available potential around the imbibing seed would be beneficial in stand establishment. Coating seeds with hydrophilic compounds maintain a high water potential around the seed, thereby ensuring that the soil water content does not fall below critical levels before germination. An increase in terms of kapas and seed yield was observed in the preset study due to seed polymer coating. The increase in physiological parameters might have greatly influenced the yield. Increase in number of bolls per plant due to higher chlorophyll content and better translocation of nutrients from source to sink might have resulted in the yield increase. Correlations between final harvested boll numbers and parameters of vegetative and reproductive growth indicate that the differences in drymatter partitioning play a key role in determining final boll number and thereby increased yield. The promotive effects of polymers have been reported in groundnut (Bhatnagar and Porwal, 1990), in mustard (Arvindkumar ***et al.***, 1991) and in cereals (Anon. 1986), which are in accordance with the present findings in cotton. Pelleting of seeds with nutrients may overcome the soil deficiency (Khan, 1980). No plant can produce its best growth if any one or more of the elements lacking or limiting. Bapat and Umale (1973) reported increased biomass production in crop plants when seeds were pelleted with phosphate fertilizer. Sathiyamoorthy and Vivekanandan (1990) found that treatment of seeds with zinc salts was effective in increasing plant growth of tomato.

In tomato, KH_2PO_4 (2%) hardened and 2g carbendazim + 1ml imidachloprid + 30g DAP + 20g micronutrient mixture + 3g red polykote + 50g *Azospirillum* film coated seeds recorded more number of fruits (Table 6), fruit weight and fruit yield (Fig. 6) than untreated seeds. This treatment recorded the increased fruit yield of 152.9 per cent over control seeds. In marigold, 0.75% KNO_3 hardened and 2g carbendazim + 1ml imidachloprid + 30g DAP + 20g micronutrient mixture + 3g pink polykote + 400g *Azospirillum* film coated seeds recorded more number of flowers plant^{-1} flower diameter, flower weight and flower yield (Fig. 6; Table 6). This treatment recorded the increased flower yield of 48.7 per cent in mine spoil, over untreated control seeds. Seed pelleting is a process that creates a complete nutrient environment in the immediate vicinity of the germinating seed. This nutrient environment could be created by adding variety of substances to the seed in such a way to enhance the seed quality. Natchimuthu (1997) emphasized that external application of micronutrients through seeds not only enhanced the initial germination, but also the productivity.

4. CONCLUSION

For early and good survival and establishment crop growth and yield, it is advisable to transform the mine spoil into productive agricultural land and prevent the environmental pollution seed soaking in a solution containing 2% KH_2PO_4 for maize, sorghum and tomato, 0.1 ppm brassinolide for ragi, 100 ppm $ZnSO_4$ for blackgram and 2% KCl for cotton and 0.75% KNO_3 for marigold and followed by seed film coating with 2g carbendazim,

Table 6. The best performing seed treatment for maize, sorghum, ragi, blackgram, cotton and tomato with respect to yield and yield attributing characters under pot culture experiment

Sl. No.	Crop	Best performing treatment details	Characters	Percentage increase over: Control in enriched mine spoil	Treated in mine spoil	Control in mine spoil
1.	Maize	2% KH_2PO_4 hardened + 2g carbendazim + 1ml imidachloprid + 30g DAP + 20g micronutrient mixture + 3g pink polykote + 60g ***Azospirillum*** film coated seed so wn in enriched mine spoil (MS+RS+Sand+FYM @1:2:1:1)	Grain yield (g $plant^{-1}$)	44.6	43.9	132.0
2.	Sorghum	2% KH_2PO_4 hardened + 2g carbendazim + 1ml imidachloprid + 30g DAP + 20g micronutrient mixture + 3g pink polykote + 40g ***Azospirillum*** film coated seed s own in enriched mine spoil (MS+RS+Sand+FYM @1:2:1:1)	Grain yield (g $plant^{-1}$)	37.4	36.1	98.8
3.	Ragi	0.1 ppm brassinolide hardened + 2g carbendazim + 1ml imidachloprid + 30g DAP + 20g micronutrient mixture + 3g pink polykote + 60g ***Azospirillum*** film coated seed sown in enriched mine spoil (MS+RS+Sand+FYM @1:2:1:1)	No. of productive tillers	73.1	66.7	275.0
			Grain yield (g $plant^{-1}$)	36.7	35.6	110.5
4.	Blackgram	100 ppm $ZnSO_4$ hardened + 2g carbendazim + 1ml imidachloprid + 30g DAP + 20g micronutrient mixtu re + 3g black polykote + 30g ***Rhizobium*** film coated seed sown in enriched mine spoil (MS+RS+Sand+FYM @1:2:1:1)	Pod yield (g $plant^{-1}$)	22.2	37.9	88.6
			Grain yield (g $plant^{-1}$)	20.9	27.9	129.3
5.	Cotton	2% KCl hardened + 2g carbendazim + 1ml imidachloprid + 30g DAP + 20g micronutrient mixture + 3g pink polykote + 80g ***Azospirillum*** film coated seed sown in enriched mine spoil (MS+RS+Sand+FYM @1:2:1:1)	No. of monopodia $plant^{-1}$	7.7	27.3	40.0
			No. of sympodia $plant^{-1}$	19.4	48.1	110.5
			Kapas yield (g $plant^{-1}$)	11.7	48.0	85.8
			Seed yield (g $plant^{-1}$)	11.6	3.5	22.8
6.	Tomato	2% KH_2PO_4 hardened + 2g carbendazim + 1ml imidachloprid + 30g DAP + 20g micronutrient mixture + 3g red polykote + 50g ***Azospirillum*** film coated seed sown in enriched mine spoil (MS+RS+S and+FYM @1:2:1:1)	No. of fruits $plant^{-1}$	33.3	100.0	300.0
			Fruit yield (g $plant^{-1}$)	65.2	0.6	87.2
7.	Marigold	2% KH_2PO_4 hardened + 2g carbendazim + 1ml imidachloprid + 30g DAP + 20g micronutrient mixture + 3g pink polykote + 400g ***Azospirillum*** film coated seed sown in enriched mine spoil (MS+RS+Sand+FYM @1:2:1:1)	Flower yield (g $plant^{-1}$)	43.6	400.0	643.6

1 ml imidachloprid, 30g DAP, 20g micronutrient mixture, 3g polykote and biofertilizer *Azospirillum* and *Rhizobium* and mine spoil can be enriched with red soil + sand + FYM @ 2:1:1 with 1 pot of mine spoil can be considered as the best viable option for the revegetation of mine spoils.

The fertility level of the spoil in the present dumps has to be improved by adoption of suitable techniques such as, addition of nutrients and biofertilizers through seed enhancement techniques and addition of soil amendments such as red soil, sand and FYM to the mine spoil. Using the above reclamation procedures, the mine spoil is converted into agricultural land.

5. FUTURE STRATEGIES

The studies related to seed enhancement techniques for better survival and establishment, growth and yield of various agricultural, horticultural and tree species and mine spoil enriching amendments for mine spoil management are very scanty. So research on afore said techniques are warranted.

Physical, chemical and micro biological properties of mine spoil can be analysed by the scientist of soil science and agricultural chemistry and agricultural microbiology to know the status of mine spoil for effective reclamation of mine spoils to rehabilitate the environment.

REFERENCES

1. Afria, B.S. and R.E. Narnolia. (1999). Effect of cycocel and saline irrigation on physiological attributes, yield and its components in different varieties of wheat. *Indian J. Plant Physiol., 4(4)* : 311-314.
2. Angamuthu, K. (1991). Studies on seed pelleting, physiological maturity, sowing quality and storage in small millets. M.Sc.(Ag.) Thesis, Tamil Nadu Agricultural University, Coimbatore-3.
3. Anonymous. (1986). Polymeric material 'Jalashakthi', *Sci. Reptr., 8* : 532.
4. Aravindkumar, R., V. Prakash and R.P. Singh. (1991). Influence of starch polymer on yield attributes, yield and quality of mustard (*Brassica juncea*) under field conditions. *Indian J. Agron., 36(4)* : 612-613.
5. Asborno, M.D., A.A. Vidal, R. Bezus and J. Beltrano. (1999). Rice : Temperature and Gibberellic acid effect on initial growth stages. *Agro-ciencia, 15(1)* : 47-53.
6. Bapat, M.V. and S.R. Umale. (1973). Seed treatment studies with phosphates. 1. Wheat. *P.K.V. Res. J., 1* : 214-220.
7. Baxter, L. and L. Waters Jr. (1986). Effect of a hydrophilic polymer seed coating on the field performance of sweet corn and cowpea. *J. Amer. Soc. Hort. Sci., 111(1)* : 31-34.

8. Bhatnagar, G.S. and M.K. Porwal. (1990). Effect of organic polymeric compounds on the seed yield of chickpea. *International J. of Tropical Agri., 8(3)* : 217-219.
9. Chandel, A.S., S.K. Tiwari and S.C. Saxena. (1990). Effect of micronutrient application on soybean (*Glycine max* L.) grown in Uttar Pradesh foot hill. *Indian J. Agric. Sci., 60* : 62-63.
10. Chikkanna, C., S. Thimmegowda, R. Paramesh and N. Devakumar. (1993). Use of hydrophilic polymer on water use efficiency of different crop under rainfed farming. *Indian Agriculturist, 37(1)* : 39-45.
11. Chio, C.D., S.C. Kim and S.K. Lee. (1990). Interaction between brassinolide and auxins on bioassays. *Korean J. Crop. Sci., 35(1)* : 58-64.
12. Cornett Zane, J. (1982). Nutrient and mycorrhizal effects on the root-shoot ratio of containerized ponderosa pine seedlings. Tucson, AZ: University of Arizona. 90 p. Ph.D., dissertation.
13. Dexter, S.T. and M. Takao. (1960). Acceleration of water uptake and germination of sugar beet seed balls by surface coatings of hydrophilic colloids. *Agron. J., 51* : 388-389.
14. Dharmalingam, C., K. Paramasivam and V. Sivasubramanian. (1988). Seed hardening to over come adversity. *The Hindu, No. 16*(Wed.).
15. Durai Singh, R. (1988). Effect of plant geometry and levels of fertilizers (N and P) on the productivity of soybean (*Glycine max* L. Merrill). M.Sc. (Ag.) Thesis, Tamil Nadu Agricultural University, Coimbatore -3.
16. Durrant, M.J., A.H. Loads and P.A. Payne. (1983). The use of water and some inorganic salt solutions to advance sugar beet seed. *Ann. Appl. Biol., 103* : 517-526.
17. Farley, R.F. and A.P. Draycott. (1978). Manganese deficiency in sugar beet and the incorporation of manganese in the coating of pelleted seeds. *Plant and Soil, 49(1)*: 71-83.
18. Hayat, S., A. Ahmad, A. Hussain and M. Mobin. (2001). Growth of wheat seedlings raised from grains treated with 28 homobrassinolide. *Acta physiologie plantarum, 23(1):* 23-30.
19. Hegarty, T.W. (1970). The possibility of increasing field establishment by seed hardening. *Hort. Res., 10* : 59-64.
20. Henckel, P.A. (1964). Physiology of plants under drought. *Ann. Rev. Pl. Physiol., 15:* 363-386.
21. Iqbal, S.H. and K.R. Srinivasan. (1987). Invention intelligence, Division of Technical Services, NCL, pp.382-385.
22. Jakobsen, I., E.J. Joner and J. Larsen. (1994). Hyphal phosphorus transport, a keytone to mycorrhizal enhancement of plant growth. In: Gianinazzi, S. and Schuepp, H. (eds.). Impact of Arbuscular mycorrhizae on sustainable Agriculture and Natural Ecosystems. Birkhauser Verlag, Switzerland, pp. 133-146.

23. Jayaseelan, K. (1997). Studies on the effect of sequencing of seed treatments on the seed yield and quality in redgram cv. CO 5 (*Cajanus cajan* (L.) Mill. Sp.) and greengram cv. CO 5 (*Vigna radiata* (L.). Wilcezek). M.Sc. (Ag.) Thesis, Tamil Nadu Agricultural University, Coimbatore -3.

24. Jeyabal, A., G. Kuppusamy and A.R. Lakshmanan. (1992). Effect of seed coating on yield attributes and yield of soybean (*Glycine max* L.). *J. Agron. & Crop Sci., 169* : 145-150.

25. Kachapur, M.D., A.S. Hadimani, B.S. Vyakaranahal and A.I.S. Prabhakar. (1987). Response of grain sorghum to micronutrients in Black cotton soils of Tungabhadra Project Area. *Sorghum Newsl., 30:* 39.

26. Kalarani, M.K., R. Sivakumar, Mallika Vanangamudi and K.B. Sujatha. (2001). Effect of seed soaking treatments in late sown finger millet under rainfed condition. *Madras Agric. J., 88(7-9)* : 509-511.

27. Kannaiyan, M. (1987). Effect of date of sowing and seed hardening on yield of pearlmillet (*Pennisetum typhoides* stapf and Hubb) under rainfed condition. M.Sc. (Ag.) Thesis, Tamil Nadu Agricultural University, Coimbatore -3.

28. Kavitha, S. (2002). Seed hardening and pelleting for maximizing the productivity of black gram (*Vigna mungo* L.) Hepper) cv. Vamban 3 under rainfed conditions. M.Sc. (Ag.) Thesis, Tamil Nadu Agricultural University, Coimbatore -3.

29. Khan, A.A. (1980). **In:** The physiology and biochemistry of seed dormancy and germination (Ed. A.A. Khan), North Holland Publishing Company, Amsterdam, New York, Oxford, pp. 127.

30. Kurundkar, B.P., P.A. Thombre and B.R. Kawale. (1991). Effect of different inoculum levels of *Bradyrhizobium japonicum* on soybean. *Legume Res., 14(4)*: 201-204.

31. Lu, S.L., S.D. Min and X.X. Jian. (1987). Effect of seed pelleting on rhizobial survival on inoculated seed and yield, food legume improvement for Asian farming systems, ACLAR Proceedings No.18, Canberra Australia (Edited by Wallis, E.S.I., D.E. Byth) pp. 28.

32. Mahatim Singh, Jagadish Singh and Kalyan Singh. (1983). Effect of phosphorus and biofertilizers on chlorophyll content of leaves and Leghaemoglobin contents of fresh nodules in *Kharif* grain legumes. *India J. Agron., 28(3)* : 229-234.

33. Maitra, S., G. Sounda, D.C. Ghosh and P.K. Jana. (1997). Effect of seed treatment on finger millet (*Eleusine coracana*) varieties in rainfed upland. *Indian J. Agric. Sci., 67(10):* 478-480.

34. Mehrotra, O.N., J.P. Srivastava and S. Nath. (1968). Germination, growth and yield of wheat as influenced by pre-sowing seed treatment in phosphate solution. *Indian J. Agron., 13(2)* : 53-56.

35. Menon, K.K.G. and H.C. Srivastava. (1984). Increasing plant productivity through improved photosynthesis. *Proc. Indian Acad. Sci., (Plant Sci.,) 93:* 359-375.

36. Min, D.H. (1988). Effects of liming and seed coating on early growth, drymatter yield and nutritive value of alfalfa (*Medicago sativa* L.) and soil properties. M.Sc. Thesis. Seoul National Univ.

37. Mucke, J.D. (1987). The regulation of water transport in pelleted sugarbeet seed. *J. Agron. and Crop Sci., 161(1)* : 79-83.

38. Namdeo, S.L., S.C. Gupta and K.S. Bangar. (1989). Effect of rhizobial strains on symbiotic traits and yield of chick pea genotypes under rainfed conditions. *Indian J. Pulses Res., 2(1)* : 64-67.

39. Natchimuthu, A. (1997). Studies on the influence of pelleting and foliar nutrition on plant growth, yield and quality of seeds of gingelly (*Sesamum indicum* L.) cv. TMV 3. M.Sc. (Ag.) Thesis, Tamil Nadu Agricultural University, Coimbatore -3.

40. Nayak, B.K., S.N. Misra and L. Dixit. (1989). Response of soybean to molybdenum and fertility levels. *Indian J. Agron., 34:* 434-435.

41. Newman, E.L., W.R. Eason, D.M. Eissenstat and M.I.R.F. Ramos. (1992). Interaction between plants: The role of mycorrhiza. *Mycorrhiza, 1* : 47-53.

42. Niranjana, S.R., A. Pandit, H.S. Prakash and H.S. Shetty. (1999). Effect of triacontanol on the seed quality of maize, paddy and sunflower. *Seed Sci.& Technol.,, 27(3):* 1007-1013.

43. Padmani, D.R., B.L. Porwal and J.C. Patel. (1994). Effect of levels of irrigation, nitrogen and Jalashakthi on growth and yield of Indian mustard (*Brassica juncea*). *Indian J. of Agronomy, 39(4):* 599-603.

44. Ponnusamy, A.S. and J. Vijaya. (1997). Effect of seed fortification and pelleting on crop growth and yield in cowpea. *Madras Agric. J., 84(7)* : 393-395.

45. Prabakaran, J. and K. Srinivasan. (1995). Effect of *Rhizobium* inoculation and coir dust on growth, nodulation and yield of pigeon pea (*Cajanus cajan*). *Indian J. Agron., 40(3):* 518-519.

46. Prabakaran, J. and M. Rangarajan. (1992). Effect of Rhizobial inoculation on the nodulation, nitrogen fixation and yield of lablab (*Phaseolus vulgaris* L.). *Madras Agric. J., 79(3)* : 172-174.

47. Punithavathi, N. (1997). Seed hardening and pelleting studies in Ragi cv. CO 13. M.Sc.(Ag.) Thesis, Tamil Nadu Agricultural University, Coimbatore-3.

48. Purushothaman, D. and S. Gunasekaran. (1981). A biofertilizer for cotton. *TNAU Newsletter, 11(2)* : 4.

49. Rana, K.S., Ashok Kumar and R.C. Gautam. (2001). Effect of starch polymer (Jalashakthi) on yield attributes, yield and water use efficiency of mustard under rainfed conditions. *Crop Res. 22(3)*: 395-397.

50. Rangasamy, A. (1986). Evaluation and management of sorghum and groundnut based intercropping system in a rainfed Alfisol. Ph.D. Thesis, Tamil Nadu Agricultural University, Coimbatore -3.

51. Robani, H. (1994). Film coating horticultural seed. *Hort. Technol., 4* : 104-105.

52. Sabir-Ahamed, A. (1999). Field performance of hardened green gram seeds. *Legume Res., 22(3)*: 207-208.

53. Sandhu, A.S. and A. Hussain. (1961). Effect of seed treatment with gibberellic acid on germination and growth of bajra. *Indian J. Agron., 5* : 269-272.

54. Saraswathi, S., A.G.R. Reddy and B.G. Singh. (1988). Effect of seed treatment with fungicides and *Rhizobium* on seed germination, nodulation, growth, dry matter, nitrogen content and yield in greegram. *J. Res., APAU, 16(1)* : 14-22.

55. Sathiyamoorthy, P. and M. Vivekanandan. (1990). Pretreatments of seed for improved growth and productivity in some crop plants. *Agric. Rev., 11(3)*: 139-148.

56. Scott, J.M. (1989). Seed coatings and treatments, their effects on plant establishment. *Adv. Agron., 12* : 43-83.

57. Selvaraju, K. (1992). Studies on certain aspects of seed management practices for seed production under moisture stress condition in sorghum cv. CO 26 (***Sorghum bicolor*** (L.) Moench). M.Sc. (Ag.) Thesis, Tamil Nadu Agricultural University, Coimbatore - 3.

58. Shekhawat, A.S., K.N. Bansal and S.P. Singh. (1993). Effect of various levels of phosphorus, Jalashakthi and Plantomin on growth and yield of soybean (***Glycine max*** L.) *Annals of Agril., Res. 14(1)* : 107-108.

59. Shinde, S.S. and R.K. Bhalerao. (1991). Effect of seed treatment with plant growth regulators on yield and yield components of grain sorghum. *Journal of MAU., 16(3)* : 447.

60. Srimathi, P. and K. Malarkodi. (2000). Influence of nutrient pelleting on seed yield of soybean cv. CO 1. *Indian Agriculturist, 44(1/2)* : 79-82.

61. Thiyageshwari, S. and G. Ramanathan. (1998). Efficiency of micronutrients on the yield and quality parameters of soybean (*Glycine max* L.) *Micronutrient News. A Bulletin of Inidan Micro-fertilizers Manufacturers Association, 12(11)*: 5-7.

62. Usha, R., N. Ramamurthy, P.N. Bhavani and P.M. Swamy. (1999). Alleviation of PEG-6000 induced water stress effects by calcium. *Indian J. Plant Physiol., 4* : 95-99.

63. Vaishya, J.K. and G.R. Gajendragadkar. (1984). Response of pigeon pea to biofertilizers. *Int. Pigeon pea Newsl.,* 2: 49-50.

64. Wang, S.Y. and G.L. Steffens. (1985). Effect of Paclobutrazol on water stress induced ethylene biosynthesis and polyamine accumulation in apple seedling leaves. *Phytochemistry, 24*: 2185-2190.

65. Yopp, J.H., N.B. Mandava and J.M. Sasse. (1981). Brassinolide a growth processing steroidal lactone. *Physiol. Plant, 53*: 445.

66. Zubenko, V.K.H. (1959). The effect of preplanting hardening of seeds against drought on the productivity of corn in late plantings. *Plant Physiol., 6*: 341-343.

PART VII

SPECIAL TOPICS

FUTURE PERSPECTIVES FOR RESEARCH ON SEED HARDENING

K. RAMAMOORTHY

1. INTRODUCTION

The reduction in yield of crops due to stress is the cumulative effect of impairment of many physiological functions within the plant, since the physiological processes are the link between plant stress and yield response. The altered physiology of plant as reflected through reduction in leaf water potential, reduced photosynthesis, impaired metabolism of nitrogen and carbohydrates leads to reduced growth and yield. When subjected to water stress, plants react physiolcgically in many ways. Some of these reactions may be incidental with no significant consequences, some may be deleterious and some may be adaptive (Anon, 1980). Plants usually yield to any stress conditions and the reactions may be elastic or plastic. If elastic, the reaction is temporary and plants revert back to original state. In plastic reactions, the plants are deformed and changes are not reversible. Any hardening treatment should take into account both the reactions. The stress may be immediately in the plant or plants may become resistant, when exposed to stress. This is otherwise referred as seed hardening. The physiological approach to cropping strategy in dry farming regions therefore, can be approached in two ways (Begg and Turner, 1976).

(i) To identify plant adaptations to stress and use them as criteria for selection of crops resistant to drought.

(ii) To evolve management practices to minimize the effect of harmful reactions.

Department of Seed Science and Technology, Agricultural College & Research Institute, Madurai-625 104

2. PLANT ADAPTATIONS TO DROUGHT

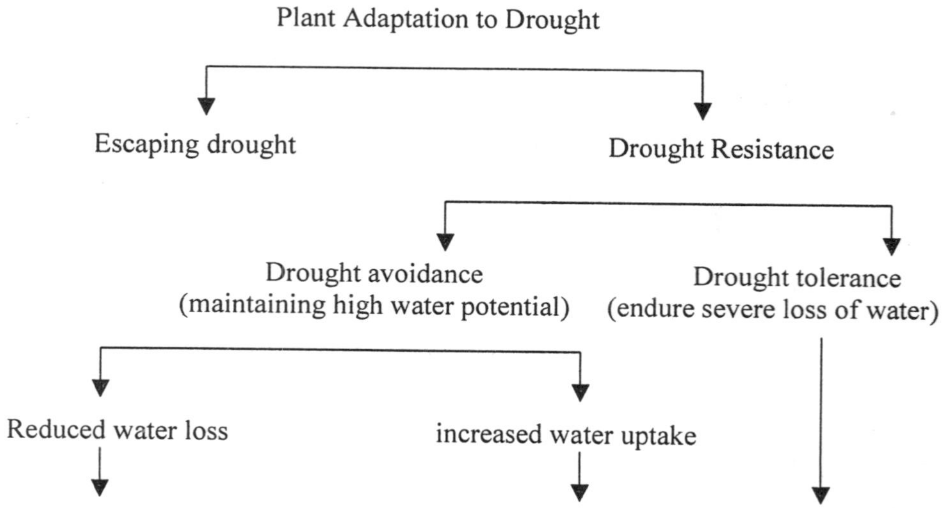

i) Early closer of stomata	i) Efficient roots	i) Resist dehydration
ii) Low transpiration	ii) High root-top ratio	ii) Prevention of cell collapse
iii) Lipid deposits on leaves	iii) Osmotic adjustment	iii) Reduction of protein breakdown
iv) Increased phytosynthesis		iv) Reduced accumulation of toxins
		v) Resuming growth after stress

2.1. Drought escape

Evading the period of drought by completing the lifecycle within the rainy period is the adaptation exhibited by desert plants. These germinate with rains, flower and set seed within 5 weeks. Otherwise, these plants have no other mechanism for withstanding water stress. In cultivated crops, the ability of a cultivar to attain maturity before the soil dries out is a desirable character while growing under drought conditions. But very few cultivars with such short duration of growth are available. Even those cultivars which mature within 60-70 days have low yield potential and should have some mechanism of drought resistance for short periods of stress.

2.2. Drought resistance

Plants that cannot escape drought develop drought resistance in two ways.

(i) Avoiding stress (ii) Tolerating stress

2.2.1. *Drought avoidance*

Stress avoidance is the ability of the plant to maintain favourable internal water balance and turgidity even when exposed to drought conditions and thus delay the onset of moisture stress within the plant. By this, the adapted plants are able to avoid the adverse effects due to soil moisture stress. Stress avoidance is mainly due to morphological and anatomical characteristics which are themselves consequence of physiological processes induced by drought. The minimum leaf water potential for survival for herbaceous crops is –15 to –25 bars for grasses from –20 to –22 bars and for desert plants, it is –55 to –80 bars. The favourable water balance within the plant can be maintained by

(a) Conserving water by reducing transpiration

(b) Increasing water uptake

2.2.2. *Drought toleraance*

Drought tolerance refers to the capacity of protoplasm to endure severe loss of water i.e. dehydration. This is measured in terms of critical water content, which is achieved by resistance to cell collapse and protein breakdown mechanism, for decreasing the accumulation of toxic substances eg. NH_3, ability to resume growth after stress on rewatering etc. The drought tolerance is measured by high relative water content. Dehydration causes water loss and because of this proteins are disrupted. Water molecules are responsible for keeping complex fluids in a stable configuration. Water loss causes concentration of solutes leading to higher concentration of cell sap and intercellular fluids cause greater decrease in water potential of the fluids, causing stress on the protoplasm and changes in cell pH affecting biochemical processes. Though it is difficult to exploit physiological responses of plant to water stress because it is difficult to ascertain which response is detrimental and which are adaptations (eg.) Proline accumulation is a physiological reaction to stress and it is a symptomatic rather than causative of drought susceptibility (Hanson *et al.*, 1979). Drought tolerance in plants can be achieved to a certain extent, through use of hydrophilic substances in the protoplasm (eg.) high molecular weight proteins, some carbohydrates (alginic acid). Seed hardening may enhance sugars in solution in the protoplasm which lowers the water potential of cell sap. This will help younger plants to retain water, conserve moisture and save from desiccation.

3. FUTURE APPROACHES

- The crop growth period can be divided into three *viz.*, vegetative (formation of organs for nutrient absorption) and two periods of

8. Kasam, A.H. (1975). Cereal physiology in relation to genetic improvement at ICRISAT with reference to drought endurance. Crop Phy. Abstracts, 6(1): 635.
9. Kapoor, V.P., Joshi, H. and Chaubey, M. (2001). Application of seed gums in pharmaceutical formulations. *Journal of Medicinal and Aromatic Plant Sciences, 22/4 A and 23/1 A 42-44.*
10. Nilima, S. and C.P.Malik. (1983). Physiological and biochemical changes under water stress in pigeonpea plants grown from seeds soaked in water and chemicals. *Indian J. Agric. Sci., 53(7)* : 550-6.
11. Siband, P., Dieya, I. and B.C.Baye. (1980). Evaluation of grain seedling system in pearlmillet. *Arid Land Abstracts, 2(10)*: 3740.
12. Saroja, S., Padma, P.R., Radha, P. and P.Thilagavathy. (2001). *Journal of Medicinal and Plant Sciences 22/4 A and 23/1 A*: 37-41.
13. Turner, N.C. and P.J.Kramer. (1979). Adaptation of plants to water and high temperature stress. John Wiley and Sons, New York.
14. Vasuki, K. and K.Thangaraj. (1997). Kothamalli Vidhai Nerthi muraigal. Valarum Velanmai, 23(12): 25-26.

ECONOMICS OF SEED HARDENING AND PELLETING OF BLACKGRAM

A.S.PONNUSAMY[1], J.VIJAYA[1],
V.KRISHNASAMY[2] AND K.RAJA[3]

1. INTRODUCTION

Pre-treatment of seeds with nutrient chemicals produce physiological effects on seed for increasing the yield. Seed fortification is nothing but a seed enrichment with nutrient solutions. Pelleting is the process of enclosing a seed in a small quantity of inert material to enlarge size of seed to facilitate percision planting. Seed coating has been reported to increase the yield of pulses particularly blackgram and greengram (Kuppusamy *et al.*, 1992). This paper reports the response of blackgram to seed fortification and pelleting on crop growth parameters and yield attributes.

2. SEED FORTIFICATION AND PELLETING IN BLACKGRAM

Investigations were carried out during 1995 to evaluate the effect of seed fortification and pelleting on crop growth and yield in blackgram. The following treatments were given. T_0-Control, T_1 – Fortified with micronutrient ($ZnSO_4$ 0.2%, $MnSO_4$ 0.2%, Na_2MO_4 0.1%), T_2 – Pelleted with micronutrients ($ZnSO_4$ 0.2%, $MnSO_4$ 0.2%, Na_2MO_4 0.1%), T_3 – Pelleted with macronutrient (120 g DAP kg^{-1}) of seed, T_4 – Fortified with micronutrient + pelleted with macronutrient ($ZnSO_4$ 0.2%, $MnSO_4$ 2%, Na_2MO_4 0.2% + 120 g DAP kg^{-1} of seed). The biometric observations were made on 10 randomly selected plants and the data were analysed statistically.

1 *Department of Seed Science and Technology, Tamil Nadu Agricultural University, Coimbatore – 641 003.*
2 *Centre for Plant Molecular Biology, Tamil Nadu Agricultural University, Coimbatore – 641 003.*
3 *Central Institute for Cotton Research, Coimbatore – 641 003.*

2.1. Method of seed fortification / hardening

One-third volume of the nutrient solution was added to the known volume of seeds and allowed to imbibe for five hours. The imbibed seeds were dried under shade for three days to bring back to the original weight.

2.2. Sequence of pelleting

Seed + Fungicide + coating material + nutrient mixture

2.3. Effect of fortification and pelleting and field emergence and crop growth

The field emergence percentage and plant height were not significantly influenced by the treatments. The dry matter production of the seedlings collected at 20th, 40th and 60th days after sowing (DAS) differed significantly due to the treatments. Fortieth and Sixtieth DAS showed high dry matter production in the seeds fortified with micronutrients and pelleted with macronutrient. The same has been reported for neem (Ponnuswamy, 1993). The pod production was significantly altered by fortification and pelleting treatment. T_4 recorded the maximum number of 25.5 pods plant^{-1}, and the minimum number of 23.6 pods plant^{-1} was recorded in T_0. The pod length was significantly inflouenced by all the treatments. T_4 and T_0 registered the maximum and minimum of 4.5 and 4.1 cm respectively.

2.4. Effect of fortification and pelleting on yield attributes

Number of seed pod^{-1} differed significantly due to treatments. The seed yield was significantly influenced by the fortification and pelleting treatments. T_4 and T_0 registered the maximum and minimum of 5.6 and 4.8 g of seed plant^{-1} respectively. Jayabal *et al.* (1992) and Kuppusamy *et al.* (1992) have registered the increased yield due to pelleting of soybean and greengram respectively. The increased yield could be due to enhanced seed vigour resulting better seedling establishment and activation of metabolic activities of the seed on account of fortification and pelleting treatments (Table 1).

2.5. Effect of fortification and pelleting on quality of resultant seed

100 seed weight was significantly influenced by the treatments. T_4 and T_0 registered the maximum and minimum of 3.80 and 3.60 g respectively. Similar findings were reported by Efimenko and Gnida (1975) in buckwheat. The maximum and minimum of 94 and 90 per cent of seed recovery was registered in T_4 and T_0 respectively. The germination per cent of the resultant seeds obtained from different treatments were not significant. The cost

Table 1. Effect of fortification and pelleting treatments on seed yield and resultant seed germination

Treatment	Field emergence (%)	Number of pods plant^{-1}	Seed yield (g plant^{-1})	Seed yield (kg plot^{-1})	Seed recovery (%)	Germination (%)
			40 DAS	60 DAS		
T_0 – Control	91 (72.54)	23.6	0.86	6.86	90 (71.57)	94 (75.82)
T_1 – Fortified with micronutrient ($ZnSO_4$ 0.2% + $MnSO_4$ 0.2% + Na_2MO_4 0.1%)	92 (73.57)	24.4	0.92	7.10	92 (73.57)	95 (77.08)
t_2 – Pelleted with micronutrient	92 (73.57)	24.3	0.92	7.16	92 (73.57)	94 (75.82)
T_3 – Pelleted with macronutrient (120 g DAP/kg of seed)	93 (74.66)	24.8	0.92	7.20	93 (74.66)	95 (77.08)
T_4 – Fortified with micronutrients + pelleted with macronutrient	92 (74.66)	25.5	0.93	7.25	94 (77.08)	96 (78.46)
CD (P=0.05)	NS	0.58	0.02	0.34	0.68	NS

Table 2. Cost benefit ratio of seed fortification and pelleting in black gram

Treatment	Cost of cultivation (Rs./ha)	Computed yield (kg/ha)	Total income (Rs.)	Increased yield over control (kg/ha)	% of increased yield over control	Increased income over control (Rs/ha)	% of increased income over control	Cost benefit ratio
T_0	6425	1372	24696	-	-	-	-	1 : 3.8
T_1	6476	1498	26964	126	9.18	2268	9.18	1 : 4.1
T_2	6533	1523	27414	151	11.0	2718	11.0	1 : 4.2
T_3	6534	1625	29250	253	18.0	4554	18.3	1 : 4.4
T_4	6534	1725	31055	353	25.7	6359	25.6	1 : 4.7

benefit ratio analysis also revealed that the T_4 treatment recorded the maximum cost benefit ratio of 1 : 4.7 when compared with other treatments (Table 2).

REFERENCES

1. Efimenko, D.Y. and Gnida, A.P. (1975). Effectiveness of presowing enrichment of buckwheat seed with treatment. Ukaramion SSR Nomkova Dumka. 148-153. *From Seeds Abst., 1979, 2(1)*: 218.
2. Jeyabal, A., Kuppuswamy, G. and Lakshmanan, AR. (1992). Effect of seed coating on yield attributes and yield of soybean (*Glycine max* I.). *J. Agron. and Crop Sci., 169*: 145-150.
3. Kuppuswamy, G., Lakshmanan, AR and Jeyabal, A. (1992). Effect of seed pelleting with bio-digested slurry and macronutrients on rice fallow green gram cv. CO 2. *Madras Agri. J., 79(4)*: 183-187.
4. Ponnuswamy, A.S. (1993). Seed technological studies in neem (*Eadrachta indica* A.Juss). Ph.D. Thesis, Tamil Nadu Agricultural University, Coimbatore.

CROP MANAGEMENT STRATEGIES TO SALVAGE AGRICULTURE UNDER DROUGHT SITUATIONS

S.RAMANATHAN

1. INTRODUCTION

Indian agriculture is predominantly a rainfed agriculture. Out of 143 million hectares of total cultivated area in India, about 100 mil lion hectares are rainfed. The rainfed lands of the country contribute to about 44% of the food grain production and support 40% of India's population. Unfavourable crop growth environment, limited choice of crops and varieties, poor soils, lower cropping intensities. Short growing and low-unstable productivity are the major concerns of the thousands of dry land farmers under drought situations. Despite the draw back of this fragile situation, prospects of boosting the crop yield are attractive. There are unlimited yield potential hidden in this environment, that could be exploited through careful selection of crop management strategies and tactics. The opportunities to improve the efficiency of dry land crops through adoption of improved crop management practices and some other factors are to be considered while determining the crop management options for a given production situation are presented. Crop cultivation itself is a risky phenomenon in drought situations due to several factors related to rainfall distribution and amount of rainfall. Despite many limitations, the dry land farming could be altered into an economical enterprise with due consideration on many improved technologies developed over many decades.

2. CROP MANAGEMENT STRATEGIES

2.1. Selection of Crops

Rainfall pattern, potential evapotranspiration and water retaining capacity of the soils decide the choice of crops and cropping systems under

---

Director of research, Tamil Nadu Agricultural University, Coimbatore- 641 003.

drought situations. However, selection of crops and varieties are governed by the needs of the farm, family and traditional experience rather than the crops suitability for a given environment. Therefore, under drought situation we are experiencing high variations in productivity. The crops and their genotypes differ in their yield potential and water use efficiency, some being more efficient than others in the same environment. Rainfall pattern and effective growing season are the most commonly used parameters for the selection of cropping system.

In regions receiving 350-600 mm rainfall with an effective growing season of 20 weeks, only single cropping (100% cropping intensity) is possible in alfisols, shallow vertisols, deep alfisols and entisols. In areas receiving more than 750 mm rainfall and having an effective growing season of 30 weeks or more, double cropping (200 per cent cropping intensity) is possible. This pattern is evident from crops grown in bi-model rainfall pattern regions. Calculation of length of growing period will be the major step in crop planning under drought situations. The overall ranking of crops in the increasing order of drought tolerance is: soybean < blackgram < greengram < maize < sorghum < pearl millet < cowpea < cotton. The following are some of the drought tolerant varieties:

Cotton	:	LRA 5166, KC 2, MCU 10
Pearl millet	:	Co 7
Sorghum	:	APK 1, K 8, Co 25
Greengram	:	KM 2
Blackgram	:	APK 1 and Co 5

2.2. Sowing Time

Success in crop production under drought situations depends upon proper time of seeding. Since the length of growing season is likely to be shortened by late seeding, time of sowing assumes importance. The very first shower should be utilized for crop growth. This calls for preparing land prior to probable rainfall event. Studies at Coimbatore have shown that planting red gram immediately after onset of monsoon had the highest mean seed yield compared to the planting on second fortnight of July. The reason was that the crop had the benefit of sufficient moisture through the receipt of rainfall during the early stage of the crop growth. At Aruppukkottai, pre-monsoon sowing of sorghum and cotton two weeks prior to onset of monsoon was optimum (Table 1).

Dry sowing is considered appropriate under drought situations. The most important advantage of dry sowing is that it gives an early start to the

Table 1. Effect of pre-monsoon sowing on rainfed cotton and sorghum

Treatments	Yield (kg ha^{-1})	
	Cotton	Sorghum
Pre-monsoon sowing two weeks ahead	419	2033
Monsoon sowings	210	1848
Sowing two weeks after onset of monsoon	120	629

crop. Dry seeding is possible in case of bold seeded grains like pigeon pea, sorghum and cotton. The rainfall use efficiency can be increased by pre-monsoon sowing (Table 2).

Table 2. Effect of pre-monsoon dry seeding of cotton

Sowing	Yield (t ha^{-1})	Rainfall use efficiency (kg/mm/ha)
Coimbatore	419	2033
Pre-monsoon (38th standard week)	2.9	8.9
Monsoon (40th standard week)	1.6	7.7
Kovilpatti		
Pre-monsoon (38th standard week)	2.4	8.2
Monsoon (40th standard week)	1.7	6.8

Pre-monsoon dry sowing of redgram (APK 1) during the first fortnight of September recorded highest grain yields (Table 3) at Aruppukkottai.

Table 3. Effect of pre-monsoon sowing of redgram at Aruppukkottai

Treatments	Grain yield (kg ha^{-1})
Sowing during September first fortnight	812
Sowing during September second fortnight	304
Sowing during October first fortnight	158
Sowing during October second fortnight	78
CD	34

2.3. Sowing Techniques

Sub-optimal and poor plant stand is largely responsible for low productivity of dryland crops. Proper germination and good crop stand is the first requirement for obtaining higher yield. Common broadcast method has to be discouraged on account of uneven stand. Seed drill sowing is more useful in view of precision in the depth of sowing, uniformity in spacing and above all rapid coverage that it ensures (Table 4). Tractor drawn seed drills are already popular among the farmers.

Table 4. Effect of sowing methods on rainfed sorghum and pearl millet at Kovilpatti

Sowing method	Grain yield (t/ha)			
	Sorghum	% increase over broadcasting	Pearl millet	% increase over broadcasting
Tractor drawn seed drill	2.08	23.0	2.04	10.9
Bullock drawn seed drill	1.92	13.6	1.95	6.0
Ferti seed drill	1.97	16.5	1.99	8.2
Sowing behind the country plough	1.74	3.0	1.87	1.6
Hand dibbling	1.82	7.7	1.92	4.3
Broadcasting	1.69	-	1.84	-

Depth of sowing is another non-monetary input, which decides the crop stand under drought situations. Optimum depth of sowing is very important for dry seeding. Recommended depth of sowing is 5 cm for cotton and sorghum for maximizing yields.

2.4. Seed Hardening

Seed hardening is conditioning the seed to withstand the adverse environment and edaphic conditions. It is a creation of resistance in the seed for better outgrowth of seedling. Pre-sowing hardening of seed is one of the techniques where the physiological reorganization is induced by hydration and dehydration process. Hardening process brought out physico-chemical changes by modifying the physico-chemical characters by increasing the physiological activity of the embryo and associated structures. This results in absorption of more water due to increase in the elasticity of cell wall and development of a stronger and efficient root system.

2.4.1. *Advantages of Seed Hardening*

- Accelerates rapid germination and growth of seedlings
- Induced resistance to salinity as well as to drought condition
- Seeds withstand higher temperature (80-105°C) for prolonged period
- Early emergence enable seedlings to compete more effectively with weeds
- Hardened seeds withstand drought in later stages and perform better than untreated crop

Besides the physiological preconditioning of seeds, it makes up the plants in a sort of preparedness to resist the adverse weather if any in later stages of plant growth. The seed hardening chemicals used for different crops are presented in Table 5.

Table 5. Chemicals recommended for different crops for seed hardening

Crop	Chemicals used
Sorghum	KH_2PO_4 2%
Paddy	KCl 1%
Cotton	KCl 2%, CCC 1000 ppm
Maize	CCC 250 ppm
Bajra and Ragi	Nacl (0.5%), KCl (2%)
Sunflower	GA_3 (500 ppm) (or) 1% nutrient solution (equal quantity of manganese sulphate, zinc sulphate, boric acid and ferrous sulphate)
Groundnut	$CaCl_2$ (0.5%)
Redgram, Cowpea	Zinc sulphate (100 ppm)
Greengram	Manganese sulphate (100 ppm)
Bengalgram	KH_2PO_4 (1%)
Gingelly	$CaSO_4$ (2%)

2.5. Moisture Conservation by Mulching

Mulching is the practice of maintaining farm wastes and other materials to reduce soil evaporation. Soil is also used as mulching material by creating surface soil particles, which arrests the capillary continuity by shallow tillage implements called "dust mulching". Residue mulch prevents direct impact of rain drop on soil aggregates, maintains pore space continuity and high infiltration rate (Selvaraju and Ramaswami, 1998). Runoff and soil loss decreases exponentially with increase in the mulch rate (Table 6).

Table 6. Effect of mulch rate on 10% slope when the total rainfall is 430 mm

Mulch rate (t ha^{-1})	Runoff (% of rainfall)	Soil loss (t ha^{-1})
0	17.4	9.6
2	10.0	2.3
4	3.5	0.5
6	1.2	0.1
12	0.0	0.0

Surface mulching with different organic mulches such as sorghum stubble, pearl millet straw, paddy straw, saw dust, groundnut shell and dust mulch showed that mulching increased the soil moisture content by 3 per cent. In cotton, saw dust mulch of 2.5 cm thickness recorded higher moisture content (30.9%) and was on par with groundnut shell mulch (30.4%) and stubble mulch (30.3%). Application of organic waste as surface mulch increased the yield and economics of sorghum (Table 7).

Table 7. Effect of organic wastes on grain yield of sorghum and economics

Treatment	Grain yield (kg ha^{-1})	Net return (Rs ha^{-1})	B:C ratio
Control	1096	1751	1.06
FYM (5 t ha^{-1})	1586	2550	1.18
Pressmud (5 t ha^{-1})	1375	2116	0.99
Coirpith (5 t ha^{-1})	1157	1673	0.82
Rice husk (5 t ha^{-1})	1371	2198	1.08
Bajra trash waste (5 t ha^{-1})	901	1122	0.57
Distillary waste (5 t ha^{-1})	1290	2061	1.02

The results revealed that application of farmyard manure (FYM 5 t/ha has consistently produced significantly higher grain yield, press mud and rice husk @ 5 t/ha followed the FYM.

2.6. Weed Management

Diverse weed flora are seen under drought situations. The major weeds identified in drought areas are *Cyperus rotundus, Cyanodon dactylon, Chenopodium album, Ageratum conyzoides, Convolvulus arvensis, Cleome viscosa, Argemone mexicana, Digera arvensis, Echinochloa sp., Eleusine indica, Digitaria sp., Trianthema sp.* etc. Incessant rains during initial stages not only hinder conventional weed control measures but also it provides ideal condition for weed emergence over an extended period of time. The critical period of crop-weed competition vary with crop duration, plant population, geometry of the crop and weed growth habits. often this period is stated to be approximately the first one-fourth to one-third of crop life without any reference to the growth stages (Table 8). The yield loss due to weed is greater (15-67%) as compared to yield loss during a favourable season (21-35%). Although complete weed control through adoption of any single method is doubtful, combination of suitable techniques normally have complementary effect. Cultural, mechanical, chemical methods are the components of integrated weed management under drought situations. Cultural methods of weed management can be regarded as an indirect method as their primary objectives are to provide optimum environment for crop growth.

Table 8. Crop yield losses and critical period of competition in dryland crops

Crop	% yield loss	Critical period (days)
Maize	57	15-30
Sorghum	60	0-30
Pigeon pea	40	50-57
Soybean	51	0-30
Groundnut	50	15-60

Studies on groundnut have revealed that application of Glycel (1%) followed by summer ploughing 15 days after during fallow period and application of Metalachlor 1.0 kg ai ha^{-1} + one hand weeding on 40 DAS for *kharif* groundnut recorded maximum yield of 1500 kg/ha over 818 kg ha^{-1} with control treatment.

2.7. Low Cost Inputs

The attempts to study the influence of non-monetary low-cost inputs on the yield of redgram under rainfed condition at Coimbatore revealed that the choice of improved variety Co 5 had resulted in the increased yield (717 kg ha^{-1}) as against conventional variety (local) SA 1 which had recorded the mean seed yield of 625 kg ha^{-1}. One of the major constraints for the low productivity in most of the pulses was low plant population per unit area. It had been reported that the crop yield increased with the corresponding increase in the plant population. Increased plant population of 100,000 plants ha^{-1} had recorded the highest mean seed yield as against the low plant population of 5000 plants ha^{-1} (Arunachalam *et al.*, 1995).

2.8. Nutritional Management

The nutrient such as potassium helps the plant to maintain tissue water potential by osmo-regulation mechanism and increase the water use efficiency. The translocation and partitioning of dry matter from source to sink is an important factor in determining crop yields. The potassium nutrition enhances the translocation, besides improving the water retention capacity of crops. Delayed flowering, spikelet sterility and incomplete emergence of panicle are the common symptoms of drought occurring at reproductive phase. In addition, water stress will reduce the availability of nutrients further delaying panicle emergence and increasing the sterility. The work on rainfed rice indicated that foliar application of N as urea at 2% was found to be superior and effective than soil application. Under water stress conditions during reproductive stage, foliar application of diammonium phosphate (DAP) at 2% (20 g of DAP per litre of water) was also equally effective since it contains, both N and P and elements necessary for root growth (Table 9).

Similarly, seed setting in rainfed cumbu can be improved by the combined foliar spray of 0.5 per cent Zinc sulphate, 0.5 per cent copper sulphate and 0.2% boric acid at flowering stage and one week after the first spray. This treatment recorded 16.8 per cent increased yield over control. Pod setting is a serious problem in rainfed groundnut. Foliar spray of DAP (0.5%) along with planofix (40 ppm), ammonium sulphate (0.2%) and boric

Table 9. Migration of water stress in crop through foliar spray of nutrients

Crop	Recommendation	Yield improvement
Rice	i) Foliar spray of muriate of potash 2.0% at flower initiation ii) Foliar spray of 0.3 ppm of Brassinolide at panicle initiation and flowering	1. 400 kg ha^{-1} increased yield 2. 6.5 t ha^{-1}
Sorghum	Kaolin 3% foliar spray	-
Cumbu (Pearl millet)	Combined sprays of 0.5% $ZnSO_4$ and 0.5% $CuSO_4$ together with 0.2% boric acid	1. 16.8 per cent yield increase
Soybean	a) Kaolin 3% or liquid paraffin at 1% as foliar spray b) Foliar spray of salicylic acid @ 100 mg/litre on 30th and 45th day	• – • 16 per cent increased yield
Blackgram	Foliar spray of KCl at 0.5% during flowering phase	100 kg ha^{-1} increased yield
Groundnut	Spraying of KCl at 0.5%	-
Cotton	Foliar spray of Kaolin 3% and liquid paraffin 1% spray	27.4 per cent yield increase

acid (0.1%) alleviated this deficiency and increased the yield by 30 per cent. Foliar spray of DAP at 2% with 0.5% zinc sulphate at flowering improved the pod set and yield in soybean. Under severe water stress condition, foliar spray of salicylic acid 100 ppm on 30th and 45th day enhanced the grain yield by 16 per cent in soybean through increased flower production and source activity. In rice fallow cotton, foliar spray of 20% DAP and 1.0% potash on 45th and 55th day after sowing resulted in an additional yield of 1.7 q ha^{-1}.

2.9. Use of Antitranspirants

Antitranspirants is a substance that is applied to transpiring plant surface with the aim of reducing water loss due to transpiration. Approximately 99 per cent of water taken by plant root is transported to atmosphere through stomatal pores in the leaves. Film forming antitranspirants like waxy or plastic emulsion produce an external physical barrier to retard the escape of water vapour. Another approach that reduce the transpiration rate is by coating the leaf surface with white reflecting material. Thus, white wash at 2% (20 g of calcium oxide per litre of water) helps to lower the leaf temperature and reduce the transpiration when applied on foliage. Studies have shown that about 27 per cent of normal irrigation water in sorghum crop could be saved with two per cent white wash or three per cent kaolinite sprays. Similarly, in soybean 3% kaoline or

1% liquid paraffin wax as foliar spray can be used to alleviate the effects of water stress. In an antitranspirant trial, kaoline three per cent and liquid paraffin one per cent foliar spray recorded 27 per cent (@ 127 kg ha^{-1}) and 19 per cent (1995 kg ha^{-1}) increased yield over control respectively. Transpiration suppressants such as atrazine and CCC have also been found effective (Table 10).

Table 10. Effect of transpiration suppressants on soybean (Co 1)

Treatments	Increased grain yield (kg ha^{-1}) over water stress control		
	Vegetative stage	Flowering stage	Pod filling stage
CCC (250 ppm)	234	223	230
Kaoline (3%)	125	126	44
Atrazine (25 ppm)	339	397	437
Liquid paraffin (1% v/v)	168	168	81

2.10. Use of Growth Hormones

Water stress results in reduced cytokinin production in roots and its supply to shoots. This shortage of cytokinin hastens the leaf senescence. Leaf senescence due to water stress can be postponed by the application of cytokinin or benzyl adenine. Yet another approach for enhancing drought tolerance is to slow down the rate of vegetative growth by spraying specific chemicals. Growth retardants like CCC (2 dichloroethyl trimethyl ammonium chloride) have been found to slow down the vegetative growth and enhance the root growth and activity to reduce the plants to survive the dry period especially at the time of ear emergence, when drought is most critical. The growth retardants in the treated plants results in reduced stomatal opening and hence lowered transpiration. Growth retardants like CCC also delays leaf senescence and increase the ability of the plant to survive drought. The studies on water stress management in rice revealed that seed hardening with potassium chloride one per cent and CCC 50 ppm followed by foliar spray of 1000 ppm at critical water stress stage recorded the maximum grain yield of 5720 kg ha^{-1} as against 4080 kg ha^{-1} in control. In Brinjal, foliar spray of CCC at 5 ppm concentration improved the fruit set and fruit yield by 24 per cent under water stress condition at flowering stage.

3. CONCLUSION

A closer look of the agricultural scenario of the developng country like India brings to light the challenging task of achieving food grain target of 250 million tonnes to meet the demand of around one billion people.

Adequate and quality seed production is also required to be achieved in the available area. The task becomes all the more challenging in the present trend of diminishing land and water resources. The major focus has to be on rainfed areas, which are potentially productive but development of drought management technologies involving multi-disciplinary approach in the context of integrated technological package would ensure soil and moisture conservation and enhance productivity of crops, especially for seed production. This type of technology should be eco-friendly without degradation of the natural resources in the large tract of drought areas and is bound to play a vital role to the user prosperity for millions of the people toiling hard in these areas.

REFERENCES

1. Arunachalam, L., S. Purushothaman, S.P. Palaniappan and M.Markdevasahayam. (1995). Relative contribution of non-monetary/low-cost inputs in redgram production. *Madras Agric. J., 82(3)*: 179-181.
2. Selvaraju, R. and C. Ramaswami. (1998). Effect of pre-season fallow management and legume cropping on succeeding intercrops in rainfed peninsular India. *Tropical Science, 38*: 190-195.